Hathorn, Davey
of Leeds

Manufacturers of Steam Pumping Machinery
1872 to 2016

To Annika and Hazel Vernon

Robert W. Vernon

Hathorn, Davey of Leeds

Manufacturers of Steam Pumping Machinery
1872 to 2016

First published in 2021 by
Robert W. Vernon
Bredon GL20 7AZ
United Kingdom

HDRV1901@Outlook.com

ISBN 978-1-8383621-0-2

Typeset in Times New Roman

Printed in Great Britain

Contents

List of illustrations   vii

Foreward   xiii

Preface   xvii

References   xxi

Acknowledgements   xxii

Definitions and Abbreviations   xxv

Chapter 1:   Introduction   1

Chapter 2:   The Sun Foundry   9

Chapter 3:   Carrett and Marshall   15

Chapter 4:   Hathorn, Davis and Campbell: 1872   23

Chapter 5:   Henry Davey   31

Chapter 6:   The Differential Expansive (Compound) Pumping Engine   41

    The Differential Gear   42

    The Compound Engine   47

Chapter 7:   Hathorn, Davis, Campbell and Davey: 1874 to 1877   53

Chapter 8:   Hathorn, Davis and Davey: 1877 to 1878   67

Chapter 9:   Hydraulic Engines and Pumps   77

Chapter 10:   Hathorn, Davey and Company and a new Partner   85

Chapter 11:   Hathorn, Davey and Company: The International Market   97

Chapter 12:   Hathorn, Davey and Company: The European Market   105

Chapter 13:   Davey's Domestic Safety Motor   115

Chapter 14:   Hathorn, Davey: The late 1880s and more Partnership changes   125

Chapter 15:   Hathorn, Davey and the South Staffordshire Mines Drainage Commission   133

Chapter 16:   Henry Davey and the London Office   141

Chapter 17:   Triple Expansion Steam Engine   153

Chapter 18:   Compound Engines and their remains in the 1890s   165

Chapter 19:   The Inverted Vertical Compound Engine   177

Chapter 20:   The Sun Foundry: 1872 to 1900   187

Chapter 21:   Hathorn, Davey and Company, Limited and the Lupton Family   197

Chapter 22:   The Limited Company: 1900 to 1914   209

Chapter 23:   Sales of the Compound Engine: 1900 to 1914   219

Chapter 24:   Triple Expansion Engines: 1900 to 1914   229

Chapter 25:   Engines and Pumps sold in Britain: 1900 to 1914   241

Chapter 26:   Hathorn, Davey during World War I   253

Chapter 27:   Hathorn, Davey in the 1920s   263

Chapter 28:   The Staff and Workforce   273

Chapter 29:   Engine Sales in the 1920s   283

Chapter 30:   The 1930s: Lack of Orders   293

Chapter 31:   Receivership, and a possible Take-Over   303

Chapter 32:   Sale to Sulzer Brothers   309

Chapter 33:   Hathorn, Davey: Part of Sulzer Brothers of Switzerland   317

Appendix 1:   Hathorn, Davey Order Books   327

Appendix 2:   Henry Davey's Patents   329

Index   333

List of illustrations

	John Fletcher Hathorn and Henry Davey	xii
1	Location Map of Hunslet, Leeds	xv
2	Lowfield Mine, Lancashire: Horizontal Compound Engine being erected	xvi
3	Newcomen Engine - Main Components	2
4	Drawing of a 60" Cornish Pumping Engine by Henry Davey	4
5	Early 19th Century Compound Engines	5
6	The Bull Engine	6
7	Location of the Sun Foundry	9
8	Advertisement for an Auction at the Sun Foundry: 1841	10
9	Charles Todd: Advertisement for the Sale of Locomotives	11
10	Sun Foundry: Auction of Plant and Machinery dated 1852	12
11	Carrett and Marshall: Road Locomotive	14
12	Carrett and Marshall: Patent Steam Pump and Water Lift	14
13	Carrett and Marshall: Portable High-Pressure Steam Engine	16
14	Carrett and Marshall: Hydraulic Organ Blower	17
15	Carrett and Marshall: Hydraulic Coal Cutting Machine	18
16	Indian State Railways: Small Pumping Engine and Boiler	25
17	Hydraulic Engine and Pump for the Clay Cross Company	26
18	Horizontal Compound Engine with a Flywheel	27
19	New Hartley Pit: Arrangement of Pumping and Winding Engines	28
20	A Portrait of Henry Davey dated 1891	31
21	Double Action Deep Well Pump with Vertical Boiler	33
22	Milford Haven Harbour: Engine House and the Accumullator	36
23	Sudbury Waterworks: Engine and Differential Gear	38
24	Davey's Differential Valve Gear (1875 Patent)	42
25	Davey's Differential Valve Gear (1899 Catalogue)	45
26	Chichester Waterworks: Rotative Compound Engine	46
27	Horizontal Compound Condensing Pumping Engine	49
28	Horizontal Compound Engine: Arrangement of Piston Rods	49
29	Solana Mine, Spain: Pumping Arrangements for a Compound Engine	50
30	Cheddars Lane, Cambridge: Horizontal Compound Pumping Engine	51
31	Navigation Colliery, S.Wales: Compound Differential Engine and Pumps	54
32	Navigation Colliery: Arrangement of Engine and Pumps	55
33	Erin Colliery, Rhur Coalfield, Germany: Pumping Machinery	57
34	List of Machinery manufactured by Hathorn, Davis, Campbell and Davey	59
35	Advertisement for the Compound Differential Engine	60
36	Advertisement for the Separate Condenser (Davey's Patent)	61
37	Croydon: Differential Compound Engine on 60 feet deep well	64
38	Public Notice 1878: Dissolution of Hathorn, Davis, Campbell and Davey	67

39	Chiswick: Differential Vertical Compound Condensing Pumping Engine	68
40	Chiswick: Vertical Compound Engine on a well	69
41	Davey's concept of parallel motion	70
42	The Company's Model of the Differential Compound Pumping Engine	72
43	Gold Medal Awarded to Hathorn, Davey and Company	74
44	A. Clack and Door Valve B. Harvey and West Valve	76
45	Beaconsfield Gold Mine, Tasmania: Clack Valve	76
46	South Durham Colliery: Plunger Pumps with Harvey and West Valves	79
47	Griff Colliery: Arrangement of Hydraulic Engines in a dip-heading	80
48	Sir Francis Level: Arrangement of the Pumping and Winding Engines	81
49	Sir Francis Level: Two vertical Cylinderof the Hydraulic Pumping Engine	83
50	Henry Davey's Meter for Recording Engine Performance	88
51	Burnt Fen: Plan and Section of the Pumping Station	90
52	Luton Waterworks: Compound Differential Condensing Pumping Engine	93
53	Yarlside Mine: Satellite Photograph showing Compound Engine-Bed	95
54	Kai ping, China: Side Elevation and Plan of the Pumping Engine	97
55	Kai ping, China: Elevation of Double Cylinder Winding Engine	99
56	Kai ping, China: Plan of the Double Cylinder Winding Engine	100
57	Shanghai Waterworks: The Engine / Boiler House	101
58	Brillador and San Carlos Copper Mine, Chile. Capstan and Man Engine	103
59	Solana Mine, Spain: Hathorn Davey Engine House	106
60	Solana Mine: Plan of the Engine House and Associated Structures	106
61	Solana Mine: Brick Built Block for the Hathorn Davey Engine	107
62	Lhuillier Colliery, France: The Surface Pumping Installation	109
63	Lhuillier Colliery: Illustration of the Hathorn Davey Engine	110
64	Davey's Differential Valve Gear fitted to a Cornish Beam Engine	111
65	Davey's Simplex Motor	114
66	Davey's Low Pressure Steam Motor	117
67	Willard's Advertisement for Davey's Low-Pressure Steam Motor	118
68	Horsepower, Sizes, Weight and Cost of the Davey Safety Motor	122
69	A larger version of the Davey Safety Motor	123
70	Hathorn, Davey - Balance Sheet dated 31st July 1888	129
71	Plan showing the locations of the Bradley, Moat and Stowheath Engines	132
72	The Bradley Compound Engine	135
73	Advertisement 1889: Invitation to tender for engine operation	138
74	Davey's Compound Beam Engine	142
75	Simplified Diagram showing how the Weston Super Mare Engine worked	143
76	Henry Davey's Design for the 'Great Tower of London'	146
77	Henry Davey's Patent for 'Game'	148

<h1 align="center">List of illustrations (continued)</h1>

78	Two Expansive Direct-Acting Accumulator Engines	150
79	Triple Expansion Pumping Engine for Brazil	152
80	Triple Expansion Pumping Engine for Brazil: Cylinders and Valves	154
81	Leeds City Waterworks, Headingley: Vertical Triple Expansion Engine	158
82	Leeds City Waterworks: Three Throw Ram Pumps	160
83	Trent Valley Pumping Station: Horizontal Triple Expansion Engine	163
84	The Marine Colliery, Ebbw Vale, S. Wales: Horizontal Compound Engine	165
85	The Marine Colliery: Hathorn, Davey Engine in Undercroft	166
86	Two Engines had a combined water lifting potential of 1000 horsepower	167
87	Miyanohara Pit, Miike Coalfield, Japan: Remains of Engine House Wall	169
88	Charters Towers, Queensland, Australia: Vertical Compound Engines	171
89	Charters Towers: Davey's Differential Gear	172
90	Cheddars Lane, Cambridge: The Oscillating Disc that operates the Pumps	173
91	Cheddars Lane: Diagram showing Engine and Pump Arrangements	174
92	Davey's Vertical Inverted Compound Cornish Pumping Engine	176
93	Waihi Gold Mine: New Zealand: Elevations and Plan of the Engine	178
94	Marriott's Shaft, Cornwall and San Rafael Shaft, Cerro Muriano, Spain	180
95	Waihi Gold Mine: The 110" Low-Pressure Cylinder	182
96	Waihi Gold Mine: Transportation of the Low-Pressure Cylinder	183
97	Advertisement 1902: Davey's High Pressure Cornish Engine	185
98	Sun Foundry 1891: Interior of the Erecting Shop	188
99	Sun Foundry 1899: New Machine Shop	190
100	Sun Foundry 1899: The Erecting Shop	191
101	Advertisement 1897: Non-Unionist Labour	194
102	The 'New' Hathorn, Davey Offices on Jack Lane	195
103	Table Showing Company Finances 1885 to 1901	198
104	Table Showing Foundry Costs between 1885 and 1901	199
105	The Foundry	200
106	Hugh and Isabella (Ella) Lupton	202
107	Chart Showing the Lupton family, and their positions in Hathorn, Davey	204
108	Reginald Hamilton Lupton	206
109	Hathorn, Davey and Company, Limited: Advertisment 1904	208
110	A Green's Economiser	210
111	Photograph of Henry Davey dated 1904	211
112	Portable Combined Petrol Motor and Centrifugal Pump	216
113	Beaconsfield, Tasmania: Transporting the Low-Pressure Cylinder	220
114	Beaconsfield: Fitting a Low-Pressure Cylinder End Plate	221
115	Beaconsfield: Interior of Grubbs Shaft Engine House	222
116	Beaconsfield: Compound Engine, Grubbs Shaft	222

List of illustrations (continued)

117	Beaconsfield: An Underground Pump Lodge	223
118	Beaconsfield: Salvaged Pumping equipment	224
119	Miike, Japan: Interior of the Manda Pumping Engine House	225
120	Llanover Collery, South Wales: Hathorn, Davey Engine in an undercroft	226
121	Hathorn, Davey Advertisement 1905 showing the Spotswood Engine	228
122	A. The Spotswood Engine 2011 B. Counter and Gauges	228
123	Rosario, Argentina: Elevation showing the Engine and Pumps	231
124	Montevideo Waterworks, Uruguay. One of the Triple Expansion Engines	233
125	South Africa: The Zwaartkopjes Pumps	234
126	Miike, Japan: Horizontal Triple Expansion Engine and Accumulator	235
127	Miike: Horizontal Triple Expansion Rotative Power Engine	236
128	Miike: The Underground Duplex Hydraulic Pumps	237
129	Nakama City, Japan: The Onga River Engine and Boiler House	239
130	London Museum of Water and Steam: Newmarket Engine	241
131	South Field Pumping Station: Hathorn, Davey Quotation for the Engine	242
132	Twyford Pumping Station, Hampshire: The remains of the 'Pitman'	243
133	Ilford Waterworks, Essex: Spur and Pinion Gearing to work the 'Pitman'	244
134	The Kesteven Asylum Engine	245
135	West SussexAsylum Waterworks Engine	246
136	Types of Two and Three-Throw Pumps	247
137	Plymouth: Centrifugal Pumps driven by Steam Cross-Compound Engines	248
138	The Stereophagus Pump	250
139	Davey's Patent 12,224 / 1914: Improvements in and Relating to Pumps	252
140	National Projectile Factory: Shell Presses	260
141	National Projectile Factory: Lathes	260
142	Hathorn, Davey: Financial Results 1920 to 1929	263
143	Henry Davey	266
144	Horizontal version of the Gargantua Pump	270
145	Vertical version of the Gargantua Pump	271
146	Indenture of Apprenticeship 1875	278
147	Hugh Luptons Indenture of Apprenticeship 1886	279
148	Salaried Staff below Works Manager, January 1932	280
149	Ahmedabad, India: Triple Expansion Engine	282
150	Mount Crosby, Brisbane Waterworks, Queensland: The Pump House	282
151:	Singapore: Triple Expansion Engines	284
152	Mount Crosby: Installation of a Triple Expansion Engine	286
153	Walton on Thames, England: Triple Expansion Engine	288
154	Mill Meece, Staffordshire: Compound Engine	290
155	Mill Meece: The two Pumping Quadrants	291
156	Unemployment in the Leeds area 1930 and 1931	293

List of illustrations (continued)

157	The Sun Foundry Erecting Shop in the 1930s	295
158	Electrically Driven Three-throw Ram Pump	298
159	Set of Three-throw Ram Pumps	299
160	The Sutton Poyntz Triple Expansion Engine	301
161	750 tons Hydraulic Forging Press	305
162	Hathorn, Davey: List of Debtors	311
163	Hathorn, Davey: Statement showing basic finances 1935 and 1935	314
164	Announcement of take-Over by Sulzer	315
165	Terkos, Turkey: The last steam engine manufactured by Hathorn, Davey	318
166	Hathorn, Davey: Drawing Office employees in 1947	320
167	Hathorn, Davey: Advertisement for a Draughtsman	322
168	Pump at Cheddars Lane, Cambridge	323
169	Brass name plate on the Twyford Pumping Station engine	324
170	Charters Towers: probably the oldest surviving Hathorn, Davey engines	326

Front Cover. Top Left: The Erecting Shop at the Sun Foundry 1899
Top Right: Horizontal Compound Engine
Cheddars Lane, Cambridge, United Kingdom
Bottom Left: Vertical Compound Engines
Charters Towers, Queensland, Australia
Bottom Right: Triple Expansion Engine
Spotswood, Melbourne, Victoria, Australia
(The Engineer 23rd August 1903 - Supplement)

Rear Cover. Top Brass Makers Plate, Cheddars Lane Engines, Cambridge
Bottom The Author and his wife at Nakama City, Japan visiting the Onga River Pumping Station. The person on the right with the Author is Manabu Hamada, who was responsible for the inclusion of the Pumping Station in the Japanese World Heritage Sites List.

The Founders of Hathorn, Davey

John Fletcher Hathorn   **Henry Davey**

'Mr Henry Davey was a ready-made engineer when he was born.'

(Quotation from the Mining Journal. 5th December 1891 p. 1366)

Foreword

The firm of Hathorn, Davey was at the junction of Dewsbury Road and Jack Lane in the heart of the Hunslet area of Leeds, once one of the most significant manufacturing districts in the north of England.

In a vote of thanks given to the author at a meeting of the Newcomen Society, Bristol on the 18th November 2010, John Anning MA (Cantab), CEng, FIMechE, described Hunslet and his memories of the Hathorn, Davey Company. A passage from the vote of thanks is reproduced below:

HATHORN, DAVEY: '*This famous family firm had a worldwide influence but I have a particular affinity with the engineering legacy of the Hunslet region of Leeds. Let me explain:- My father knew the Lupton family [Directors of Hathorn, Davey] and also being friends with the Alcock family arranged for me to have what is now called work experience, at the Hunslet Engine Company for two weeks during most of my school holidays starting in September 1952. I used to cycle from Headingley, leaving home at 6.30am for the Hunslet works and on turning into Jack Lane passed the main offices of SULZER HATHORN DAVEY on the corner of the Dewsbury Road. No traffic to speak of then, but the numerous tram tracks along the route were a constant hazard for the unweary. Even as late as 1953 the output from the Hunslet Engine Company was more of less equally split between diesel and steam locomotives with a large portion for export markets. It was an exciting place for a young apprentice engineer particularly in the erecting shop alongside Cancel Street.*

The 600 acres or so, or roughly a square mile bounded by the Dewsbury Road - Jack Lane - Grape Street and Hunslet Lane was the home of a number of medium sized but still largely family owned engineering companies all with a world-wide reputation and which all "punched above their weight" of which Hathorn Davey was just one. Sadly the region has now been decimated by motorways and business units and is hardly recognisable. Although thankfully some of the historic offices and works buildings still exist and are listed.

To give some idea, let us imagine we are travelling down the Dewsbury Road from Leeds City Centre 100 years ago we pass:- Mann's Patent Steam Cart Wagon Company at the Pepper Road Works on our left then almost next door was the Yorkshire Patent Steam Wagon Company who were closely associated with Clayton and Co, another Hunslet based engineering company and famous for gasometers and associated equipment. Then turning left into Jack Lane almost opposite the Sun foundry, we pass the offices of Hathorn Davey, then next door is J & H McLaren on the right at the Midland Engine works, makers of road locomotives and traction engines then Manning Wardle, Locomotive and steam engine builders at the Boyne Engine Works a few yards up Jack Lane on the left and next door is the Hunslet Engine

The Author has compatable memories to John, recalled from his childhood in the 1950s, when it was not unusual to see low-loaders laden with heavy machinery or locomotives, produced by the engineering companies on the outskirts of Liverpool slowly moving along the East Lancashires Road, en route to the docks for export.

The passing of such companies marked the end of an era, when British heavy engineering had a worldwide reputation for quality, reliability and efficiency.

The history of Hathorn, Davey described in this book is a detailed historical account of just one of the many successful engineering companies in the Hunslet area of Leeds (Figure 1).

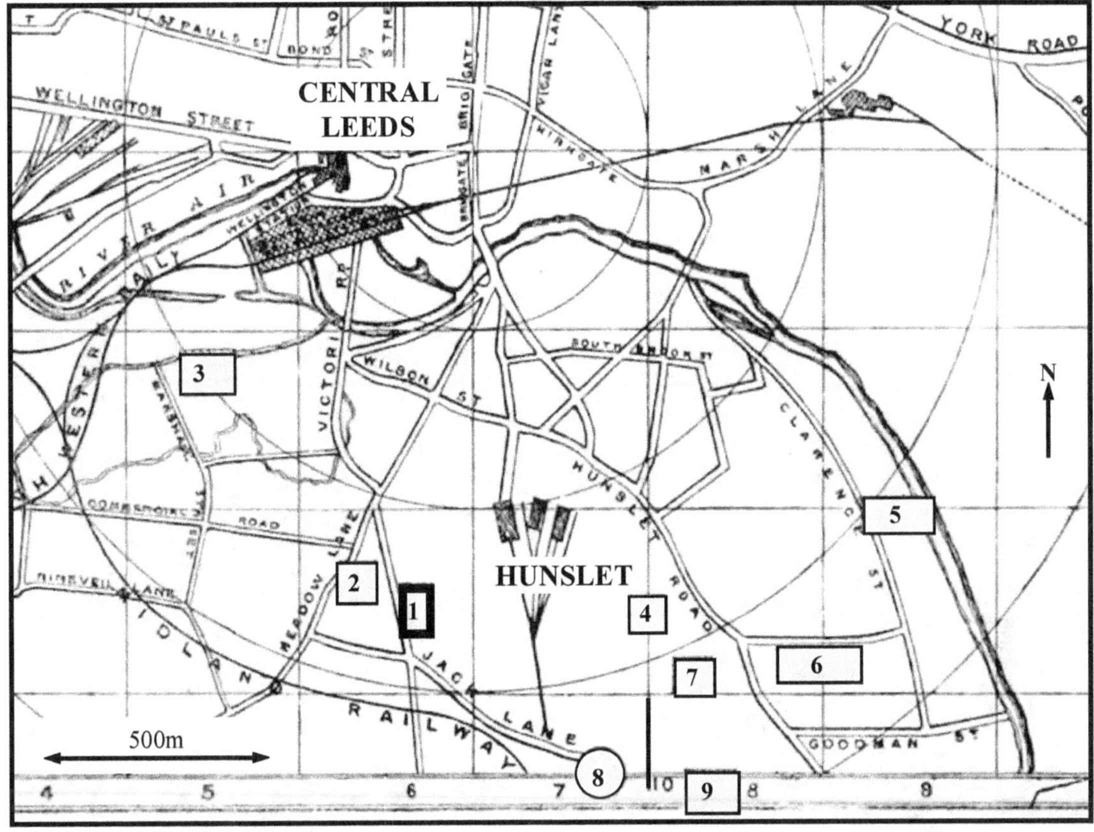

Figure 1. Plan of Leeds South, showing the location of principal engineering firms in1882. Based on a description in *The Engineer*, 21st July 1882, p.39 [GG].

Key:
1. Hathorn, Davey and Company — (Pumping machinery)
2. Joshua Buckton and Company — (Machine tools)
3. Smith, Beacock and Tannet — (Machine makers)
4. Fowler and Company — (Traction and railway engines)
5. Taylor and Brothers — (Railway equipment)
6. Tannett and Walker — (Water pumping engines)
7. Kitson and Company — (Railway locomotives)
8. Manning, Wardle and Company — (Railway locomotives)
9. Hunslet Engine Company — (Railway locomotives)

Figure 2. A Hathorn, Davey Horizontal Compound Pumping Engine being erected in 1897 at the Lowfield Iron Mine, Ulveston, Lancashire
(The Engineer, 30th July 1897 p.110 [GG])

Preface

In the 1970s I was a geologist working in the deep coal mining industry of West Yorkshire. One of my hobbies was collecting old mining postcards, and it was that hobby that indirectly awakened my interest in Hathorn, Davey.

In 1975, when rummaging through boxes under a table at a 'flea market', Queens Hall, Leeds, looking for postcards, I came across a slim book with a red cover. It was a catalogue dated 1899 for Hathorn, Davey and Company, Leeds entitled 'Pumping Engines for Waterworks.' I bought it for £4.50. I did not know at the time the journey this purchase would take me on, or the interest that journey would arouse.

Prior to this discovery, I may have unknowingly encountered Hathorn, Davey as early as 1955. Occasionally I was taken on holiday to London and like many of my generation I have happy memories of the Science Museum, South Kensington, and particularly of the myriad of detailed engineering models on display. They included working (by turning a handle) railway locomotives and stationary steam engines. I would later discover that Hathorn, Davey did manufacture several models for exhibit at the Science Museum. I was equally oblivious to the fact that I was close to a Hathorn, Davey engine every time I visited relatives in Everton, Liverpool. They lived adjacent to the Aubrey Street Reservoir, Liverpool where a Hathorn, Davey engine was located.

My first known meeting with Hathorn, Davey 'in the field' was in 1978 on a visit to Spain. I had found evidence that a large number of Cornish beam engines had been exported to the Linares lead mining area, Andalucía, and so I wanted to see if any engine houses remained. They did, and Linares is now recognised as having the largest grouping of Cornish type engine houses outside the United Kingdom. I also thought that there was a possibility that the Cerro Muriano copper mine, just to the north of Córdoba might also have an engine house as it had been worked by a British company that had connections with Linares in the early 1900s. I did find an unusual brick-built engine house there that had once housed a large Hathorn, Davey Vertical Compound Pumping Engine. However, we did not linger long, as the mine was adjacent to a large army base.

As my interest in mining history developed, and took on a more global perspective, it became apparent how significant a role British technology played in the development of the British Empire and beyond. Railway systems were important in opening up countries. They frequently functioned as supply routes to mines, and conduits for mineral export. In the 19th and early 20th centuries the British engineering base was vast, and neither distance nor remoteness were barriers for opening up an area. A fact perhaps exemplified by the steam ships, manufactured in Hull and Glasgow that are still sailing on Lake Titicaca in the Andes, the highest navigable lake in the world.

Hathorn, Davey played its part in this worldwide bonanza. In addition to a healthy home market (Figure 2) it sold pumping engines to British companies operating in India, Australia, New Zealand, Japan, and South America for example, and exported them through the ports of Liverpool, Hull and London.

By the 1980s, Hathorn, Davey had got my full attention, and any article relating to the Company found during other research was filed away for future reference, with the ultimate intention of writing this book about the Company.

In addition, I discovered that a significant amount of archival material had been deposited at the Leeds Record Office, and it was whilst looking through this material that I first came to realise just how significant the Company had been in the field of pumping technology.

The archives were then housed at Sheepscar, north Leeds. Sulzer Brothers, the company that had taken over the Hathorn, Davey Company in 1936, had deposited them there. The archives were in two parts, i) The Hathorn, Davey Collection, that consisted of documents relating to the manufacturing side of the company at their works, the Sun Foundry and, ii) The Dibb-Lupton Collection, a collection of legal documents relating to partnership agreements and finance.

For technical information about engines, the former collection was important. It consisted of a sequence of Order Books covering the period 1852 to 1920. However, as the Hathorn, Davey partnership was only established at the Sun Foundry in 1870s, the first two order books related to a company that occupied the Sun Foundry prior to to that date.

Once an order number was known, it could be cross-referenced to a sketchbook reference. The sketchbooks, also in the Leeds Record Office, provided a numbered list of all the drawings pertinent to an order. The sketchbook lists included specific drawings as well as those for standard parts. The sketch number could then be cross-referenced to an actual drawing in the archives, if it still existed. The drawings are very detailed engineering drawings, with dimensions etc, and illustrate how the various parts are related to each other, as well as their installed location and setting.

When I first went through all the Order Books, with Margaret (Boo), my wife, in the 1990s we made pencil notes of all the engine orders, which I subsequently transposed to a spreadsheet. Since then I have re-examined the Order Books several times and photographed their indexes, together with any specific orders of interest.

In the new century, I made several trips to Australia to look at Hathorn, Davey engines, and more recently to the island of Kyūshū, Japan to look at sites where Hathorn, Davey engines were employed to dewater the Miike Coalfield.

My main source of information has been the Order Books. They provide a wealth of information on the various engines, quite often with very detailed specifications, and have provided a valuable pointer towards other sources of information, local council records for example, who would have purchased engines and pumps for sanitation schemes. Mining companies were also major buyers. Initially orders were from coalmines in the north of England and eventually they came from mines as far away as Australia and New Zealand.

The history of the Hathorn, Davey Company however, can really be regarded as the story of one man, Henry Davey. His innovations maintained the Company at the forefront of British pumping technology. His numerous patents provided a rich source of information, as well as many contemporary publications, for example, The *Mining Journal* and *The Engineer*. There are also other published works that cover steam technology in detail, including two editions of a book by Henry Davey published in the early 20th century.

I presented my first paper about Hathorn, Davey at the Eighth International Mining History Congress held at Truro, Cornwall in 2009. Perhaps this was an appropriate location to declare my interest in this remarkable company, as the majority of engines produced by Hathorn, Davey superseded the less efficient Cornish beam-pumping engine. Hathorn, Davey engines were built in the twilight years of steam pumping technology. Their engines bridged the gap between the old technology, and the introduction of pumps driven by electricity and petrol, with which the company briefly flirted, but never fully succeeded, until they were taken over by Sulzer in 1936.

Following further talks given at Bristol, Manchester and Birmingham various people came forward, who had information about Hathorn, Davey, particularly members of the International Stationary Steam Engine Society. I was also put in touch with the Lupton family, who were the last Directors of Hathorn, Davey prior to it being taken over. As a consequence of this introduction I was given a number of photograph albums that I deposited in the Leeds Record Office. However, they only contained a few photographs of steam engines, some of which are reproduced in this book.

The London Museum of Water and Steam also hold important archival material relating to Hathorn, Davey and this proved to be an invaluable source.

I have had two fundemental dilemas whilst researching the history of the Company.

The first was with the Company's name. Initially, the Company evolved through a series of Partnerships. In those instances the use of commas in the Partnerships title was straightforward. Once the Company was titled Hathorn, Davey it was the usual practice to show a comma between the two names, as in Figure 109. However, in later useage of the name, the comma was often missing, as in the nameplate shown in Figure 169, and it is now common practice to refer to the Company as 'Hathorn Davey.' For the sake of consistency I try and refer to the Company as 'Hathorn, Davey.'

The second was at what year to take any history up to. There were various options:

      i)   The year when Sulzer Brothers took the Company over (1936).
      ii)   The year when the last steam engine was manufactured (1937).
      iii)   The year that Sulzer wound-up Hathorn, Davey (2016).

Although Hathorn, Davey was mainly a dormant Company in its later years, for historically consistency, I have chosen the year 2016. This was the year when the name 'Hathorn, Davey' was removed from the Register retained by Company House, London.

Over ten years after my declared interest in the firm's history, with further talks, more papers, and numerous site visits under my belt, I now feel it it time to put that knowledge into print. This book predominently covers the period from when John Fletcher Hathorn and Partners took over the Sun Foundry in 1872, until the year the Company was wound-up, 2016. It is a history of the Hathorn, Davey Company of Leeds and its contribution to the age of steam. It is published at the end of a year that will be remembered for so many reasons both personally, and as a nation.

Robert W. Vernon
Bredon, Tewkesbury, England
Christmas 2020

HDRV1901@Outlook.com

References

The primary sources of material for this book are:

1. The collection of documents lodged as the Hathorn, Davey Collection [HDC], and the Dibb Lupton Collection [DLC], held by the West Yorkshire Joint Services - Archives, Nepshaw Lane South, Gildersome, Leeds LS27 7JQ.
2. The Hugh Ralph Lupton Collection lodged at the London Museum of Water and Steam at Kew. [LMWS]
3. A Brief History of Hathorn, Davey by Hugh Lupton [HL] given to the author by Robert Cox, Bridport, Dorset.
4. Records relating to John Fletcher Hathorn at the Scottish National Archives [SNA], Edinburgh.
5. The *Mining Journal* [MJ], held at the National Coal Mining Museum, Wakefield.
6. *The Engineer*, uploaded to the Internet by *Grace's Guide to British Industrial History* [GG] has been an invaluable source of information and can be accessed at:
https://www.gracesguide.co.uk/The_Engineer

Sources are identified and referenced in the text or in illustration captions by the appropriate initial shown in the square brackets above and, together with other references, will be numerically listed at the end of each chapter.

Acknowledgements

There are many people who have helped in this work. I sincerely apologize if I have missed out anyone.

Firstly, I would like to thank the Archivists at West Yorkshire Archives Services, Sheepscar, Leeds, and more recently at the West Yorkshire Joint Services - Archives, at Gildersome, Leeds.

Archivists, Librarians, or similar administrators, at Companies House, London; the National Coal Mining Museum for England, Wakefield, Yorkshire; the Northern England Institute of Mining and Mechanical Engineers, Newcastle upon Tyne; the National Archives, Kew, London; and the National Archives for Scotland at Edinburgh, Dumfries and Galloway Local Archives, and Mike Gill, Recorder, Northern Mine Research Society, England

The archives at the London Museum of Water and Power, at Kew were also very informative, and I would like to acknowledge the help provided by their Registrar Toby Sleigh and John Porter, and previously Bryce Caller.

In addition to the pumping engine at Kew, elsewhere in the UK, I visited the Mill Meece pumping station, Staffordshire; Cambridge Museum of Technology Cheddars Lane, Cambridge where Alan Denney has been most helpful in supplying information, and Twyford Pumping Station, Hampshire (Alan Down).

In Australia, I visited Spotswood (Scienceworks Museum) at Melbourne, Victoria, and would like to thank Matthew Churchward, Rohan Lamb (Spotswood) and Karen Jakubec, Loans Manager, Museum Victoria; at the Beaconsfield Mine and Heritage Centre, Tasmania. I would like to thank the help provided by various curators including: Prue Wright, Sally Kleine, Julieanne Richards, and the Board of Trustees. In addition I received help from Annaliese Claydon Archivist, State Library and Archive Service, Libraries Tasmania. At Charters Towers, Queensland, I would like to thank officials of the Town Council for their assistance in taking me to see the engines.

For New Zealand (not visited) I am pleased to acknowledge help provided by Keith Rimmer and Peter Short, who provided information about the Waihi engines.

Barry Gamble, who I would like to gratefully thank, initiated my Japanese visit. In Japan, I would like to acknowledge the help provided by Koko Kato and Noriko Hanada of the National Congress of Industrial Heritage for my visit to the Miike Coalfield, and to Manabu Hamada, Nakama City, on the Island of Kyūshū. I would also like to mention the late Stuart Smith with whom I first had discussions about the Hathorn, Davey links with Japan.

Assistance on the Hathorn, Davey links with South American orders have been provided by Frank Vicencio López (Brillador Mine, Chile) and Dennis Cahill (Montevideo Pumping Station, Uruguay).

In France I recognise the assistance provide by Lucile Decombe, Manager, Musée de la Mine, Puits Hély d'Oissel and Bernard Deschamps, for their knowledge about the Charbonnage des Bouches-du-Rhone Coalfield, near Marsailles.

In Spain, my good friends Antonio, Pepe and Paco of the Colectivo Proyecto Arrayanes, Linares, (Jaén), and Juan Manuel Cano Sanchiz (Córdoba) have always found time to respond to my queries, and offer assistance. I would also like to especially thank Antonio Cabrera (Córdoba) for arranging a visit to the Solana mine site, and later our collaboration that produced a joint prize-winning paper on the subject.

In the United Kingdom, I would like to thank Linda Reed who provided an insight into Henry Davey's associations with George Bower of St.Neots.

Members of the International Stationary Steam Engine Society, kept me informed of matters relating to Hathorn, Davey, including Chris Hodrien, Paul Stephens, Brian Hillsdon and John Hole, for observations and information. I am indebted to Brian for taking the initiative in the 1970s of copying information from the now missing Hathorn, Davey Order Books, and to John for listing Henry Davey's patents. I must also give special thanks to Chris Allen who meticulously read the book and applied his extensive experience and knowledge to produce a list of very constructive comments.

There are many others who must be thanked, including relatives of former Hathorn, Davey employees. Christine Hart (née Auty) provided information about her father who worked in the drawing office, and Mike Beevers whose father retained copies of Hathorn, Davey catalogues and other memorabalia, and of course Paul King and Alan Lupton, members of the Lupton family.

Others who have assisted along the way include Robert Cox, Graham Roberts, John Anning, John Wellington and Jon Knowles.

I must also thank my wife, Margaret (Boo) who has been my companion on this adventure, and who has helped me in so many ways.

Last but not least, I must thank, posthumously John Fletcher Hathorn, for taking the initiative to purchase the Sun Foundry, and Henry Davey for the many innovations he brought to the age of steam, without which I wouldn't have had the enjoyable experience of researching this book.

Definitions and Abbreviations

There are various expressions used in this book to describe different types and components of a steam engine

1. Terms used to describe a steam engine

Engines can be:
Single-acting engine: Steam acts on only one side of the piston.
Double-acting engine: Steam acts alternately on both sides of the piston.

The number of cylinders can vary;
Single Cylinder or 'Simple' engine: The steam enters the cylinder at high pressure, expands as it pushes the piston through its stroke, and is then exhausted to atmosphere as the piston returns,

Two cylinders or 'Compound' engine: The exhausted steam from the first, or 'high pressure' cylinder is reused in a second 'low pressure' cylinder where it expands further as it pushes the low pressure piston through its stroke.

Types of cylinder configuration for a Compound engine include:

Cross-compound:	the cylinders are side by side
Tandem compound:	the cylinders are end to end, driving a common connecting rod
Telescopic-compound:	the cylinders are one inside the other
Angle-compound:	the cylinders are arranged in a vee (usually at a 90° angle) and drive a common crank

Three cylinders or 'Triple Expansion' engine: The steam will pass into an intermediate cylinder between the high and low-pressure cylinders. The cylinders increase in size; and the steam pressure decreases.

In all types of engine, the steam from the final cylinder exhausts to a condenser or to atmosphere.

Horizontal engine:	The long axis of the cylinders is horizontal
Vertical engine:	The long axis of the cylinders is vertical
Inverted:	The cylinders are vertical and the piston rod comes out at the bottom, usually to work pumps in a pit beneath the engine.

2. Engine Parts

Air Pump: A pump for removing condensed steam / water from a condenser.

Barring Engine: A small engine that is used to turn the main engine to a favourable position from which it can be started. Sometimes the barring engine engages with teeth on the rim of the flywheel.

Condenser: A device to convert steam to water.

Con Rod: A rod that connects the crosshead to a crankshaft or other mechanism.

Crank Shaft: Connected to a con rod, it converts the reciprocating motion to circular motion.

Crosshead: A sliding metal block connected to the end of a piston rod.

Cylinder: A cylindrical chamber for a piston.

Flywheel   A heavy revolving wheel in a machine used to increase the machine's momentum.

Gudgeon: A socket-like, cylindrical fitting attached to one component to enable a pivoting or hinging connection to a second component.

Pinion: A gear with a small number of teeth, especially one engaging with a rack or larger gear (spur wheel).

Piston: Circular moving component that slides up and down in a cylinder.

Piston Rod: A rod that joins a piston to the crosshead.

Plummer Block: A mounting block for a quadrant.

Quadrant: Connected to a con rod, it converts a horizontal to vertical reciprocating motion.

Valve: A device for controlling the passage of steam in an engine, or fluid in a pump.

3. Abbreviations

All costings are pre-decimal and are given in Pounds (£), shillings (s) and pence (d)
12 pence = 1 shilling. 20 shillings = £1 (Sterling).

Cylinder, stroke, pump and boiler sizes are given in feet (') and inches (")
12 inches = 1 foot

Depths are given in feet.

rpm	=	revolutions per minute
psi	=	pounds per square inch
gpm	=	gallons per minute
gph	=	gallons per hour
gpd	=	gallons per day

1. Introduction

Many books have been written about steam pumping technology, and the use of steam power is well documented. Currently, research on the subject is tending to concentrate on the early phases of its development in the 18th century, and rightly so, as this is a period that is poorly documented, and scarcely examined. There are numerous books that relate to later periods in the development of steam power particularly the Cornish beam engine.[1] There are also a variety of publications about the companies that manufactured them, for example Harvey's of Hayle, and the Williams' Perran Foundry, etc.[2] In view of those facts, the early development of steam power will get the briefest of mentions, but sufficient to place the Hathorn, Davey Company in the time line for the 'Age of Steam.' Similarly, any comment on steam technology will be limited to those that may be pertinent to the engines that the Hathorn, Davey Company manufactured, and mostly designed by their innovative engineer Henry Davey.

There were many contenders in the race to produce the first steam engine. However, it was Thomas Newcomen's collaboration with Thomas Savery (inventor of the 'miners friend - a simple method of pumping mine water) in the early 1700s that became the frontrunner.[3] Together they produced an atmospheric engine that revolutionised steam-pumping technology, and is usually referred to as a Newcomen engine. Despite its inefficiency, the engine worked, and it's use spread rapidly throughout the British mining industry. The technology was simple and is illustrated in Figure 3. Steam was introduced into a cylinder below a piston, which was then condensed by a jet of cold water. Low pressure was thus created below the piston that was sufficient for the piston to be forced downwards by atmospheric pressure. A vertical piston rod from the top of the piston was connected by chain to an overhead beam balanced on a fulcrum and that end of the beam was pulled down. The other end of the beam was connected to weighted pump rods that ran vertically down a mine shaft. The pump rods were raised by the downward action of the piston. The pressure in the cylinder was then normalised and the weight of the pump rods (or spears) pulling on the beam raised the piston, ready for the cycle to be repeated. This produced an up-down motion that worked the pumps in the shaft.

A major difficulty that Newcomen had to overcome was to ensure that the piston rod stayed vertical. Obviously if the piston rod had been connected directly to the beam, the rod would oscillate from side to side. This would create wear on the piston and break the piston's seal with the side of the cylinder. Water on top of the piston maintained the seal. To alleviate this problem, a chain was arced (or curved) over the end of the beam and the chain was connected to the end of the piston rod. A similar arrangement at the shaft end of the beam maintained the verticality of the pump rods.

At this time shaft pumps consisted of a column of simple valves in a vertical pipe that raised the water in stages. Prior to steam technology this type of pump was often operated by a waterwheel, a method used to dewater mines since the late medieval period, and in use into the late 19th century. Whilst inefficient, the source of power, i.e. water, was free, unlike the expenses incurred by steam technology fuelled by coal.

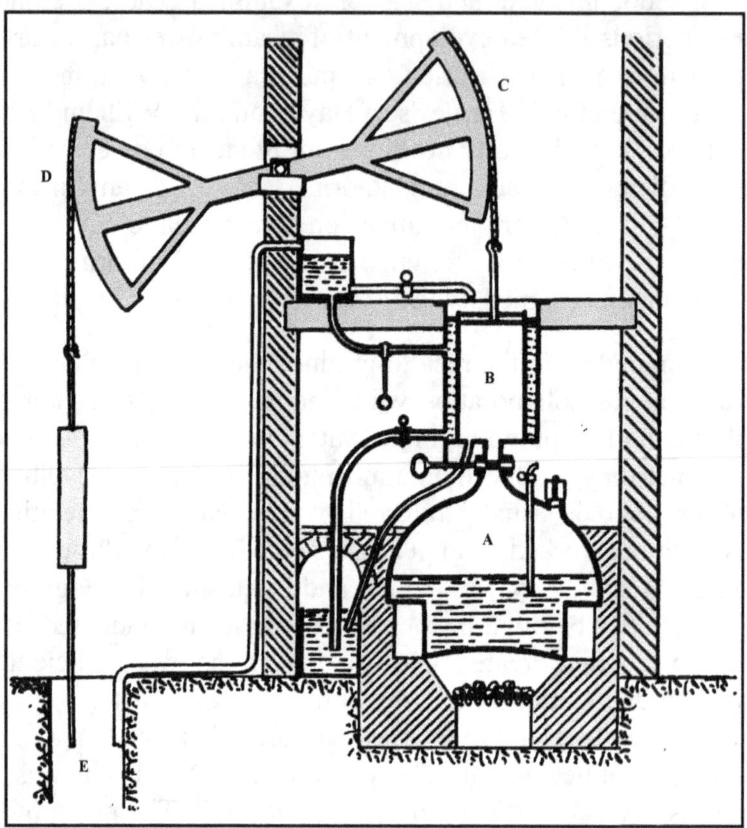

Figure 3. Newcomen Engine: Main Components. A= Boiler; B= Cylinder with piston; C - D Beam. Arced ends to the beam ensure that the chain connection is aligned vertically with the piston rod and the shaft pumps. E = Shaft
(Davey, H. 1900. Principles, Construction and Application of Pumping Machinery. 1st Edition. p.6)

The general arrangement first seen in the Newcomen's engine formed the basis of pumping technology for the next 200 years. The Newcomen engine did undergo various modifications. In the early engines, the steam cylinder was fixed immediately on top of the boiler within the engine house. In later models, steam was introduced into the cylinder via a pipe from an external boiler.

It was some 50 years after Newcomen's invention that James Watt successfully made the engine more efficient, e.g. he used steam pressure to activate the piston. In addition, condensing the steam in the cylinder meant that energy was being lost to reheat the cylinder. Watt overcame this problem by adding an external condenser. The condensed water was removed by an air pump. The combination of condenser and air pump was later to become components of the Compound engines designed by Henry Davey.

Watt had previously experimented with double-acting engines, which introduced steam under pressure alternatively on either side of the piston. He patented his innovations in 1781 -1782.

Other significant modifications made by Watt included the parallel motion in 1774, which eliminated the wooden arc and chain system used to keep the piston and pump rods vertical. Watt also devised a sun and planet gear system in 1781, which converted the reciprocating motion of the beam to circular motion that was used primarily in single cylinder beam winding engines. Both inventions were used in the engines that Watt manufactured with his business partner Matthew Boulton. These are usually referred to as 'Boulton and Watt' engines.

By the time the firm of Hathorn, Davey was established in 1872, the number of single cylinder beam engines being manufactured in Britain was starting to wane. In addition Hathorn, Davey specialised in pumping engines, rather than engines used for winding. Henry Davey, however, devised his own form of parallel motion, for use with vertical compound engines pumping directly above a well.

The valve system employed in both the Newcomen, and Boulton and Watt engines was relatively similar, and provided an open and close sequence for introducing steam and exhausting 'spent' steam to the condenser. The various rods, which operated the valves, hung vertically from the beam, and therefore moved in time with the movement of the piston. They were adjustable and the operation of the valves could be finely tuned. They incorporated a cataract valve, a form of governor, which could be used to regulate the speed of the engine. In theory such arrangements should operate 'like clockwork' but the author knows from his own supervised experience of operating the 200 year old Boulton and Watt engine at Crofton, Wiltshire, that the system required constant visual and manual attention, to ensure that everything ran smoothly.

Further improvements were made to the beam engine in the 19th century. Richard Trevithick, introduced the concept of using high-pressure steam (above atmospheric pressure). Initially, high-pressure steam was introduced above the piston, and on exhaust it was returned to the other side of the piston, below atmospheric pressure. Eventually, this method became known as the Cornish cycle. The cycle, and its variations were utilised by Henry Davey in some of his engine designs.

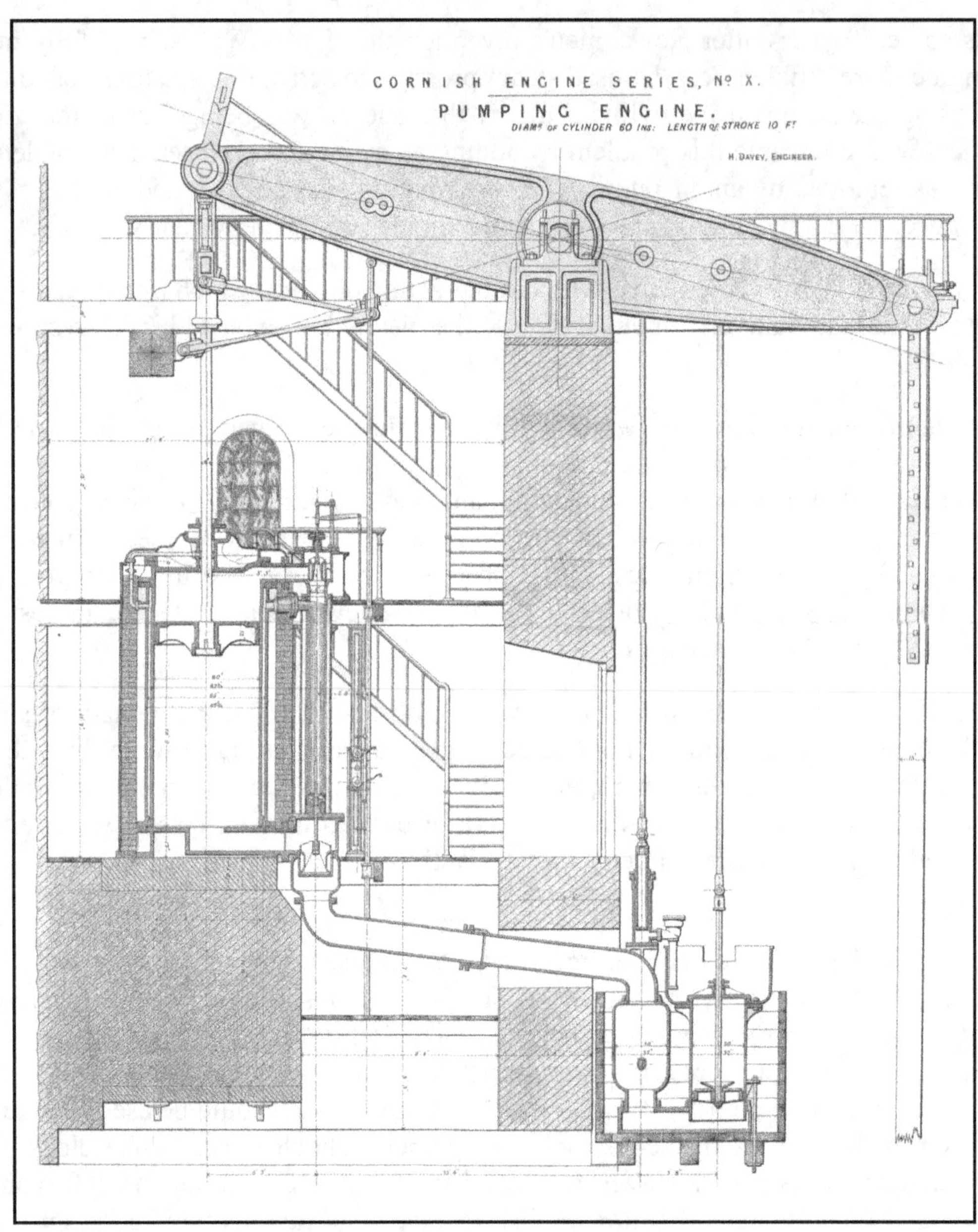

Figure 4. Drawing of a 60" Cornish Beam Pumping Engine by Henry Davey
(The Engineer, 1st December 1871. (Courtesy of Robert Cox))

Figure 4 shows a drawing of a Cornish Beam Engine by Henry Davey, published in *The Engineer* 1871. One costly problem associated with the traditional Cornish beam engine was the damage that could arise if the load was suddenly removed. This usually occurred if the joint connecting the rods to the end of the beam broke, causing the piston under steam pressure to accelerate quickly and smash down into the base of the cylinder. The result would frequently damage the engine. In the 1870s Henry Davey patented the Differential Gear, which if the engine behaved erratically, allowed steam to cushion the piston, and switch off the steam.[4] Not only did Davey fit the Differential Gear to the Compound Engines he designed, the gearing was purchased by others and fitted to other types of stationary steam engine, including the Cornish Beam Engine.

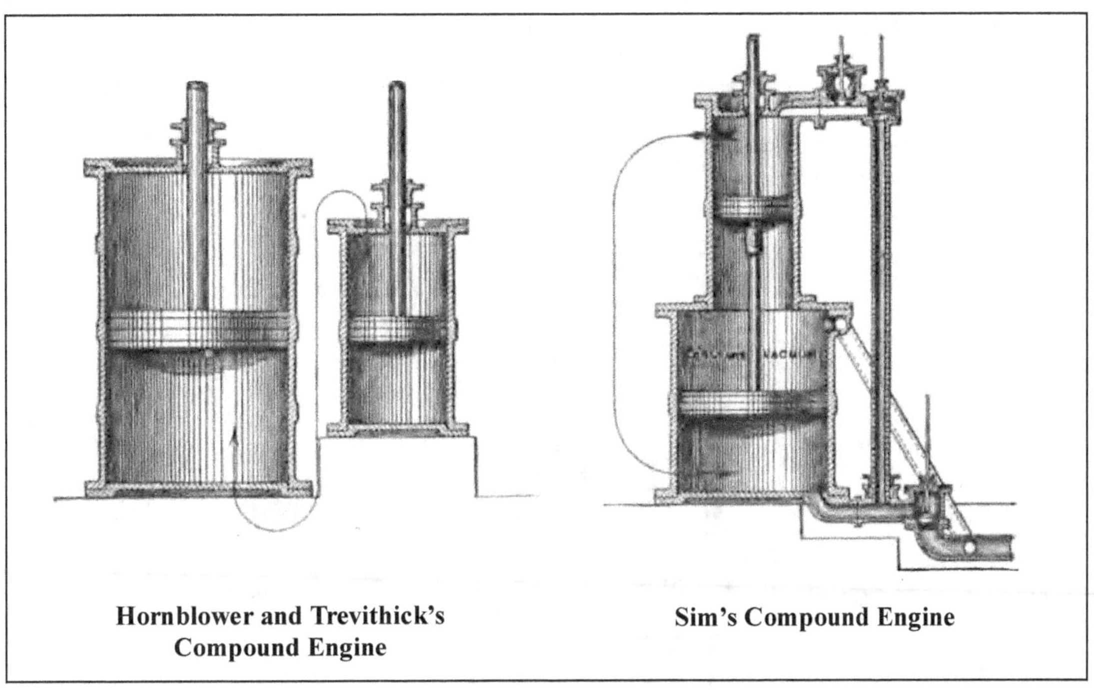

Figure 5. Early Nineteenth Century Compound Engines
(Davey, H. 1900. Principles, Construction and Application of Pumping Machinery.
1st Edition. p.20)

Other ideas were applied to the beam engine to try to improve efficiency. Earlier in the 19th century, experimentation by Jonathan Hornblower and Richard Trevithick had resulted in a two cylinders engine with the steam expanding successively in each cylinder. The steam was compounded, a principle pursued further by James Sims. The engines are shown in Figure 5. Both engines had two cylinders, high (small cylinder) and low pressure, but worked on different cylinder configurations. However, neither engine was as efficient as the single cylinder Cornish Beam Engine so they never gained popularity. Later Henry Davey improved the concept, and patented his own efficient, and very successful, version of the Compound Engine using both cylinder configurations.

Edward Bull examined cylinder configuration further and inverted the pumping engine. The Bull Engine (Figure 6), as it became known, had an inverted steam cylinder over the shaft top to operate the pump rods. However, such a method could clutter a shaft, so one variation was to have the pump rods operated by a beam located in a pit beneath the engine house. It was then possible for the inverted engine cylinder to be positioned so that the piston rod was connected to the beam away from the shaft. This was also a method that could be used to alter the stroke of the shaft pumps.

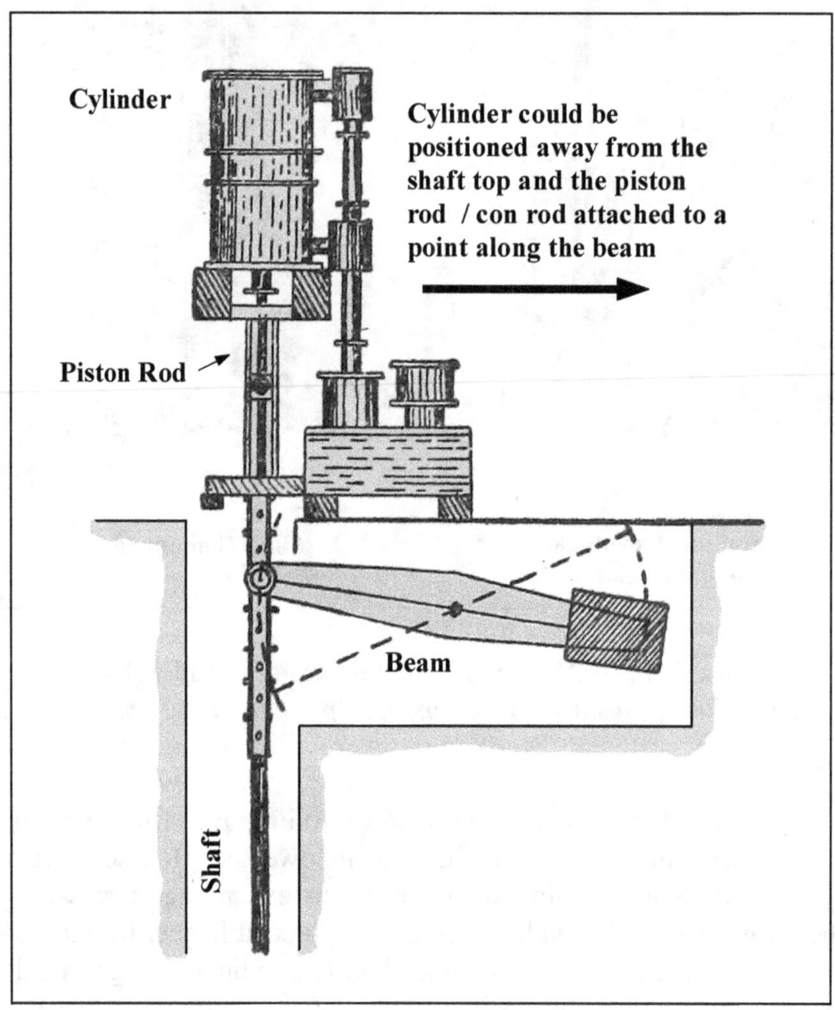

Figure 6. The Bull Engine
(Modified from: Kerr, G. 1905. Practical Coal Mining. Charles Griffin & Co. p.319)

Henry Davey combined both the concepts of the Compound and Bull Engines and produced the Vertical Compound Engine that consisted of two inverted cylinders located above, and at each end of the beam. A Vertical Compound Engine with a 110-

inch low-pressure cylinder for the Waihi Gold Mine, New Zealand was the largest engine produced by the Hathorn, Davey Company.

As previously mentioned, the success of Hathorn, Davey, certainly in the late 19th Century, can be attributed to the innovations of Henry Davey who joined the partnership in 1873. In his formative years Davey had gained much experience in engine design and manufacture, and their applications. The combination of the Horizontal Compound Engine with Davey's Differential Gear was a recognised success that endured well into the 20th Century. The Vertical Inverted Triple Expansion Engine, much favoured by water companies, superseded the Compound Engine. After World War I, fewer engines were made, and the last order for one was placed with the Company in 1937.

Hathorn, Davey also manufactured pumps. In the 20th Century, despite trying to operate pumps with other power sources, for example the petrol engine, the Company never quite established a large market for those products. After World War I, they specialised in several types of sewage pumps that had the ability to disintegrate larger particles. When Hathorn, Davey got into financial difficulty, partly the result of the great depression, their specialisation in pumps, made them an attractive proposition for a take-over by the Swiss firm of Sulzer Brothers in 1936. Sulzer retained the name of Hathorn, Davey until 2016.

Every story needs a framework, and for the Author, that proved to be the Hathorn, Davey Order Books that Sulzer deposited in the Leeds Record Office sometime in the 1970s. Unfortunately, only the Order Books that cover the period up to 1920 were deposited. However, thanks to the foresight of Brian Hillsdon who made notes from some of the post-1920 Order Books in the 1970s, together with contemporary reports from newspapers, it has been possible to determine what post-1920 engines were manufactured by the Company. The known and unknown Order Books are listed in Appendix 1.

The Order Books are important, not only because they are a chronological record of the Company's history, but for the fact that they generally give a very detailed specification, as well as costing and delivery dates for engines. Where orders are mentioned in the following chapters, the order number will be shown together with the date of the order (where known), followed by the type of engine and cylinder sizes and stroke. Pump diameters and stroke will also be given where possible.

Throughout his lifetime, Henry Davey filed many patents (Listed in Appendix 2) and they are cited where appropriate, in addition to references to the many papers that Davey published in the journals of professional bodies.

Many of the illustrations of engines have been taken from contemporary sources, predominantly *The Engineer* (Courtesy of Grace's Guides), and the *Mining Journal* and other similar weekly publications.

Hugh Lupton, one of Directors of Hathorn, Davey wrote a brief history of the Company in 1940 that outlined technological changes, and covered other aspects like wages, job functions, etc.

The material for the period after World War I comes from two sources. i) Information deposited in the archives of the London Museum for Water and Steam, Kew relating to the time when Hugh Ralph Lupton was a Director of Hathorn, Davey. ii) Photograph albums compiled by Reginald Hamilton Lupton (Courtesy of the Lupton Family) who was once the Commercial Director of Hathorn, Davey.

The Hathorn, Davey Company was at its peak in the transitional period between the waning of the Cornish beam engine and the introduction of electric pumps, and many examples of their engines still survive today. The Author has visited some of the surviving engines and sites.

It must be emphasised that this work is not intended to be a detailed account covering steam-pumping technology. There are many contemporary works on that subject. Rather, it is the Author's intention to produce a meaningful work that details the history of the Hathorn, Davey Company, and in addition provides details of some of the many fine Hathorn, Davey mine and waterworks pumping engines that once operated in Great Britain and elsewhere in the world.

Chapter 1 References

[1] Barton, D.B. 1965. *The Cornish Beam Engine*. D. Bradford Barton Ltd. Truro, Cornwall

[2] Williams' Perran Foundry Co., 1974 *Illustrated Catalogue of Pumping and Winding Machinery.* (Reprint) Trevithick Society and Ironbridge Gorge Museum Trust

[3] A lithograph dated 1712 of the Steam Engine near Dudley Castle states that it was invented by, 'Capt. Savery and Mr. Newcomen.' *The Engineer*, 28th November 1879. p.400

[4] Patent 277. Dated 1st February 1871. Steam and Pumping Engines. Filed by Henry Davey, St. Neots, Huntingdon

2. The Sun Foundry

The partnership that became Hathorn, Davey established their business at the Sun Foundry, Leeds in 1872. The foundry was located on the northeast side of the intersection of the Dewsbury Road and Jack Lane, Hunslet, to the south of Leeds City centre.

Up until the 1820s the land on which the foundry was to be built was under two ownerships, William Smithson, (Esquire) of Heath, near Wakefield, and the Milnes family of Fryston Hall, near Castleford.[1]

It is known that Smithson sold an estate known as Dunwell Ings, Hunslet, Leeds to James Leather (Gentleman) of Beeston, Leeds, and Robert Ritchie, also of Leeds on the 10th and 11th June 1823. It is believed that this land formed part of the site later to be occupied by the Sun Foundry.

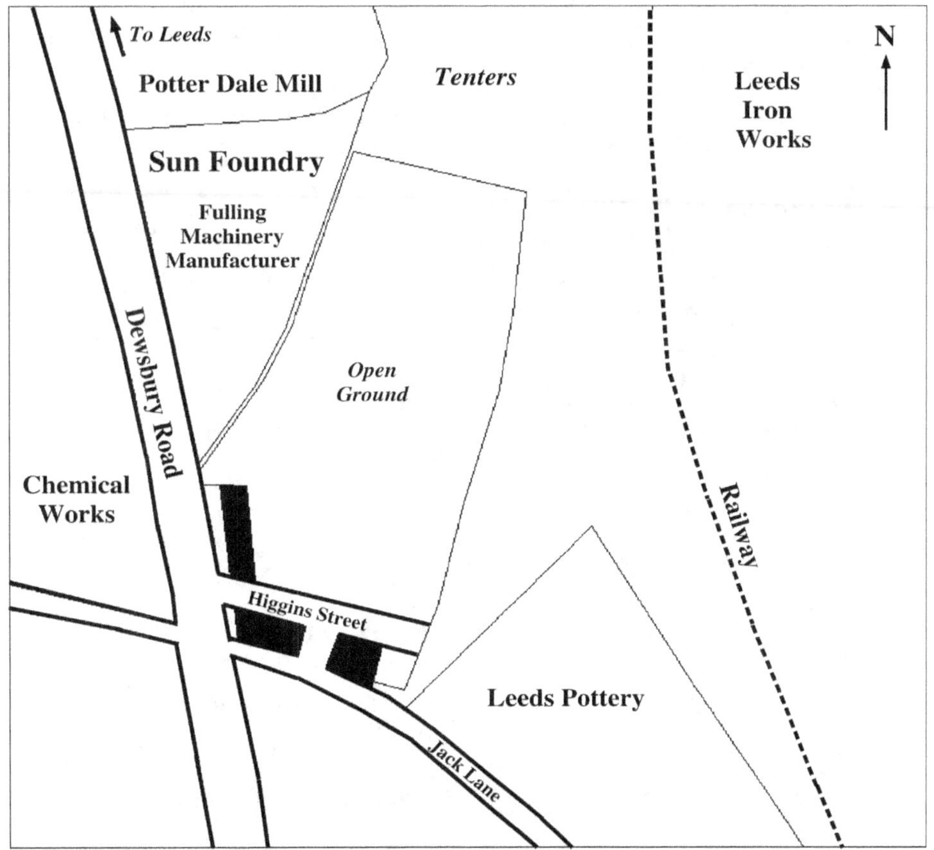

Figure 7. 1850 Survey: Location of the Sun Foundry
(Based on Ordnance Survey Town Plan: Leeds. Published 1850)

On the 8th and 9th September 1828, Rachel Milnes (widow) and Robert Pemberton Milnes (of Thorne and MP for Pontefract) sold portions of the land to William Sayers (accountant), John Brooks (publican), and James Brook (innkeeper) all of Leeds. By the middle of 1829, Sayers had acquired more land from the Milnes and had split, and sold, some of the previously purchased land. It is believed that this became the Higgins Street development that was later demolished for an expansion of the foundry.

SUN FOUNDRY, DEWSBURY ROAD, LEEDS.—EXTENSIVE SALE OF MACHINE MAKERS' TOOLS, FOUNDRY IMPLEMENTS, WOOD AND IRON MODELS, JOINERS' TOOLS, AND OTHER EFFECTS.—DUTY FREE.

Mr. THOMAS HARDWICK begs to announce that he will SELL BY PUBLIC AUCTION, on *Monday, Tuesday, Wednesday, and Thursday, the 7th, 8th, 9th, and 10th Days of June next,* upon the Premises of Mr. John Sugden, the Sun Foundry, Dewsbury Road, Leeds;

THE Whole of the Valuable and Extensive STOCKK-IN-TRADE and EFFECTS, including One Screwing Machine, with Counter Shafts; One Boring Lathe, with Side Rest and Gantry; One small Planing Machine, by W. Collier and Co. Manchester; One Self-acting Slide Lathe, 14 Feet Bed and Gantry; One Boring Lathe and Gantry; One Upright Drilling Machine.

ONE LARGE PLANING MACHINE,
31 Feet Bed, 4 Feet 2 Inches wide, self-acting, for Horizontal and Vertical Planing, by W. Collier and Co., Manchester; One small Self-acting Slide Lathe, 10 Feet Bed; Two Upright Drilling Machines; Three Throw Lathes and Gantries; One Slide Lathe, 12 Feet Bed;

FIRST MOTION WHEEL AND LINES OF SHAFTING AND APPENDAGES;
Two Glaziers and Frames; large Crane and Fixings; Iron Turning Frame, Joiners' Benches, Vices, Throw Lathe, 14½ Inches high; One Ditto, 8 Inches Ditto; One Ditto, 6½ Ditto; Circular Saw and Frame;

TWO LARGE FOUNDRY CRANES AND FIXINGS;
Cross Shafting; Two Cupolas, Blowing Machine and Apparatus; a Quantity of Seasoned Wood; large Scales and Weights;

ALL THE SMITHS' TOOLS,
comprising Fly Punch, Four Anvils, Swages, Bellows, a large Quantity of Bar and Rod Iron.

An extensive Stock of WOOD and IRON MODELS for Woollen and other Machinery; small Wherry, Two Carts with Patent Arms, One Horse, a Quantity of Ironmongery and Brass Work, Files, together with the

COUNTING-HOUSE FURNITURE,
Including Double Desk, Deal Cupboard, and an Assemblage of other valuable and important Property.

The above Machines are chiefly of the most modern Manufacture, and in the best working Condition.

Catalogues to be had of the Auctioneer, at his Office, in Trinity Street, One Week previous to the Sale.

The Sale to commence each Morning at Eleven o'Clock.

Figure 8. Advertisement for Auction at the Sun Foundry dated 1841.
(Leeds Mercury. Saturday 29th May 1841. p8.c.2.)

By the 1830s this area of Hunslet was being developed, with a mixture of woollen mills, machinery manufacturers and potteries. The area is still known today as Pottery Field. A survey of the area dated 1846-47 indicated that a fulling machinery manufacturer occupied the site, although by this date the Sun Foundry was being used for heavy engineering. A later survey, dated 1850, places the Sun Foundry at the northern end of the site (See Figure 7). Between the Sun Foundry and Jack Lane, there were three groups of terraced houses centred on Dewsbury Road and Higgins Street.

It is believed that John Sugden had established a firm for manufacturing fulling machinery at the Sun Foundry in the 1830s. By 1833, Sugden was certainly associated with that type of business, as he was involved with an insolvency auction of woollen machinery.[2] There was considerable competition in the Leeds and Bradford area for the production of woollen manufacturing machinery and eventually Sugden also succumbed to this commercial pressure and put his plant and the Sun Foundry up for auction. The sales advert (Figure 8) provides some information on the scale of his business and confirms that the foundry accommodated lathes, gantries and other plant.[3] Sugden eventually filed for bankruptcy[4] in May 1841, and after several years, the Sun Foundry had a new proprietor.

Hugh Lupton was to later mention (1940) [5] that the Sun Foundry had been, *'run by a Co-operative Society of working engineers,'* but no evidence has been found to support this assertion.

TO be DISPOSED OF, Two powerful and well built NEW LOCOMOTIVE GOODS ENGINES and TENDERS. These Engines have each six 5-feet wheels all coupled; 16 inches cylinders, and 2 feet stroke; copper fire box, brass tubes, the link motion; and are in every other respect of the most modern and approved build. Several goods engines of the same construction are now at work on the York and North Midland Railway, where an account of their efficiency may be obtained.

For further information apply to CHARLES TODD, Locomotive Engine Builder, &c., Sun Foundry, Leeds.

Figure 9. Charles Todd: Advertisement for the sale of locomotives.
(The Economist. 19th August 1848. No 260. Front)

Sun Foundry, Leeds.

Very Important to Engineers, Railway Companies, Machine Makers, Millwrights, Boiler Makers, Smiths, Brokers, and Others,

MR. WHEATLEY KIRK is honoured with instructions from Charles Todd, Esq., the eminent Engineer, &c., of the Sun Foundry, Dewsbury-road, Leeds, who is declining the business on account of ill health, to SELL BY AUCTION, *on Monday, the Tenth of May*, 1852, *and Three following days*, on the premises of the Sun Foundry aforesaid, all his truly valuable PLANT, MACHINERY, Tools, Utensils, Stock-in-Trade, &c., among which may be enumerated—large back-geared wheel or crank lathe, with 10½ feet face-plate, with two slide rests, &c.; back-geared wheel lathe, with two face-plates, slide rest, and traverse motion; back-geared face lathe, with two face-plates, traverse motion, &c.; powerful double back-geared lathe, for turning two railway wheels on the same axle, with four face-plates, two slide rests, traversing motion, &c.; one 16-inches back-geared slide lathe, with screw cutting apparatus, &c., with 18 feet bed; one 10½ inches back-geared slide lathe, also for screw cutting; one 5½ inches do., for sliding and screw cutting; together with other valuable back-geared and single-speed lathes, making 14 lathes in number; boring machine, for locomotive cylinders; slotting machine, will slot 8 inches, with circular table, &c.; two upright back-geared drilling machines, three upright single power drilling machines; power planing machine, 6½ feet bed, will plane 5 feet long and 2 feet 6 inches wide, and 2 feet 6 inches deep; do. 4½ feet bed, will plane 3 feet long and 1 foot 6 inches square; back-geared screwing machine, to screw from ¼ inch to 1½ inch; one single power ditto, will screw from ¼ inch to ¾ inch; one shearing, two punching, and one boiler plate bending machine, double wheeled furnace, very complete sets of engineers', boiler makers', smiths', and foundry tools; wheel, pulley, and a large quantity of other engineering models, in wood and iron; together with the vices, files, turning tools, &c.; stores of malleable iron, pig iron, steel, new files, &c. &c.; a model girder or vertebral arch, as exhibited at the Great Exhibition, 30 feet span, showing a new and extraordinary principle in the construction of bridges and roofs, for spans of large dimensions, being originally designed by Mr. Todd for a bridge across Runcorn Gap, near Liverpool; new boiler makers' and erectors' shed, covered with blue slate; one excellent four-wheeled phaeton; together with other valuable effects, full particulars of which will be given in descriptive catalogues, which will be ready on the 3rd of May, and may be had on the Premises of the works as above; or at the offices of the AUCTIONEER, 24, Princess street, Manchester; or will be sent by post on receipt of four stamps.

N.B. The Auctioneer begs to intimate, that the concern having been recently constructed, the whole of the plant is in first-rate order, and the engineering tools were chiefly made by the eminent firm of Smith, Beacock, and Tannet, of Leeds.

Sale to begin each day at eleven o'clock in the forenoon.
Refreshments provided for purchasers.

Figure 10. 1852 Sun Foundry: Auction of Plant and Machinery
(Leeds Mercury. 8th May 1852. p2.c.4.)

Charles Todd eventually purchased the Sun Foundry. He was described as an engineer, and had previously been in partnership with John Shepherd.

However, this partnership had been dissolved by mid-1844, due to Shepherd's debt, and Todd seems to have purchased the Sun Foundry after the dissolution of the partnership.[6]

Certainly by 1846, Charles Todd, now described as a locomotive manufacturer, soon turned his business into a profitable concern.[7] He manufactured locomotives for several local railway companies, as indicated by an advertisement from *The Economist* (1848) (Figure 9) that confirms his proprietorship of the Sun Foundry.

Todd started to suffer from bad health and in May 1852, the plant and machinery were put up for auction at the Sun Foundry. The sales advertisement (Figure 10) shows that Todd had made a significant investment in plant and other facilities, that included cylinder boring equipment and other items associated with engine manufacturing.[8] Todd, then aged 57, died a few months later on the 30th September 1852.[9]

Chapter 2 References

[1] DLC Box 2. Reconveyance of the Sun Foundry. 7th January 1878.
[2] Woollen Machinery, School Close Leeds. *Leeds Intelligencer*. 23rd March 1833. p.1
[3] Sales by Auction - Sun Foundry. *Leeds Mercury*. 29th May 1841. p.8
[4] Declaration of Insolvency. *The Standard* (London). 29th May 1841. p.1
[5] HL. 1940. *Hathorn, Davey and Company, Sun Foundry, Leeds.* p.1
[6] Public Notice. *Leeds Intelligencer*. 8th June 1844. p.2
[7] HL. 1940. *Hathorn, Davey and Company, Sun Foundry, Leeds.* p.1
[8] Sales by Auction - Sun Foundry. *Leeds Mercury*. 8th May 1852. p.2
[9] Deaths. *Bradford Observer*. 7th October 1852. p.8.

Figure 11. Carrett and Marshall: Road Locomotive used as an illustration by Hathorn, Davey in their *1933 Catalogue of Pumping and Hydraulic Machinery*

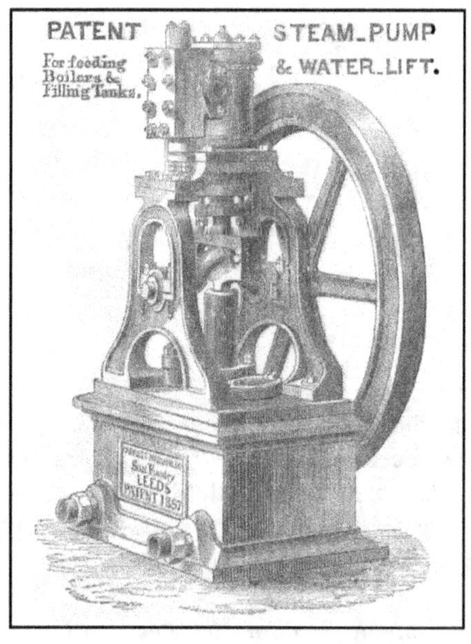

Figure 12. Carrett and Marshall: Patent Steam Pump and Water Lift.
(Illustrated Guide to the North West Railway 1861. p.379)

3. Carrett and Marshall

The new proprietors of the Sun Foundry were the partnership of William Elliot Carrett, William Ebenezer Marshall and John Telford and traded under the name of Messrs. *Carrett, Marshall and Company.* They took over the foundry in 1853, and four years later in 1857 the partnership purchased the freehold of the site, from the then owners, George and William Leather, the original purchasers of the land having died in the intervening years.[1,2]

The partnership soon expanded the range of products at the Sun Foundry. Some were exceptional, and included a steam carriage, or road locomotive, for use on public roads. (See Figure 11) One such carriage, manufactured for the Yorkshire industrialist Titus Salt, was demonstrated at the International Exhibition held at London in 1862. It was reported to have successfully proceeded under steam from the Exhibition to Blackwall via Knightsbridge and Piccadilly.[3]

The company manufactured machine tools and steam engines, including Horizontal Compound Engines. They also won various awards for their innovations, including silver medals at the Paris Universal Exhibitions of 1855 and 1867.

A description of the Company in an 1861 Railway Guide provides some detail of the range of products manufactured by the Company.[4]

'Besides the ordinary branches of engine and boiler making, iron founders Messrs. Carrett, Marshall and Co. are celebrated for their Patent Steam Pumps and Water Lifts, of which we are enabled to present to our readers illustrations; which are all of the plunger construction, and adapted for forcing the water to any distance or height, besides raising it by suction, which is limited in practice to 29 feet deep, or what is equivalent to the usual pressure of the atmosphere. (Figure 12) These simple and compact machines are entirely portable and complete in themselves, being, in fact, a combined engine and improved pump in one apparatus.'

The feed pumps were primarily used for supplying marine and locomotive boilers, but clearly they had other uses, as they,

'Form at the same time a complete steam-engine, which can be detached from the pumps, and used as a motive power for actuating machinery.'

The various sizes of these steam-pumps and water-lifts are constructed to raise and force from 3,500 to 100,000 gallons, fifty to one hundred or more feet high, in ten hours.

They are made also of the piston and bucket description, for drainage and mining purposes, and are in operation even so large as five feet diameter of pump, raising three gallons per minute.'

Figure 13. Carrett and Marshall: Portable High-Pressure Steam Engine
(Illustrated Guide to the North West Railway 1861. p.382)

'We are further enabled to furnish a class of vertical direct-action engines, constructed from 1 to 20-horse power, which is peculiar to these makers, and meets with very high commendation; and also a new arrangement of these engines upon a vertical internal fire-box and water-chambered boiler, which forms all the foundation that is necessary for engines of two, three, four, and five horse-power (Figure 13). *Here all fixing is dispensed with, enabling the whole apparatus to be readily set to work in a few hours, and working with about five pounds of coal per horse per hour.'*

Whilst a majority of the products were steam related, they also produced a range of specific hydraulic equipment.

'Another unique and most useful invention, now applied in all the cathedrals and principal churches in the kingdom, and constructed by these makers on Joy's Patent, is the Hydraulic Organ-blower for developing power from the pressure of Town's water, or other suitable pressure; and is applied also in another form to produce rotative as well as reciprocating power.' (Figure 14)

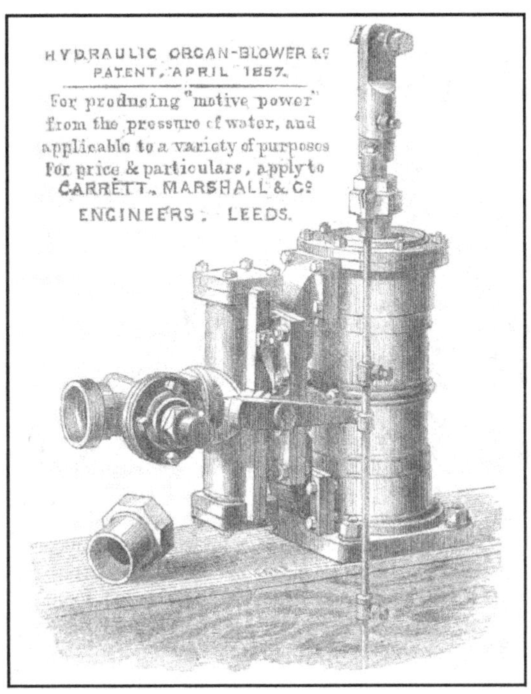

Figure 14. Carrett and Marshall: Hydraulic Organ-Blower
(Illustrated Guide to the North West Railway 1861. p.383)

The Organ (and Harmonium) blower, originally patented by Messrs. Joy and Holt, Leeds in 1857, was probably one of the company's regularly sold products.

'The cylinder is similar to that of an ordinary steam engine, the piston having a reciprocatory motion given to it by the pressure of a column or stream of water admitted alternately to the top and bottom of the piston by suitable passages. The admission of the water to the cylinder is regulated by a slide valve, the back of which is attached to a small double piston, that working in the cylinders formed at the top and bottom of the valve-box.' The reciprocating motion operated the organ bellows.[5]

The simple principle of the Joy engine would be later adapted for use in collieries in the form of a pumping engine to drain water from a sump in dip workings.[6]

One item of hydraulic equipment manufactured by the company at this time was most certainly used in coalmines. Carrett, Marshall and Telford together with Messrs. Lock and Warrington, colliery owners, patented one of the first coal cutting machines. (Figure 15) It was successfully demonstrated in November 1864 at Kippax Colliery, Leeds, and continued in operation for a number of years.[7,8] News of this innovation spread widely, and in 1870 two machines were ordered for the coalfields of Nagasaki, Japan.[9]

Figure 15. Carrett and Marshall: The hydraulic coal-cutting machine was used to undercut a coal seam. Carrett referred to it as 'an iron man.'
(The Engineer, 31st August 1866, pp.149 to150 [GG])

Carrett died from '*paralysis*' on the 25th September 1870, and Marshall took over Carrett's share from his widow.[10] Later valuations of his assets in the company seem to suggest that Carrett had suffered from a degenerative mental condition as a Court of Lunacy had placed a valuation on his share.[11] Carretts death combined with a trade dispute in 1871 placed the business in difficulty and so the surviving partners, Marshall and Telford put the company up for sale.[12,13]

In addition to being a partner, Telford was a foreman pattern maker but he didn't seem to feature in Marshall's decision to sell the business.[14] Marshall was later to state that, '*My reason for wishing to retire from business is that I find it too much work for one person to carry it out and do not care to take a partner.*' The partnership evidently still did survive at this time, as it wasn't dissolved until late August 1872.[15]

How the sale of the foundry was advertised in 1872 is not known, and perhaps it was word of mouth? It did however come to the attention of Colonel John Fletcher Hathorn who had just retired from the Army and was looking for a suitable business to purchase.

Obviously, a price had to be agreed, and the sales documentation based on a valuation made in July 1868 gives some indication as to the scale of the business.[16] There are actually two valuation documents with slightly different figures. The more coherent of the two is detailed below:

'General Statement of Carrett and Marshall and Co's Business
May 1852 to July 1868

We started our business in 1852 with a capital of £3000

We estimated our concern when we took stock in July 1868 as follows:-
(We have not taken stock since)
Land and Buildings   *£10,000*
'Fixed Stock' consisting of engines, boilers, shafting
tools, models, drawings   *£7,300*
'Floating Stock' consisting of boiler makers blocks,
moulding boxes, cranes, ladles, smiths tools,
tools for lathes, bar iron, pig iron, steel, wood, all stores,
blocks, cranes, etc.   *£4,700*

                                                                £22,000

It is to be borne in mind in the 'floating stock' there is all necessary for carrying on the
business and consisting of a large amount of material that would represent convertible
capital.

In the land and buildings are 12 cottages worth £800 paying about 7%

The estimate we had in the business in 1868 about   *£33,000*
       Less capital we started with   *£3,000*

                                                            £30,000
We had drawn out in the 16 years   *£14,300*

Equal to about £3000 a year profit   *£44,300*

You must bear in mind how small our business was for several years.
I should have no objections to lease £5,000 on the mortgage of the buildings to be paid
off at the rate of £500 per year or even £1000
As to goodwill our connection to inventions would have to be valued, but whatever the
goodwill has taken I should leave on to be paid in time.

We should finish our orders and collect our debts or pay for them.'

The second document contains further clauses about the state of the company

'Next with regard to goodwill. The Court of Lunacy fixed Mr. Carrett's share at £1500, so that his share and mine would represent £3000 thus I should wish to ask someone who knows the custom, I do not, and would be guided by custom.

As to capital required to carry on the business, you want money for credit and work in progress. We estimate we have at present in our concern about £35,000 but this we know is much more than is required to carry on the business.

I should have no objection to leave £6000 in the concern on security of the freehold at 5% to be paid off in 6 years at £1,000 per year.

As to the returns of profit of the concern its history is in little rooms. We started our business in 1852 with a capital of £3050. In 1853 we put in £370 more. From July 1852 to July 1868 we drew out £14,000 and have now as estimated capital of £35,000.

As to the future, I believe our concern was never in a more prosperous state. You must bear in mind the new buildings which were in hand when Mr. Carrett was incapacitated ....some of them were never filled with machinery so that without any more outlay for buildings, the production could be nearly doubled.'

Eventually a Memorandum of Agreement was drawn up for the sale of the business to Messrs. Hathorn, Davis and Campbell. A draft survives.[17]

'Memorandum of Agreement between Messrs. Carrant Marshall and Co., Sun Foundry, Leeds of the one part, and Colonel Hathorn, Alfred Davis and Hugh Campbell of the other part for the purchase of the whole of the lands, buildings, fixed and loose stock, exhibition medals and goodwills of the business now carried out by the said Carrett and Marshall and Co. in Leeds.

1. The freehold property. Fixed and loose stock to be valued by two valuers (one to be named by each side), and the amount agreed on to be the price paid.

2. The price to be paid as follows. Six thousand pounds in cash on obtaining possession, and the balance by capital instalments extending over twelve years from date of possession, with power to purchasers to pay off whole or larger instalments at any time on giving one months notice.

3. A mortgage over the whole of the freehold property and fixed stock to be given by the purchasers in security of the balance of the price over the £6000 to be paid down. The purchasers also to give their permission notes for the instalments. The unpaid instalments to bear interest at the rate of five per cent per annum payable half yearly.

4. The work in progress at date of sale of the business to be valued by the valuer is finished under the superintendence of Mr. Marshall and paid for by the purchasers when the cash received by the purchasers from the cashier (?) for above the work as being done. The vendors to guarantee the purchasers against loss on the work in progress.

5. The Goodwill and Business books of the vendors to be included in the purchase, but nothing to be paid on account thereof. None of the vendors to enter into the same or similar business within 200 miles of Leeds, but the vendors to be a liberty to act as consulting or civil engineers. The vendors to have access to their business books during business hours, for the purpose of the collection of their outstanding debts only.

6. Prior to entering into final agreement for purchase, the purchasers to have liberty to inspect the business books of the Vendors and the inspection to be to the satisfaction of the purchaser's accountant.

7. The business to be taken over by the purchasers as at midnight on 30th June 1872.'

The person chosen by the purchasers to independently assess the value of the company was Edward Davis, Albion Street (Leeds), who was a maker of mining instruments, and may have been a relative of Alfred Davis. He suggested a number of alterations to the agreement that included,

a) The cottages associated with the works that should be included as an option.
b) A longer period of time should be allocated for paying off the balance.
c) The initial payment at time of sale should be 5% of the total value.
d) Marshall to hold the mortgage of the Freehold of the works and the cottages.
e) Payment for goodwill not to be entertained.
f) Freehold, plan and work in progress to be valued by joint Valuers, and their decision to be final. It was suggested that this value was limited to £22,000 with the addition of a discretionary £1,000.
g) Marshall shall only assist for nothing in the completion of any work, and guarantee against any loss.

The business was sold to the Hathorn Partnership in two stages. Firstly, the land and premises were sold to them for £11,000 on the 17th August 1872, and secondly, the personal chattels were sold to them for £16,690 on the 19th August. The partnership was also legalised on the latter date.[18]

On 17th August 1872 the Partnership of John Fletcher Hathorn, Alfred Davis and Hugh Fletcher Campbell acquired the Sun Foundry for a total sum of £27,690. The new partnership had purchased a complete business, a foundry, plant and equally important, orders for machinery to complete, and a number of innovations to modify. It is unclear whether all the land was sold to the Partnership at this time as a further payment was made a few years later.

Chapter 3 References

[1] DLC Box 2. Reconveyance of the Sun Foundry. 7th January 1878.

[2] HL. 1940. *Hathorn, Davey and Company, Sun Foundry, Leeds.* p.1

[3] Mr. Salts Steam Carriage. *Kendal Mercury.* 13th December 1862. p.3.

[4] Measom, G. 1861. *Illustrated Guide to the North West Railway and all its branches.* W.H. Smith, London. pp.378 to 383

[5] Hydrostatic Organ-Blower. *The Practical Mechanics Journal* (London). 1st July 1857. p.108

[6] Bjorling, P.R. 1903. *Water or Hydraulic Motors.* Spon, London. p.140

[7] Locke W., Warrington J., Carrett W.E., Marshall W.E. and Telford J., 'Patent No. 2630. Improvements in the working and mining of coal, minerals, and earthy matters, and in the machinery, apparatus, and means to be employed therein', *The London Gazette.* 23rd February 1864. p.847

[8] Trial of a new machine for cutting coal. *Westmoreland Gazette.* 28th November 1864. p.8

[9] *The Leeds Times.* 14th May 1870. p.3

[10] HL. 1940. *Hathorn, Davey and Company, Sun Foundry, Leeds.* p.1

[11] LMWS. A19932.1

[12] Announcement of Carrett's death. *Sheffield & Rotherham Independent.* 28th September 1870, p. 3

[13] *Sheffield Independent* .1st October 1870. p.12

[14] HL. 1940. *Hathorn, Davey and Company, Sun Foundry, Leeds.* p.1

[15] Gazette News, Partnerships Dissolved. *Lloyds List.* 24th August 1872. p.iv

[16] LMWS. A19932.1, 2 and 3

[17] LMWS. A19932.2

[18] DLC. DB168.1

4. Hathorn, Davis and Campbell: 1872

The purchasers of the Sun Foundry were a strange alliance. Neither Hathorn nor Campbell had any direct experience of engineering and it is believed that Alfred Davis's engineering qualifications, if any, were minimal. Certainly the new Partnerships' ability in running such a business could be open to question.

Until 1872, John Fletcher Hathorn had been in the British Army. He was born on the 22nd August 1839 at Boquhan House, Stirling, Scotland into a family of wealthy landowners. He was educated at Harrow[1] and began a career in the army on the 19th December 1856, at the age of 17, as an Ensign in the Coldstream Guards.[2] He became a Lieutenant on the 18th July 1862. Some of his time seems to have been spent in India. Eventually he gained the rank of Captain and Lieutenant Colonel, by purchase on 23rd November 1870.[3] He retired from the regiment on the 3rd August 1872, and received the value of his commission. [4] John Hathorn did take up a Directorship with the *Central Swedish Iron and Steel Company, Limited* in early 1872, but by 1876 that company had gone into liquidation[5]

Throughout much of his involvement with the Sun Foundry, John Hathorn lived at Castlewigg House, a 3582 acres country estate at Whithorn, Wigtownshire, southwest Scotland.[6] He had moved there after the 1st March 1875, following his marriage to Charlotte Dick Lauder who came from a prominent Edinburgh family.[7] There is no doubt that John Fletcher Hathorn was the main financier of the Leeds Partnership.

Alfred Davis is described as an engineer, with an address at 31, Duke Street, Westminster, London. Despite extensive searches, very little has been discovered about him other than in May 1872, he is recorded as being the consulting engineer for the newly registered *North of Ireland Iron Ore Company, Limited.*[8]

Hugh Fletcher Campbell was a half-brother of John Hathorn, and it was John's intent to give Campbell a worthwhile occupation by bringing him into the Partnership.[9] This was a good intention that was relatively short-lived.

The first order placed with the new company is probably number 2115 dated 20th August 1872, for a 6" brass-lined hydraulic organ blower costing £40, (20% reduction applied).[10] The order was placed by John Nicholson, Organ Builder of Palace Yard, Worcester. On completion the order was to be sent directly to the Reverend Lord A. Compton, Castle Ashby, Northampton. In effect this was a continuation of a line of business previously carried out by Carrett and Marshall. They were indeed small beginnings for a Company that would soon become leaders in their field, and build some of the largest pumping machinery manufactured in England.

Contrary to popular belief, Henry Davey was not a partner at the onset of the Company although he was occasionally residing in Leeds. Certainly for the first year of operation, the Company functioned as Hathorn, Davis and Campbell. There is one person not mentioned at this time, John Telford, who appears to have remained at the Sun Foundry and provided continuity by overseeing the work previously taken on by Carrett and Marshall, and the orders placed with the new firm.

The exact date of Davey's appointment as works manager of the Sun Foundry is not known, but the position arose probably from the lack of engineering knowledge in the partnership, and the possible limitations of John Telford's experience. Certainly the first order that refers specifically to a Compound Engine designed to a Davey's patent is order 2157 dated 26th December 1872 for the *Seaton Delaval Coal Company, Limited* (Northumberland Coalfield).

For the first few years, further organ-blowers were manufactured, but by 1874, orders had reduced considerably. Later work was confined to repairing them, and to several for Edmund Schulze an organ builder from South Shields. But as the business gained momentum, some of the largest orders were for engines, both hydraulic and steam, for colliery companies based in the East Midlands, Yorkshire and the Durham and Northumberland coalfields.

Five months after the partnership was formed, business was evidently thriving, and there were plenty of orders on the books.[11] *'Messrs. Hathorn, Davis and Campbell (late Carrett, Marshall and Co) have some very important work on hand. They have just completed ten small pumping engines for the Indian State Railways, with self-containing vertical boilers. They are also engaged upon pumping engines for colliery work. These pumps are placed underground, and force the water to the surface - a special feature lately introduced to collieries. The old system was a cumbersome arrangement of spear rods, the engine and pumps being at the bottom, the steam being carried down. The engines introduced by Messrs. Hathorn and Company for this purpose are of entirely new construction and design. One of these is going to Clay Cross, and will raise water at one lift 1000 feet. They are making a similar pump for a colliery at Newton Capp [sic], near Darlington, to pump 1,300 gallons per minute. For colliery work above ground the same firm are constructing a large direct acting compound condensing engine, which is to work with a consumption of not more than four pounds per horse per hour, and the economy in the use of fuel by this machine will be an important item. Besides these they are constructing a very large pair of winding engines also for colliery work, with 14ft drums, to wind 800 tons of coal in ten hours from a depth of 70 fathoms; and a large Guibal ventilating fan, of 36ft diameter and 12ft. wide. Some idea of the size of the shaft may be judged by the fact that it weights 4½ tons, driven by a pair of 30in. horizontal engines. Following the example of their*

Under the terms of the sales agreement between the Carrett and Marshall, and the new Partnership, orders placed prior to the sale of the Sun Foundry had to be fulfilled. Consequently, Hathorn, Davis and Campbell never started a new order book. They simply continued with the order numbering system devised by the previous company. This accounts for the two Carrett and Marshall's order books surviving in the Hathorn, Davey archives. Some of the orders mentioned above were actually placed with Carrett and Marshall, but completed by Hathorn, Davis and Campbell.

Figure 16. Indian State Railways: Small Pumping Engine and Boiler
(Scientific American. April 19th 1873, p.239)

The orders for the ten small pumping engines and boilers (Figure 16) for the *Indian State Railways*, for example were placed in May 1872 (Orders 2078 & 9). They were simple Vertical Engines, with 6" cylinders and an 8" stroke used for pumping water from shallow wells. The engines were fixed upon stone foundations at the top of the well, and the pumps were located on a wood beam fixed a little above the water level. The pumps were worked by means of a spur pinion on the crankshaft, and spur wheels on the countershafts; the pump rods were fixed to the spur wheels. The stroke of the pumps, two in number per well, could be varied, and were capable of delivering eighty gallons of water per minute, thirty feet high. '*The workmanship and materials are alike excellent, and well maintain the reputation of the firm.*' [13,14]

The engines for Clay Cross (Derbyshire) and Newton Cap (County Durham) Collieries are referred to in an article in *The Engineer* that is based on a paper Henry Davey gave to the Society of Engineers.[15]

Davey's paper titled, '*On Recent Improvements in Pumping Engines for Mines,*' described a variety of engines.[16] The Clay Cross order, was for a pair of hydraulic single-cylinder engines for underground use, to pump water directly to the surface. (Figure 17) '*They pump against 1000 feet head of water, whilst a similar pair at Newton Gap have 240 feet.*' The construction of these engines was similar to that of a Compound Engine, minus the high-pressure cylinder. The engines were placed underground, and were self-contained with the pumps on one bedplate. Both the *Clay Cross Company*, and the owners of Newton Gap placed various orders with the company over the coming years. Ultimately, other colliery companies would form a high percentage of the orders received in the first ten years of trading.

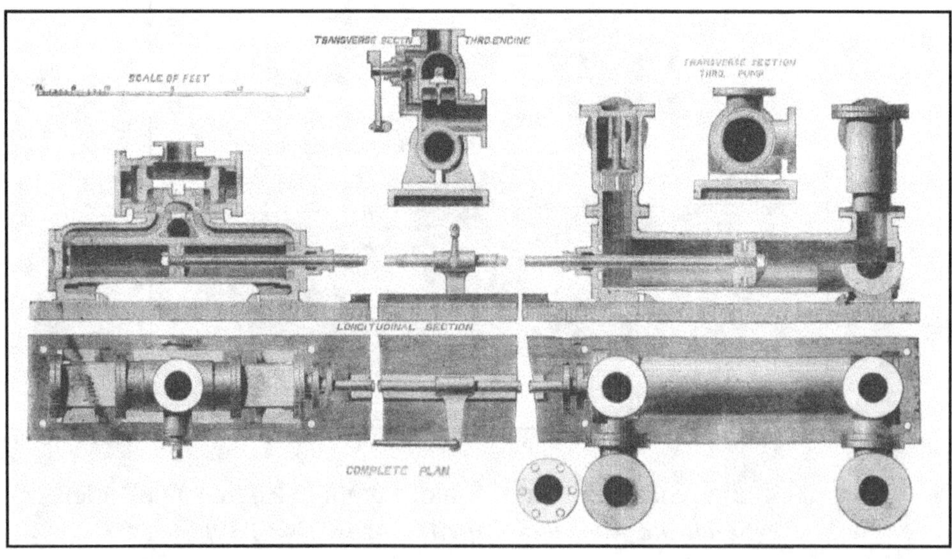

Figure 17. Hydraulic engine and pump supplied to the Clay Cross Company
(The Engineer, October 24th 1873. p.266 [GG])

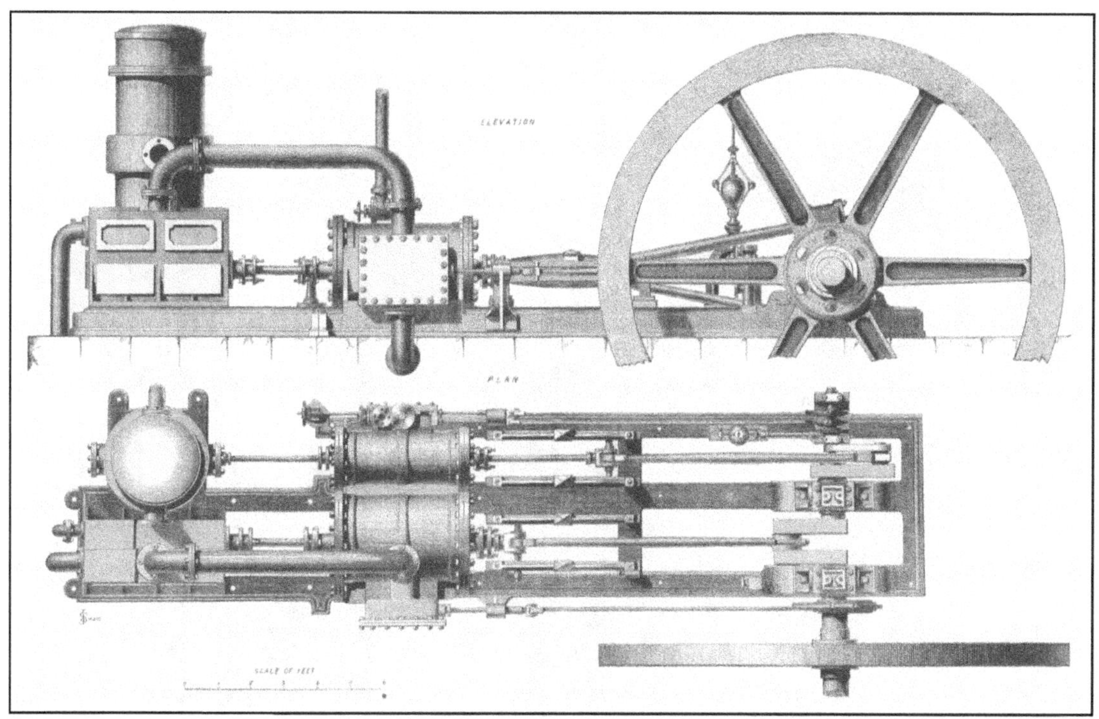

Figure 18. Horizontal Cross Compound Engine with a flywheel. The two cylinders placed alongside each other was probably a Carrett and Marshall design.
(The Engineer, July 3rd 1874. p.10 [GG])

The orders for the pair of colliery winding engines (30" cylinders and 30" stroke) and a large Guibal ventilating fan, 2100 and 2101, respectively were placed with Carrett and Marshall by the *Wheldale Coal Company, Limited,* Castleford on the 25th July 1872. The total cost for both items was £1,456 plus £125 for girders and bolts, and were delivered on the 11th June and 9th October 1873.

Two other orders, for Compound Engines, were also mentioned in the list of work in hand by the Partnership. The first, for the Admiralty waterworks at Chene, Portland (Order 2155) had cylinders of 9¾" and 16½", to be delivered by 31st March 1873. The second Horizontal Compound Engine was for the cement works of *Brooks, Shoobridge and Company*, at Grays Thurrock, Essex (Order 2130 dated 30th October 1872) and had 12" and 28" cylinders.[17] It had a condenser and 8 ton flywheel and cost £850 (Figure 18).

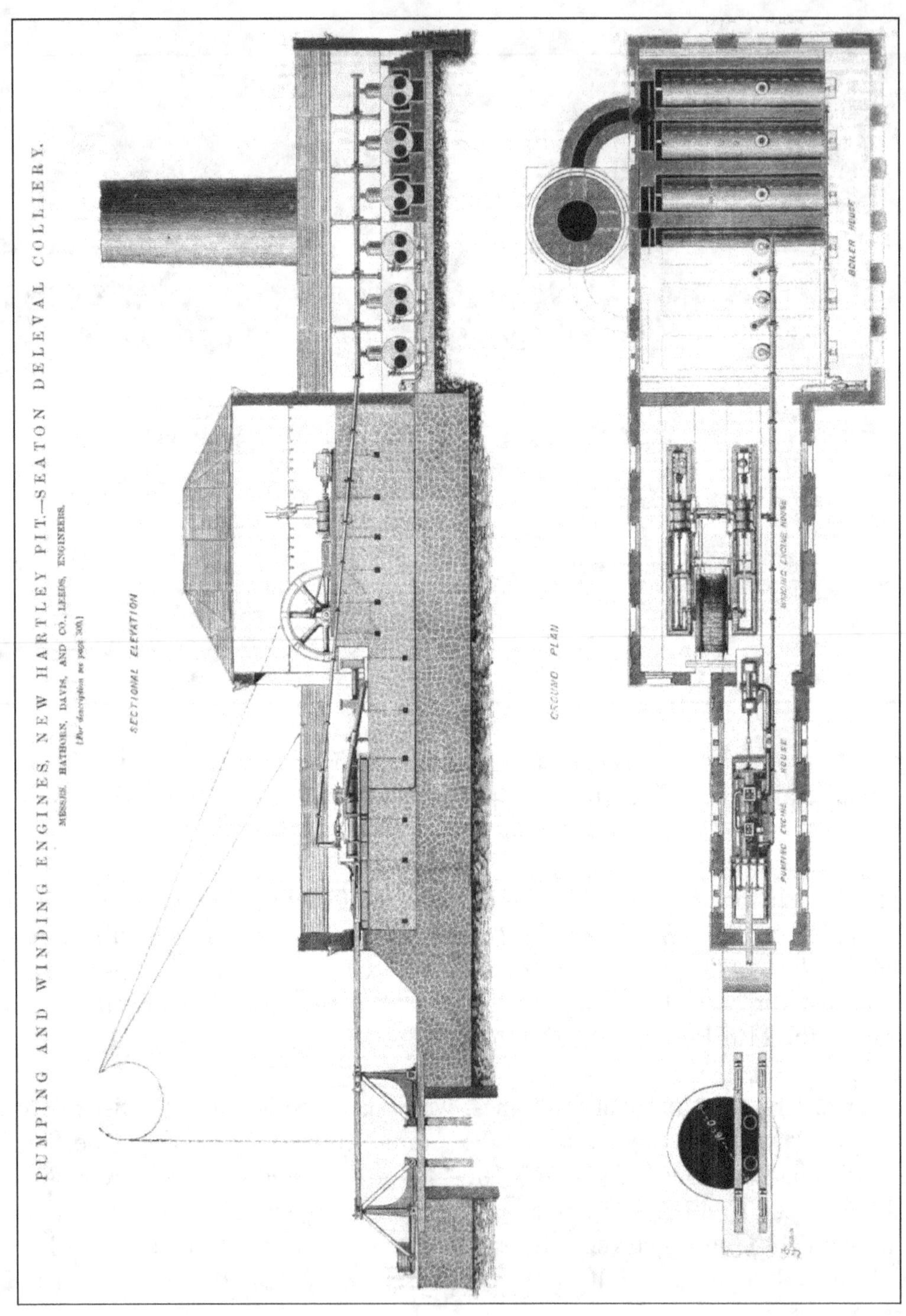

Figure 19. New Hartley Pit: The Arrangement of the Pumping and Winding Engines
(The Engineer, April 21st 1876. p.298 [GG])

The Horizontal Compound Engine was soon to become popular with coal owners. It was not necessary to site the engine immediately on top of the shaft like the Cornish Beam Engine, and in later examples it could be installed in an undercroft immediately below the surface. The engine had two cylinders. The low-pressure cylinder was nearly double the diameter of the high-pressure cylinder. Both cylinders were in line with the high-pressure cylinder mounted at the pump side of the engine.

In the first ten years of trading Hathorn and Partners sold numerous Compound Engines to colliery companies in the Northumberland-Durham Coalfield. It is not surprising that the coal companies of that area wanted a change from the traditional Cornish Beam Pumping Engine as the breakage of a cast iron pumping beam from a Cornish type engine resulted in one of Britain's worst mining disasters. The accident occurred at the Hartley Pit on the 16th January 1862. The beam had snapped at a weak point on its axle, and it fell down the shaft, blocked it, and trapped 204 miners. It was a single shaft mine with a furnace at the bottom of the shaft. The entombment resulted in asphyxiation of the trapped men and boys. The consequence of the accident was a change to British mining legislation requiring all coalmines to have a second means of egress. [18]

One of the largest orders in this period was for the *Seaton Delaval Coal Company* for the New Hartley Pit, Northumberland (Orders 2157 to 2160, dated 26th December 1872). It was for a Horizontal Compound Pumping Engine with quadrants, a pair of winding engines, and six boilers. Figure 19 shows the layout of the machinery.[19]

The Compound Engine (2157) had 24" and 44" cylinders, with a 7' stroke, and an air pump and condenser, operated by a horizontal tail rod coming out of the end plate of the low-pressure cylinder. Henry Davey's Differential Gear used for regulating steam flow to the engine had not yet gone into production, so a governor was ordered for the engine which was to be fitted with a wheel to regulate the speed of the engine. The cost of the engine was £1,815.

The pair of quadrants are listed as order 2160 and cost £175.

Order 2158 was for one pair of Horizontal Condensing Engines, with two 28"diameter cylinders, with linked motion for reversing, located on each side of a 15' diameter by 4' wide winding drum with brake. It was employed to wind from a depth of 420 feet, and cost £1,815.

In additions to the engines, six boilers (30' long by 6' diameter) were ordered (2159) and cost £2,190.

Hathorn, Davis and Campbell entered their machinery for awards in both national and international competitions. They received their first award in August 1873 when the firm was given an 'Honourable Mention' for their machinery at the Vienna Universal Exhibition. They entered two items, one a 2 horsepower high-pressure Vertical Engine with a 6" cylinder and an 8" stroke, and secondly, a hydraulic organ blower with a 3½" cylinder. (Orders 2142 and 2143, dated 22nd November 1872). There is a note in the Order Book for both engines to be, 'good bright jobs.' It is believed that the two-horsepower engine was later sold to Sir Titus Salt, the Saltaire mill owner, West Yorkshire.

Sometime during early 1874, Henry Davey was made a Partner, and the business became known as Hathorn, Davis, Campbell and Davey.[20]

Chapter 4 References

[1] The late Col. Hathorn. *Wigtown. Free Press.* 13th December 1888. p.4

[2] Ross, J. 2019. *A History of the Coldstream Guards from 1815 to 1895.* Goodpress. Item 1100.

[3] *The Edinburgh Gazette,* 29th November 1870. p.1471

[4] *Late Madras. Naval & Military Gazette and Weekly Chronicle of the United Service* - 3rd August 1872. p.6

[5] Central Swedish Iron and Steel Company, Limited. *Cork Constitution.* 1st March 1872. p.1

[6] Bateman, J. 1883. *Great Landowners of Great Britain and Ireland.* Harrison, London. p.212

[7] *The Dundee Courier & Argus and Northern Warder.* 5th March 1875. p.2

[8] Prospectus for the North of Ireland Iron Ore Company Limited. *The Belfast Newsletter.* 29th May 1872. p.1.

[9] HL. 1940. *Hathorn, Davey and Company, Sun Foundry, Leeds.* p.1

[10] The Joy Organ Blower was originally described as a 'New Hydraulic Engine'. See: Joy.1857. Description of a new hydraulic engine. *Proceedings of the Institution of Mechanical Engineers,* pp. 184 to 188

[11] West Riding Iron and Coal Trades. *Leeds Mercury.* 4th January 1873. p.6.

[12] *Leeds Mercury.* 4th January, 1873.

[13] Steam Pumping Engines for India State Railways. *Scientific American.* 19th April 1873. p.1

[14] Steam Pumping Engines for India State Railways. *The Engineer.* 7th March 1873. pp.137 to 138

[15] On Recent Improvements in Pumping Engines for Mines. *The Engineer,* 24th October 1873. p..268

[16] Davey, H. 1873. On Recent Improvements in Pumping Engines for Mines. *Transactions of the Society of Engineers* pp.182 to 195

[17] Horizontal Compound Engine and Condenser. *The Engineer,* 3rd July 1874. pp. 6 and 10

[18] https://en.wikipedia.org/wiki/Hartley_Colliery_disaster

[19] Pumping and Winding Engines, New Hartley Pit - Seaton Delaval Colliery. *The Engineer,* 21st April 1876. pp. 298 and 300

[20] HL. 1940. *Hathorn, Davey and Company, Sun Foundry, Leeds.* p.1

5. Henry Davey

There is no doubt that Henry Davey was the chief contributor to the Hathorn, Davey Company's success. It was written that, *'Mr Henry Davey was a ready-made engineer when he was born.'* This is not surprising considering the many technological innovations that would ultimately be attached to his name.[1]

Henry Davey (Figure 20) was born in 1843, at Lewtrenchard, Devon to Jonathan and Mary Davey. He would have come into contact with engineering at an early age, as his father was a mine captain and surveyor, as well as the schoolmaster at Lewdon, near Tavistock, Devon.[2] Henry often remarked that it was the illustrated catalogue for the 1851 Great Exhibition, brought home by his father that inspired him to embark on a career in engineering. He was also fortunate, as he learnt surveying skills helping his father, and also went to his father's school. He would later attend school in Tavistock under a Mr Loam, and not surprisingly showed an aptitude for geometry.[3]

Figure 20. Henry Davey: A Portrait of Davey dated 1891
(Mining Journal, 5th December 1891, p.1366)

In 1860, at the age of seventeen, Davey became articled to *Nicholson and Mathews*, later to become known as *Joseph Mathews and Company*, at the Bedford Foundry, Tavistock. They manufactured Cornish Pumping Engines and mining machinery, and specialised in the production of horizontal engines.[4]

During Davey's apprenticeship, there was another pupil at the foundry, a Mr Ingram who he befriended. Ingram possessed money, but lacked a passion for engineering. The two young men went into an informal partnership and set up a business in Hampshire manufacturing agricultural implements. Davey's role was probably to design portable and horizontal engines.[5] After Davey left the business, Ingram may have gone into a more formal partnership with a Mr. Sheppard, as in 1864, there is a 'Sheppard and Ingram', advertised as Engineers and Millwrights, based at the Stuckton Ironworks, Fordingbridge, Hampshire and they used Messrs. Howard, Clayton and Shuttleworth to supply steam engines.[6] Henry Davey got little satisfaction from the partnership with Ingram but did gain valuable business experience!

Perhaps during a period of unemployment, Davey registered the first of many patents on 17th December 1864, when he was residing at Gunnislake, Cornwall. Patent 3139 was simply titled, '*Improvements in Steam Engines.*' The patent offered four improvements to engine design that included: (i) Valve arrangements for both steam introduction and exhaust from engine cylinders. (ii) A condenser and air pump. The condenser had a cold-water cistern above, and holes in its base that allowed water to descend as a shower into the condenser. The cistern was supplied with water via a double-acting air pump. (iii) A slide valve arrangement that removed a large portion of the steam back-pressure. (iv) A modification to reversing gear. However, perhaps because of a lack of capital, Davey never proceeded with the patent, but some of the concepts were almost certainly employed in his later engine designs.[7,8]

About 1865, Davey next took up employment as a draughtsman at the *Fox and Walker Company* at Bristol, manufacturers of steam engines.[9] Not only did the company build locomotives for the Great Western Railway, they also specialised in the construction of portable steam engines for agricultural use. They exhibited, and had stands, at various agricultural shows in the West Country, something that was no doubt remembered by Davey when he later designed an engine for domestic purposes, that was exhibited at the Royal Windsor Show.

He soon tired of Bristol, and set up practice as a consulting engineer in London where he established an office at several locations that included, the Strand, and later Victoria Chambers, Westminster. By 1869, he had settled at John Street, Adelphi. He took on various contracts. One involved the manufacture of Bessemer steel, for which he made visits to both Sheffield and Glasgow.

During 1867, Henry Davey took out three patents. The first patent, 1008 dated 3rd April 1867, was for *'Improvements in obtaining motive power, and utilising the products of combustion or liquid and other fuels.'* The other two were taken out with David Davy of Sheffield. Patent 1936, dated 3rd July 1867 was for *'Improvements in centrifugal governors, and means of governing steam and other engines'* and patent 2430, dated 1867, *'Improvements in apparatus for supplying or feeding water to steam boilers and other vessels.'* The last two patents probably resulted from Henry Davey's consultancy work in Sheffield, as David Davy was almost certainly the head of the Sheffield firm, Davy Brothers, engineers and boilermakers.[10,11,12]

The two patents with Davy were probably continued for a number of years, as articles on both innovations appeared in *The Engineer*. Described as Davey's Isochronal Governor it was designed to maintain a constant engine speed under varying loads, as well as incorporating a means to stop the engine racing in the event of breakages in the system.[13] It was probably this latter patent that provided Davey with the inspiration to design, and later patent, the Differential Gear.

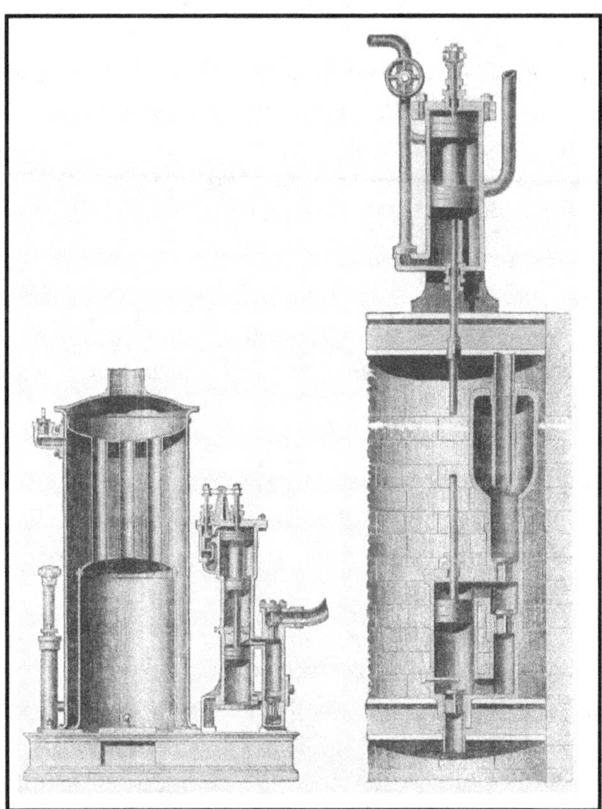

Figure 21. Double Action Deep Well Pump with Vertical Boiler
(The Engineer, 5th June 1868. p.419 [GG])

Patent 2430 for a feed pump was examined in the context of supplying water to a boiler, and also for use in extracting water from a well. The pumps are described as direct-action and of very simple and compact construction. Figure 21 illustrates a double-action pump with a vertical boiler and a deep well pump. In this example, the up-stroke is produced under water pressure, while the down-stroke is produced by the weight of the pistons and the attachments, the piston then being thrown in equilibrium by admitting steam on the topside of the piston.

It was on the 1st April 1867 that Davey presented to the Society of Engineers the first of many papers.[14] Entitled, *'Pumping-Engines for Town Water Supply'* the paper examined the economics of pumping by comparing the efficiency of the Cornish or single-acting engine with its rival types, for example the rotatory engine. *'It is often found that a well-constructed Cornish engine does a higher duty* [a measurement of engine efficiency related to the amount of fuel burnt to generate steam] *when expanding the steam nine times than a rotatory engine expanding the steam five to eight times; and as an obvious consequence, it follows that the peculiar element of economy existing in the single-acting engine is not dependent on the ratio of expansion employed, but in the mode of developing and applying the power. Notwithstanding the advantages which it possesses over its rival, the Cornish engine has not received the amount of attention of late years which has been bestowed on the rotatory; and in order that the latter engine may produce the maximum economy, recourse has been had to high expansive working, to obtain which the crude ideas of Hornblower and Wolf have been developed, a second cylinder has been applied, and the ratio of expansion increased with very good results accruing. It appears rather strange to the author* [Davey] *that the inferior type of pumping-engine should have been selected as the type to receive a higher state of development, and it is this, together with the dictates of experience, which he has had with single-acting engines, which induced him to prepare this brief paper for discussion.'*

What Davey went on to describe in his paper was the basis of his Differential Compound Engine. This is covered in the following chapters. Eventually the engine would gain him worldwide recognition, and for the future Hathorn, Davey Company, a reputation for efficient machinery.

Davey took out further patents the following year, all of which he had to finance. This undoubtedly contributed to 'cash-flow' issues between Davey's expenditure and the consultancy payments he received. Inevitably he owed money to various parties. This seems to have reached crisis point when he was adjudged bankrupt under a petition for *'Adjudication of Bankruptcy'* filed on the 1st May 1869. He was required to surrender himself to the Registrar of the London Court of Bankruptcy at the first meeting of his creditors held on the 31st May 1869.[15]

A further sitting was arranged for 3rd July 1869 for a further examination of Davey's financial situation. The meeting enabled him to make an application for his bankruptcy to be discharged.[16] There was a further adjournment of the proceedings and the whole matter of Davey's bankruptcy was settled in his favour on 2nd October 1869. [17]

'Court of Bankruptcy Before Mr. Commissioner Winslow. This was an adjourned examination and discharge sitting under the bankruptcy, upon his own petition, of Henry Davey, described as of No 20 John-street Adelphi and Victoria-street Westminster, civil engineer, who attributed his failure to want of capital. Mr W. W. Aldridge represented the official assignee, and Mr Linklater supported the bankrupt whose statement of accounts were thus summer up, viz; To creditors unsecured £487. 3s 2d. By debtors doubtful., £18 19s; deficiency £468. 4s. 2d. The bankrupt stated that his expenditure had been £100 per annum. His creditors chiefly reside at Sheffield, Limehouse, Southwark and Blackfriars, &c.

In the absence of any opposition the Court passed the bankrupted and granted his order of discharge, and the sitting ended.'

During his consultancy work Davey had gained some experience in hydraulic engineering, and had taken out a patent for *'Improvements in hydraulic machinery and in means of actuating the same, part of which is applicable to steam engines.'* on the 31st March 1869 (Patent 971).[18] At about this time he made the acquaintance of Mr George Bower, a skilled engineer, who had an interest in supporting the development of new inventions. Bower owned the Vulcan Foundry at St. Neots, Huntingdonshire (now Cambridgeshire) and had established a business manufacturing and installing equipment, which included small pumping engines, for gasworks. Altogether, Bower was to manufacture over 1,000 gasworks, which was quite an achievement for a relatively small provincial enterprise. The making of steam engines was only a tiny part of his business. He used steam pumps in his exhausters in the gasworks.[19]

In 1869, Bower acquired, through debenture holdings, a large interest in the Milford Haven harbour extension scheme in Pembrokeshire. The scheme was in financial difficulty, and Bower became the lessee of the pier, and also the main contractor to complete the extension scheme. The contract work included the installation of hydraulic hauling and hoisting equipment on a new pier. Knowing about Davey's experience in hydraulics, it was not surprising that Bower appointed him to design the hydraulic machinery for the new harbour facility.

The pier extended 600 feet from the shore before it turned at right angles for a further 180 feet, so in plan it was 'L' shaped. At the end of the main pier there was a hydraulic coal hoist and a 3½-ton hydraulic crane. There was also the provision for the erection of a further six cranes.

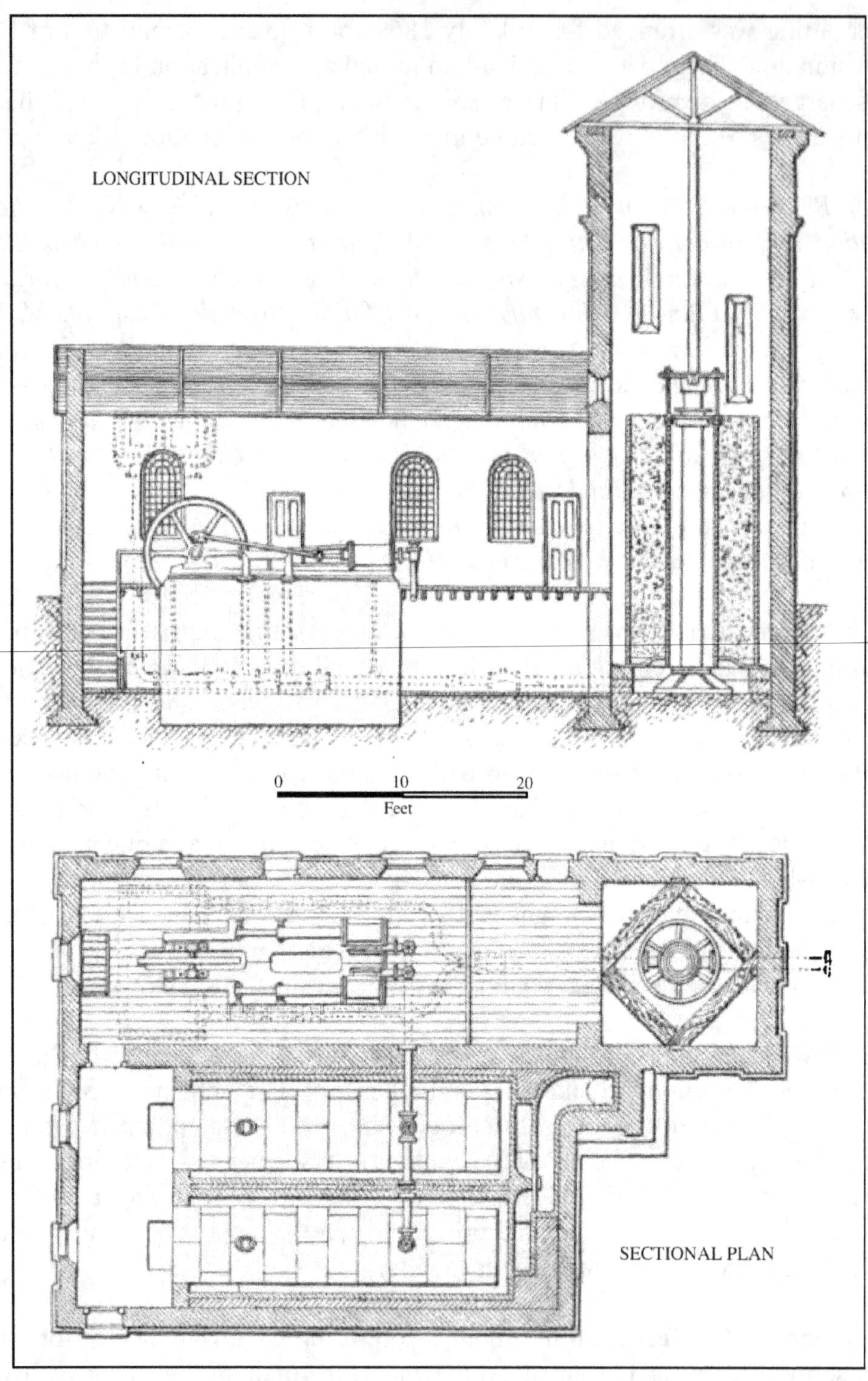

Figure 22. Milford Haven Harbour: Engine House and the Accumulator
(Transactions of the Society of Engineers 1872. p.97)

A pair of coupled horizontal high-pressure steam engines (12" cylinders with 2' 6" stroke) worked a pair of 3½" diameter piston and plunger pumps. The hydraulic circuit fed to a 15" diameter ram accumulator. It had a 17' stroke in a vertical cylinder loaded to a pressure of 800 pounds. Together with two boilers, they were all housed in a building (Figure 22) located 450 feet inland from the shore end of the pier.

The power was transmitted via a line of 3" pipes that ran from the accumulator along the pier. Branch pipes were taken off *en route* to connect to the cranes etc. At the rear of the engine house a well had been sunk for fresh water that was pumped with a direct-acting pumping engine, designed by Davey, into a wrought iron reservoir tank. The engine automatically kept the tank topped up. Davey estimated that the total cost of the hydraulic machinery amounted to about £16,000.[20]

Davey seemed to have worked on other projects for Bower. For example, during October 1870 the Jersey Water Company asked Bower, '*when shall we have your design and estimate for pumping engine particulars of which we gave your Mr Davey?*' The reply the following month was sent with letters and plans with regard to the Engine costing of £1,040 and continued '*I have a pair of Engines to erect for Sudbury Waterworks like this I enclose. It is an engine for high service. Enclosed a copy of the tracing of the engine*'. [21]

Davey must have explained to Bower the principles of the differential gear. At last his dreams were realised when he designed a small engine fitted with differential gearing for the new waterworks at Sudbury, a small market town in Suffolk.

This milestone in Davey's life is only briefly mentioned in his obituary, but the construction of the Sudbury engine was recorded in the contemporary engineering press. Davey also filed his patent (277) for the engine '*Improvements in steam engines, parts of which are particularly to pumping engines.*' on the 1st February 1871. The patent was given six-months protection on the 3rd March 1871. The 1871 census records at the time that Davey was lodging in the household of Daniel Sewell, East Street, St Neots, and that he was an unmarried civil engineer from Devon.

The Sudbury engine, detailed in the *The Engineer* is referred to as, '*A peculiar direct-acting expansive working pumping engine, designed by Mr.Henry Davey, for the Sudbury Waterworks...* ' , and was the first recorded use of Davey's Differential Gear. [22]

The engine is described as double-acting with a steam-jacketed 16" diameter cylinder with a 3' 6" stroke, performing at 12 to 20 strokes a minute. (See Figure 23) The engine actuates two pumps - one at the bottom of the well (16" plunger) and the other at the top (bucket and plunger with a 13" barrel).[23]

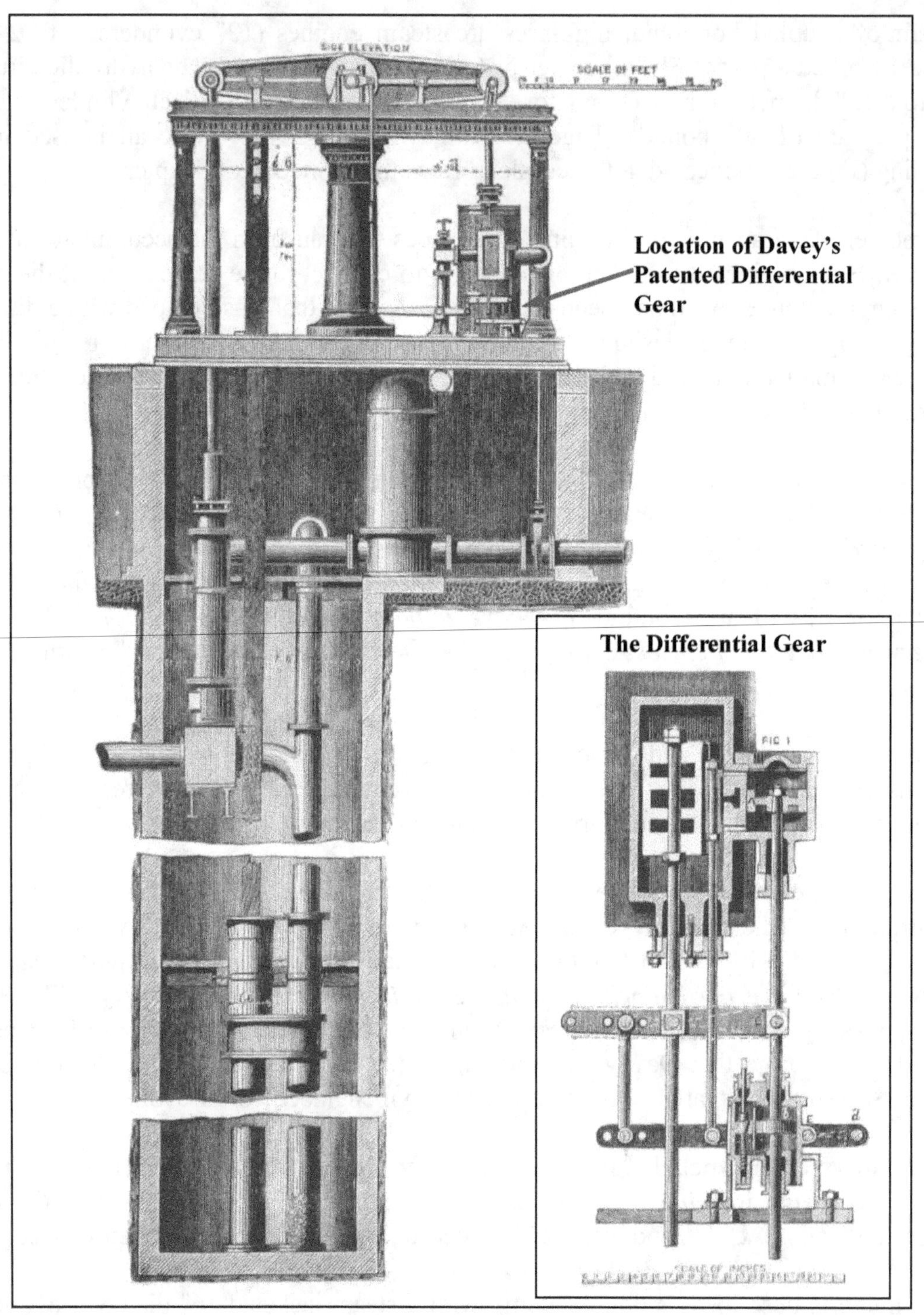

Figure 23. Sudbury: Waterworks Engine and Differential Gear
(*The Engineer.4th October 1872.p.232 [GG]*)

The water was very pure and was obtained from a large boring 300 feet in depth into chalk. A steam engine of 16-horsepower pumped the water from a deep shaft into a reservoir whence it is supplied to the town. It was capable of supplying 16,000gph and there was another engine that could supply an equal quantity if necessary. The waterworks were completed by February 1872. [24]

'The Water Works - Sudbury
These works may now be said to be completed, the reservoir having been filled, the main charged and some few of the houses supplied with the water. The steam engine, which is of 15 horse-power, on the Cornish beam principle, has been found to work exceedingly well; it is fitted with patent slide valves and a patent motion for regulating the supply of steam; it has also a double iron bed-plate for a duplicate engine should one ever be required. The maker is Mr Geo Bower, of St Neots, Hunts, and it would seem to be both a very powerful and very superior engine. The two boilers are by Garrett and Sons, Leiston Works, and are fitted with all model improvements, including a steam feed double acting pump for supplying water. A pipe from the supply main has been, however, found sufficient, as it gives a pressure of 25lbs and readily forces the water into the boiler. The pumps throw 280 gallons per minute and are estimated to fill the reservoir in eighteen hours. The reservoir itself holds 300,000 gallons and will probably give an ample present supply to the town for a week.'

On completion of the work with Bower in 1872, which had been described as a complete success, Davey was asked by Messrs. Abell's, who operated a foundry on Brook Street, Derby, if he would like to join their company. He refused to do so, but was advanced some money by Abell's, which he used to establish himself in Leeds. His address on the patent (3208) for *'Improvements in means of evacuating condensers of steam engines and apparatus therefore, part of which is applicable to the purpose of exhausting gases'* confirms that on the 15th November 1872, he was living (or lodging?) at Tanfield Street, Leeds.[25]

His appointment as Works Manager of the Sun Foundry was really the start of Davey's successful career and it gave him the freedom to promote his many ideas. It no doubt enhanced the reputation of the Hathorn, Davey Company and it was not surprising that he was made a partner in 1874. *'Here* [The Sun Foundry] *the differential pumping engine was rapidly developed and many orders were received. So many they were in fact that the use of additional capital became essential. It was provided by Colonel Hathorn, Davey's lifelong friend and partner, with whom the firm of Hathorn, Davey and Co., of the Sun Foundry, Leeds, was founded.'* [26]

Chapter 5 References

[1] Henry Davey Esq., Familiar Faces' No. 49. *Mining Journal*. 5th December 1891. pp. 1366 and 1367,

[2] Mr Henry Davey. *Pages Weekly*. 4th November 1904. p.540

[3] Obituary. Henry Davey. *The Engineer*. 20th April 1928. p.437

[4] Mr Henry Davey. *Pages Weekly*. 4th November 1904. p.540

[5] Obituary. Henry Davey. *The Engineer*. 20th April 1928. p.437

[6] Advert for Sheppard and Ingram. *Salisbury and Winchester Journal*. 25th June 1864. p. 1

[7] Class 1 - Prime Movers. *The Engineer*. 7th July 1865. p.13

[8] *London Gazette*: 30th December 1864. Issue: 22925. p.6826

[9] Obituary. Henry Davey. *The Engineer*. 20th April 1928. p.437

[10] *The Engineer*. 3rd May 1867. p.401

[11] *London Gazette*. 15th July 1870. Issue: 23634. p.3403

[12] Yorkshire Wills - Mr David Davy. *Sheffield Leeds Mercury*. 22nd July 1903. p.3

[13] Davey's Isochronal Governor. *The Engineer*. 20th March1868. p.202.

[14] Davey, H. 1868. Pumping-Engines for Town Water Supply.' *Transactions of the Society for Engineers*. Spon, (London). pp.88 to 108

[15] *London Gazette*.7th May 1869. Issue: 23495 p.2718

[16] *London Gazette*. 4th June 1869. Issue: 23504 p.3241

[17] Court of Bankruptcy. October 2nd. *Clerkenwell News and London Times*. 6th October 1869. p.6

[18] Patent number 971. Henry Davey. Improvements in hydraulic machinery and in means of actuating the same, part of which is applicable to steam engines. *The London Gazette*. 23rd April 23 1869.

[19] I am grateful to Linda Reed, St. Neots for sharing with me her own research about George Bower.

[20] Davey, H. 1874. On Milford Haven and Its New Pier Works. *Transactions of the Society of Engineers* 1872, Spon, (London) pp.89 to 108

[21] Archival information from Linda Reed, St. Neots

[22] Davey's Patent Pumping Engine at the Sudbury Waterworks. *The Engineer*. 4th October 1872. pp. 228 and 232.

[23] Davey's Patent Pumping Engine at the Sudbury Waterworks *The Engineer*. 4th October 1872. p.228

[24] Archival information from Linda Reed, St. Neots

[25] Patent 'Improvements in means of evacuating condensers of steam engines and apparatus therefore, part of which is applicable to the purpose of exhausting gases' *The London Gazette*. 15th November 1872. Issue: 3920 p.5353

[26] Obituary. Henry Davey. *The Engineer*. 20th April 1928. p.437

6. The Differential Expansive (Compound) Pumping Engine

During the 1870s Davey filed the two patents that would bring him fame. One was for the differential gear and the other for the expansive, or compound, pumping engine. Both patents would eventually become part of the Hathorn, Davey Company's assets, and that would produce extra financial gain for Davey. The combination of both patents produced a relatively simple, reliable and efficient pumping engine.

The first patent, '*Improvements in the method of working the slides of steam or water-pressure engines and apparatus for that purpose,*' dispensed with the tappet motion of the Cornish gear arrangements and introduced a governing element into the system.[1] The second patent for '*Compound pumping engines*' used both high- and low-pressure cylinders and incorporated the differential motion.[2] The steam first released part of its energy into a small diameter (high pressure) cylinder, and then passes into a larger diameter (low pressure) cylinder, where any remaining energy is released. It was a system that proved to be energy efficient and resulted in a saving in fuel consumption.

Hugh Lupton (1940) was later to summarise the two achievements.[3] '*Hathorn, Davey's engines were less liable to accident than the Cornish engines had been, both because expansive use of the steam was attained by compounding (between a high and low pressure cylinder), instead of by the rapid indoor stroke of the Cornish engine and also because the differential gear introduced an automatic throttling action in case of excessive speed were therefore able to deal with higher steam pressures than had therefore been usual in mining practice.*'[4] The Compound Differential Pumping Engine was to be one of the company's most successful products. The pumping engine, constructed to the patents of Henry Davey, would ultimately become one of the company's main products for the next thirty years, and would be used widely both nationally and internationally. The last Compound Engine to be manufactured by Hathorn, Davey was in the 1920s.

Davey gave many papers about the engine during it's first years of production, and due to his patent's uniqueness, many of the orders placed in the 1870s specified Davey's patent gearing as a requirement. Ultimately the gearing would become a standard component. Numerous sets of the gears were also fitted to existing Cornish Beam Engines, both in Britain and abroad. One particularly large regular order was for pumping engines on the Rhur Coalfield, Germany.

The Cornish Beam Engine was particularly vulnerable, because any small variation in load brought about by breakages in the shaft pumping mechanism usually meant disaster. The result from such a calamity could cause the main beam to fracture under the shock, or a cracked cylinder, or a piston smashed to pieces.

The Differential Gear

There are many descriptions of the Differential Gear that range from simple, to very complex. Perhaps Davey's earliest description of the patent was in a paper he gave to the *North of England Institute of Mining and Mechanical Engineers*, on the 13th September 1873. Davey used the Newton Gap and Clay Cross engines, as examples. The paper was later published in the Institute's Transactions together with comprehensive illustrations.[5]

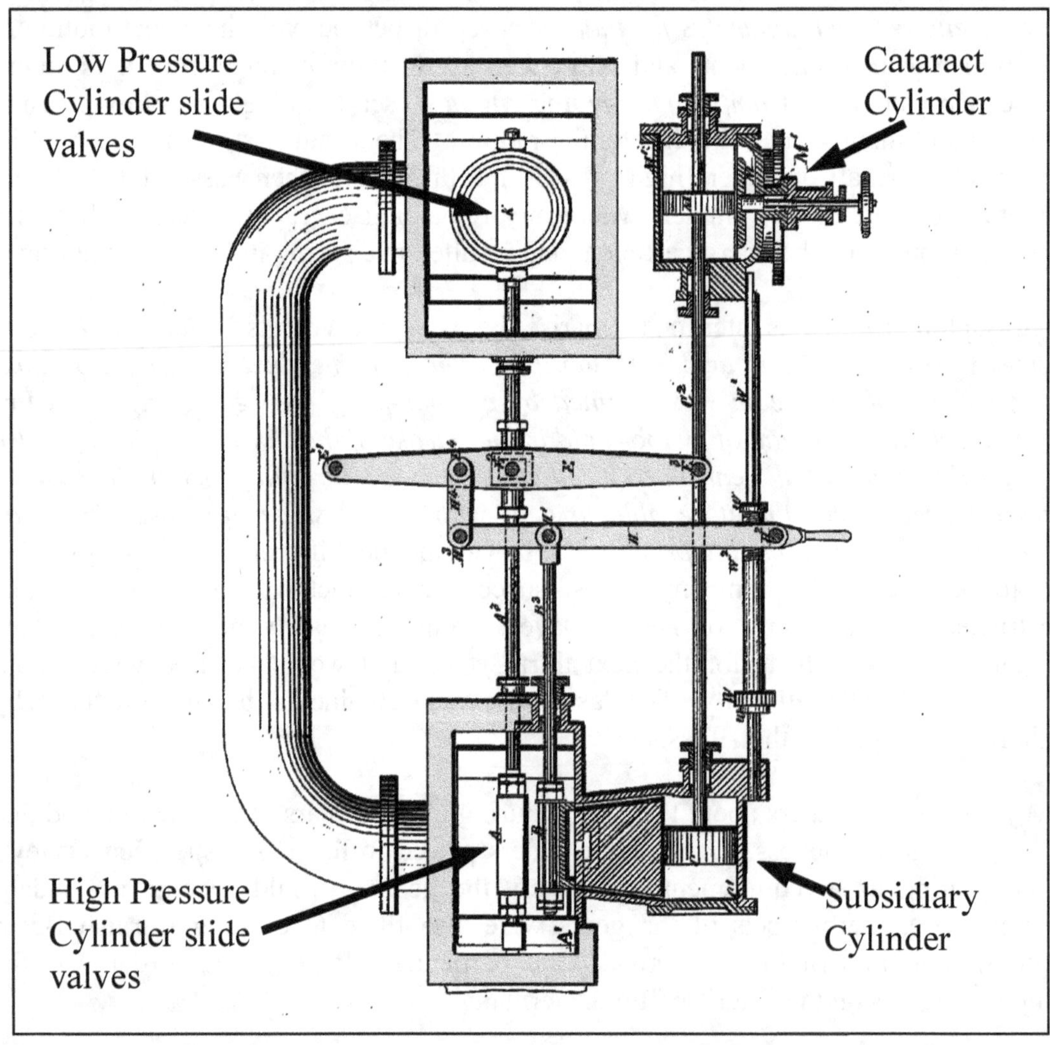

Figure 24. Davey's Differential Valve Gear originally designed to fit horizontally between the two cylinder valves on top of a horizontal compound engine
(Modified from: United States of America Patent 167,509 dated 7th September 1875)

It is not proposed to describe, or examine, the Differential Gear in detail, as there are many contemporary articles that do this. In essence, the Differential Gear consisted of a subsidiary motor that regulated the steam input that was synchronised with the main engine. Should the main engine become unsynchronised, the gear would throttle the steam going into the cylinder, or shut it off completely.

The main principle of the gear is that the valves of the engine have a motion that is the resultant of two other motions. The first is a constant and independent motion, and the second, the dependent variable motion of the steam engine. Erratic movement of the engine that resulted from a change in load produced a corresponding variation in the distribution of steam, through the resultant motion of the valves. A sudden change in this motion would administer steam on the reverse site of the piston, thereby dampening the motion. The application of differential gearing made the engine self-regulating. Steam would be shut off when the load was removed. [6]

In addition to the British patent, Davey also applied for one in the United States of America that was filed on the 1st August 1874. Simply titled, '*Improvement in Steam Engines'* the patent, number 167,509, was accepted on the 7th September 1875. Figure 24, is a modified drawing of the Differential Gear taken from the patent. [7]

Initially the Differential Gear was fixed horizontally on top of a Compound Engine between, and adjacent to, the slide valves of the high and low-pressure cylinders. Later, the gearing could frequently be found sited in a vertical position, fixed to the floor, at the side of the engine.

In his American patent Davey describes the Differential Gear as, '*a method of working the slides of steam or water pressure engines, without the use of eccentrics, cams, or other rotating parts, the movement of the main piston being employed to move the main slide through a part of its stroke, and also to move the slide of a subsidiary cylinder and the rest of the stroke of the main slide being effected by the movement of the piston of the said subsidiary cylinder. The speed with which the stroke of the main slide is completed or effected is governed by a "cataract' that is, the resistance of a liquid in a cylinder to a piston moving therein. This mode of working the slide, although peculiarly applicable to direct-acting engines, such as are employed to work pumps without connecting-rods, cranks, or fly-wheels, is applicable also to engines working rotating shafts, and either to simple or compound engines, and either in combination with or without expansion valves or slides.'*

In time, the Differential Gear became more complex and was still being advertised in the Hathorn, Davey catalogue at the end of the 19th Century, and where it was described as follows:

'Davey's Differential Valve Gear

Davey's Differential Gear is designed to regulate the admission and release of steam used in Direct-Acting Pumping Engines.

It has control of both the steam and exhaust valves throughout the stroke, and is therefore enabled to adjust the amount of steam admitted, so as to exactly suit any variations which may take place, either in the steam pressure or in the load against the engine..........

Description of Differential Gear

Davey's Differential Gear consists essentially of a small subsidiary engine, the speed of which can be regulated by means of a cataract cylinder [a cataract is a device were water or other fluid is forced from one end to the other through an adjustable opening which can be used to regulate speed] to any desired rate, and of a pair of links having no fixed anchorage, but attached at one end to a rocking shaft (or other convenient part) driven from and moving with the main engine, and at the other to the small subsidiary engine above referred to. It is evident that, supposing the subsidiary engine and the main engine be moving in opposite directions, a point at the centre of the links will not be moving in accordance with either of them, but will have a mean or differential motion between the two. It is from this central point that the valves receives their motion, and the arrangement is such that the valves are opened when the links are moved in the direction in which the subsidiary engine tends to move them, and closed when the links are moved in the direction in which the main engine tends to move them. As before noted, the subsidiary engine is controlled by a cataract cylinder, and can, therefore, be set to move at the rate which will give the necessary valve opening when the main engine is working at any required speed, but directly this speed is exceeded, either from increase in steam pressure or decrease in load, the main engine gains upon the subsidiary engine, the closing of the valves is accelerated, and steam either throttled or entirely cut off. The gear, therefore, forms a sensitive governor acting directly upon the steam valves. The illustration [Figure 25] shows a subsidiary engine of the kind above described, together with a second small subsidiary engine also regulated by a cataract. This second small engine is used to reverse the steam valve of the Differential Gear Engine, which, therefore, pauses, and with it the main engine also, until the secondary has made its stroke. The cataract regulation enables this pause to be adjusted to a nicety, from a mere dwell at the end of each stroke to a pause of 10 to 15 or more seconds, when for any reason it is desired to run the main engine dead slow.'

Depending on the use, some engines were designed so that the gearing would not put steam on the opposite side of the piston if the load were removed. The Cheddars Lane sewage pumping engines are such an example where experiments have shown that the system was designed to do this. [8]

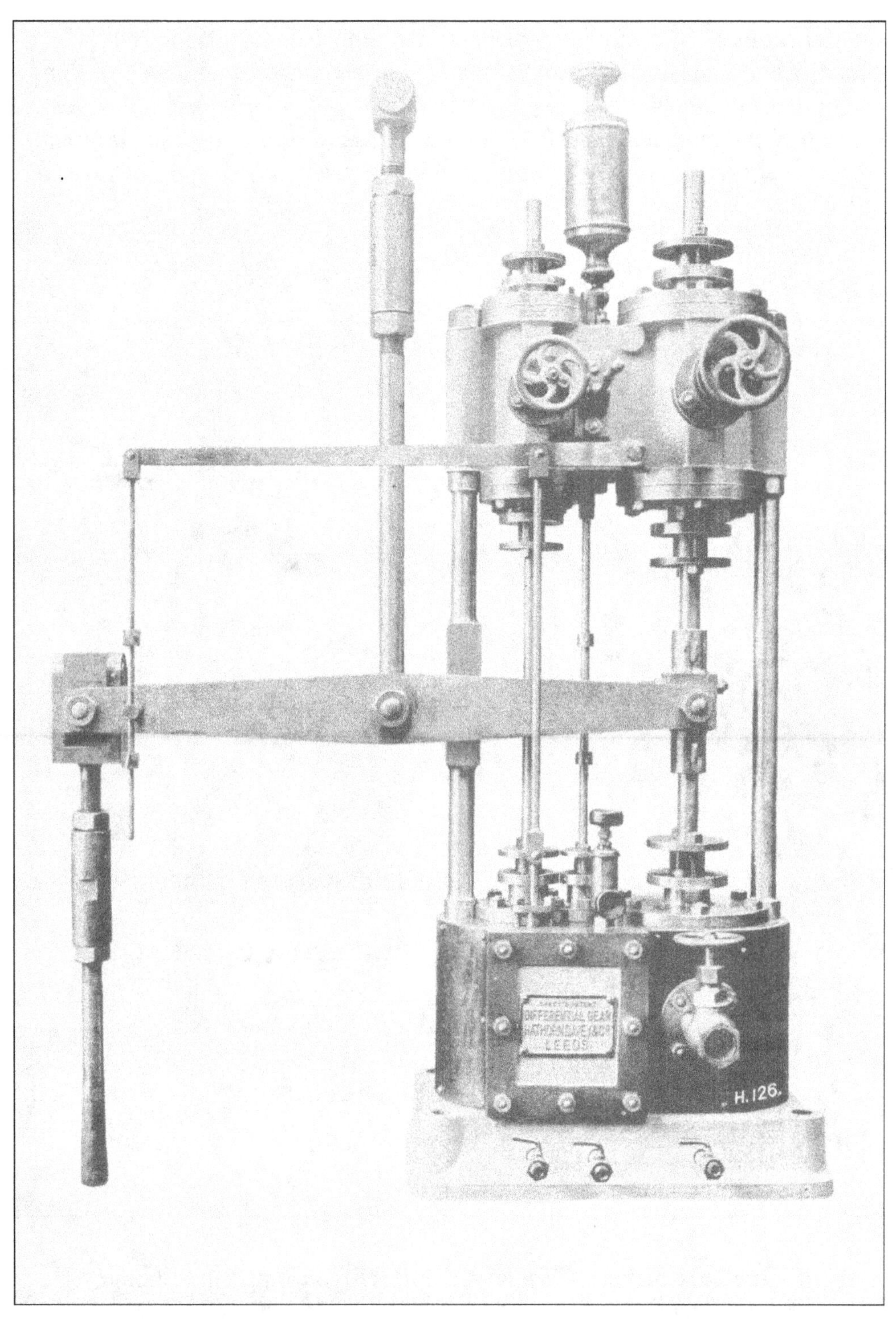

Figure 25. Davey's Differential Valve Gear
(Hathorn, Davey & Co. Catalogue: Pumping Engines for Waterworks, 1899)

An editorial footnote in a more recent paper by John Hole (2005) on the 'The Davey Differential Engine' described the gearing as, *'an elegant mechanical solution to the problem of proving a self-regulating function for the potentially unstable system that is a non-rotative pumping engine. It is so named as it relies on the difference (the differential) between a constant motion and a variable motion to provide self-regulation.'* [9]

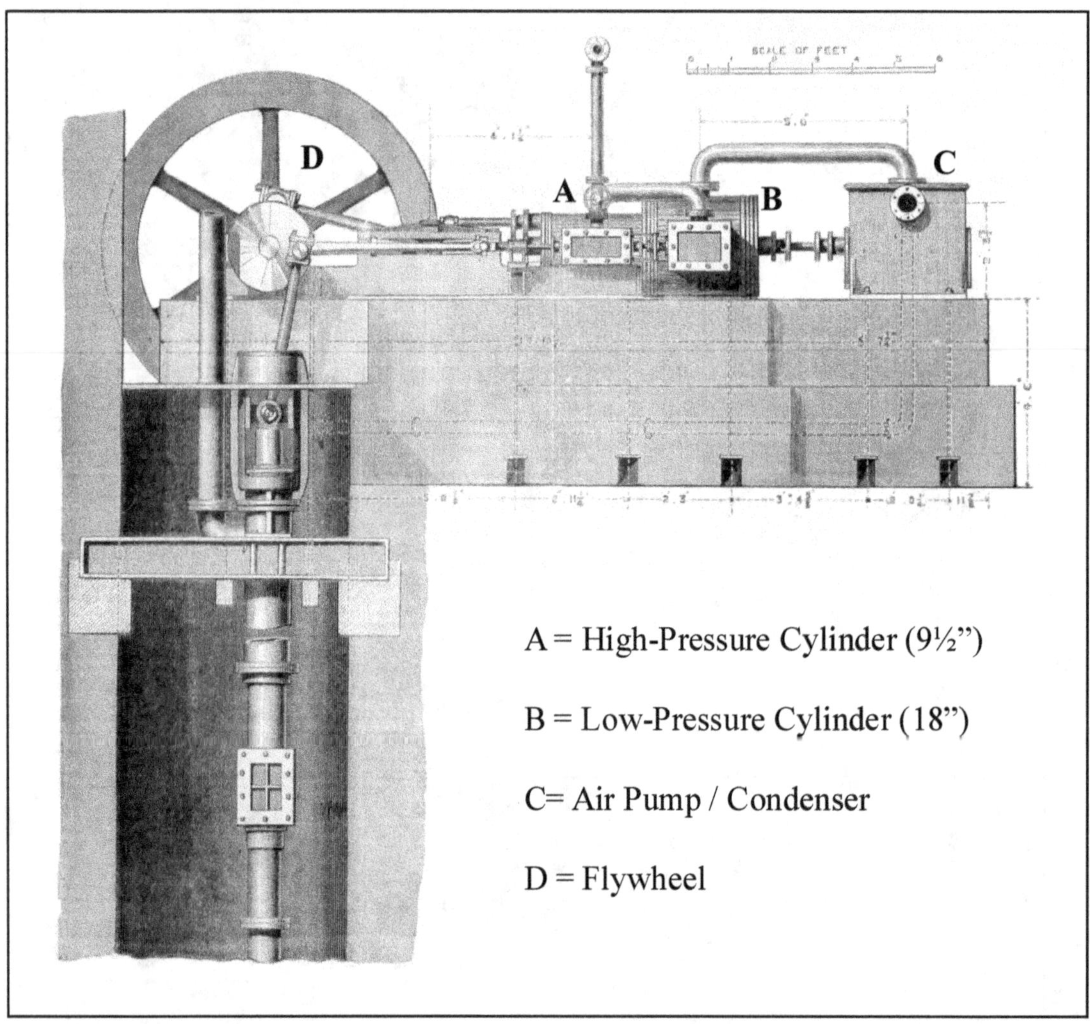

A = High-Pressure Cylinder (9½")

B = Low-Pressure Cylinder (18")

C= Air Pump / Condenser

D = Flywheel

Figure 26. Chichester Waterworks: Rotative Compound Engine
(The Engineer. 8th November 1878. pp.336 and 340 [GG])

The Compound Engine

By the 1860s, there seems to have been renewed interest in the Compound Rotative Engine, and that may well have influenced Davey in his formative years as an apprentice in Tavistock. Certainly by the 1870s there was significant evidence for Davey to conclude that, '*Compound engines are, as a rule, far more economical than single cylinder engines.*'

In the early years of the Partnership of Hathorn, Campbell and Davis, several orders were taken for Rotative Compound Engines. The engine illustrated in Figure 18 (Chapter 4) is a Rotative Cross-Compound engine that has the high and low-pressure cylinders side by side, and was almost certainly based on a Carrett and Marshall design.

Henry Davey modified the designs of the Rotative Compound Engine to incorporate some of his own ideas. The engine (9½"+18", with a 2' stroke) shown in Figure 26 was ordered for the Chichester Waterworks on the 7th September 1876. A flywheel is attached to one end of the crankshaft. There is a disc at the other end of the shaft, to which a con rod (connecting rod) is attached to work the well pumps. A tail rod from the low-pressure cylinder operates the air pump. Con rods ran from two cams located on the crankshaft to operate the valves. [10]

However it was Davey's intention to produce a double-acting non-rotative Compound Engine, as illustrated by his early installation at the Seaton Delaval Colliery.

The early attempts by Messrs. Hornblower, Trevithick, Woolf and Sims failed to produce a significant Compound Pumping Engine. However, Sims did at least have the satisfaction of knowing that in 1844 one engine (22"+44") manufactured by Harvey's of Hayle to his specification, was exported to Linares, Spain. [11]

Davey had noted that such engines had been single-acting, with steam only being applied to one side of the high-pressure cylinder only, and that on the return stroke, the expanded steam was just applied to the opposite side of the low-pressure piston. He remarked that such engines had three serious defects. [12]

(i) There was great thermal loss, due to the cooling influence of continuous exposure to the condenser.
(ii) It was difficult to maintain a uniform length of stroke.
(ii) The engines were costly to operate because both cylinders were single-acting.

While much of this early experimentation was being conducted, the double-acting principle was slowly making the transition to the Cornish Beam engine, where steam was applied to both sides of the piston, a principle patented by Watt.

Clearly Davey did not think this to be adequate, and considered that even greater efficiency could surely be obtained from the Compound Engine.

'Efforts have been made with excellent results, to obtain the greatest economy of fuel with Cornish engines; but higher results have been secured with compound rotative engine. This naturally suggests the question, whether the direct-acting engine is not capable of accomplishing all that has been attained by the rotative engine?'

Davey believed that there were four propositions that must be considered if he was to construct a reliable and efficient Double-Acting Engine.[13]

i) To expand steam to a moderate extent only in a single cylinder, the maximum strain on the piston and its attachments must greatly exceed the main strain.
ii) The maximum speed must greatly exceed the mean speed.
iii) The weight of the moving mass must greatly exceed the water load. (He defined the water load as the actual weight of water lifted by the engine at the speed of the piston.)
iv) The maximum strain and speed can be reduced by the employment of a second cylinder, even when the initial pressure and ratio of expansion are greatly increased.

By considering the above, the criteria were established for introducing a second cylinder, and the Davey's Compound Engine was created, combined with plunger pumps the engines would raise water from great depths in one lift.

Based on one of several published descriptions of the engine by Davey, the compound engine consisted of a pair of horizontal cylinders (See Figure 27), placed end to end. [14] The bottom of the high-pressure cylinder formed one of the covers of the low-pressure cylinder. There were two piston rods to the low-pressure piston, which pass through tubes cast on the jacket of the high-pressure cylinder, so that they ran in the same plane with the rod of the high-pressure piston (Figure 28). Those three rods are coupled to one crosshead, to which the connecting rod for working the pumps was attached. The cylinders were firmly secured to a strong girder bed, and the condenser was carried on a separate bed at the rear of the engine. The air pump was worked by means of a tail rod from the low-pressure piston (shown in Figure 27). This was the Compound Engine in its simplest form, and in time, Davey was to modify his design of the engine. There was, for example a single piston rod passing in the central plane of both the high and low pressure cylinder to replace the three rods coupled to the crosshead.

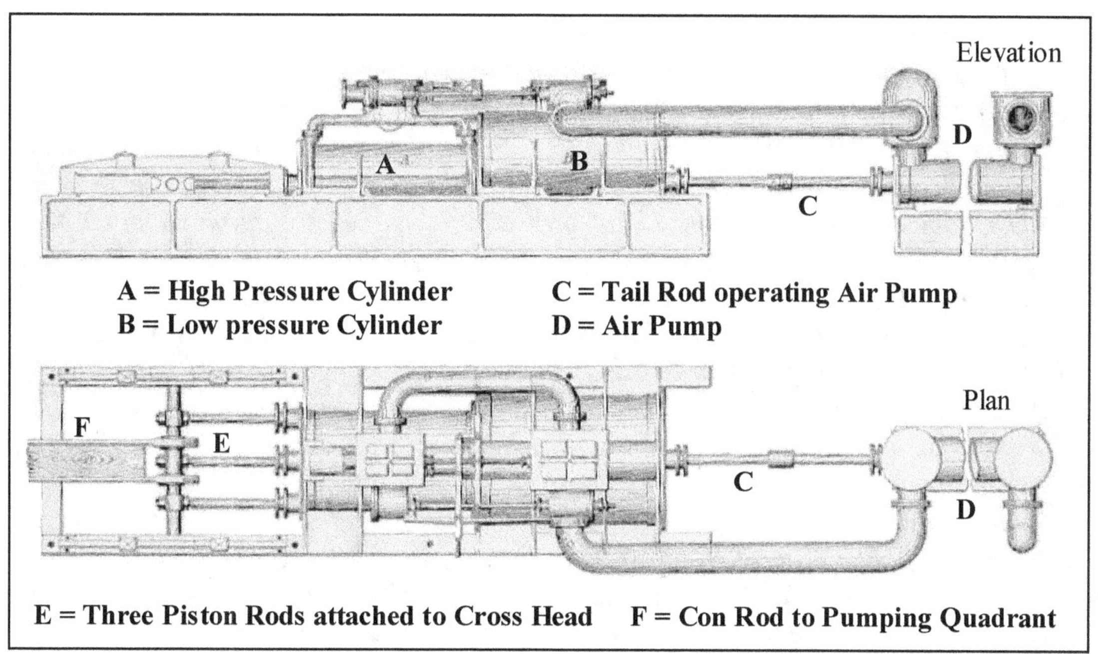

Figure 27. Horizontal Compound Condensing Pumping Engine
(Modified from: Davey. 1873-74. TNEIMME. v.23. Plate 1)

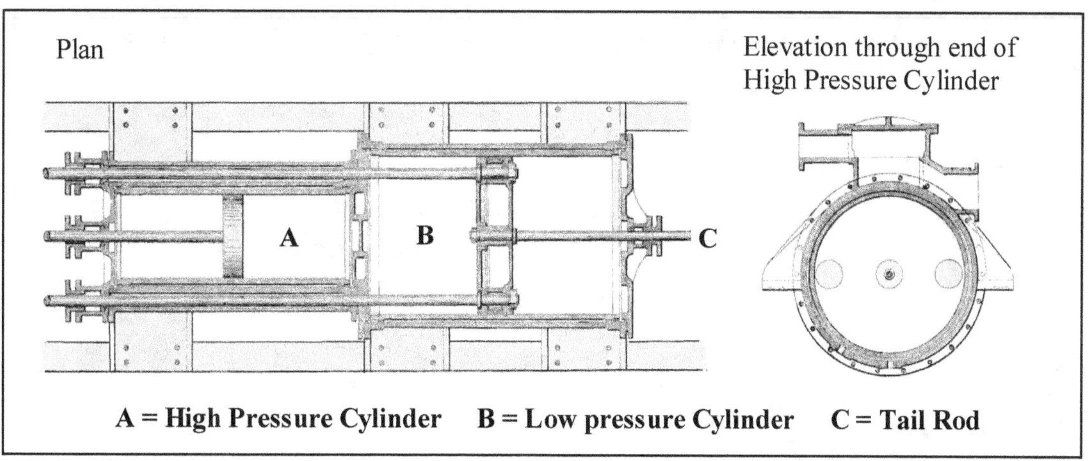

Figure 28. Diagram showing the arrangement of piston rods in the two cylinders
(Modified from: Davey. 1873-74. TNEIMME. v.23. Plate 2)

In his paper Davey also explains the practicalities of the two-cylinder compound engine.[15] *'Judging from the demand which there exists for compound engines, there is every reason to believe that they possess certain advantages, but as far as economy of fuel is concerned it is well known that this is governed by the pressure and ratio of expansion employed and the introduction of the second cylinder is therefore purely a practical question. By the introduction of a second cylinder the maximum strain thrown on the piston and its attachments is greatly reduced, and a greater uniformity of motion secured. In the present example the variation of force between the commencement and end of the stroke is about 3 to 1 with an eight-fold expansion, whereas in a single cylinder engine it would be as 8 to 1. In Cornwall during the race, for economy of fuel, some years' ago, the engineers carried expansion in a single cylinder to its utmost practical limit. Taylor's engine, at the United Mines, worked with a cut-off of one-tenth the stroke, and did a duty of 112 millions. Experience has proved, however, that such high grades of expansion cannot be safely employed in a single cylinder; and we find that the custom is now to work Cornish engines with a three-fold, or at most a five-fold, expansion; this partly accounts for the falling off in the duty reports. A high degree of expansion cannot be employed with safety in the single cylinder engine, because the variation of force during' the steam stroke is too great—therefore the reason for the use of the second cylinder. The economy in the construction of this engine and the buildings for it is very great as will be readily seen by an inspection of the model before us. Power for power, this engine does not weigh two-thirds that of a Cornish engine, although the ratio of expansion employed is far greater, whilst the gear is of the most simple description, and the safety secured in working by the peculiar action of the gear must effect considerable economy in maintenance.'*

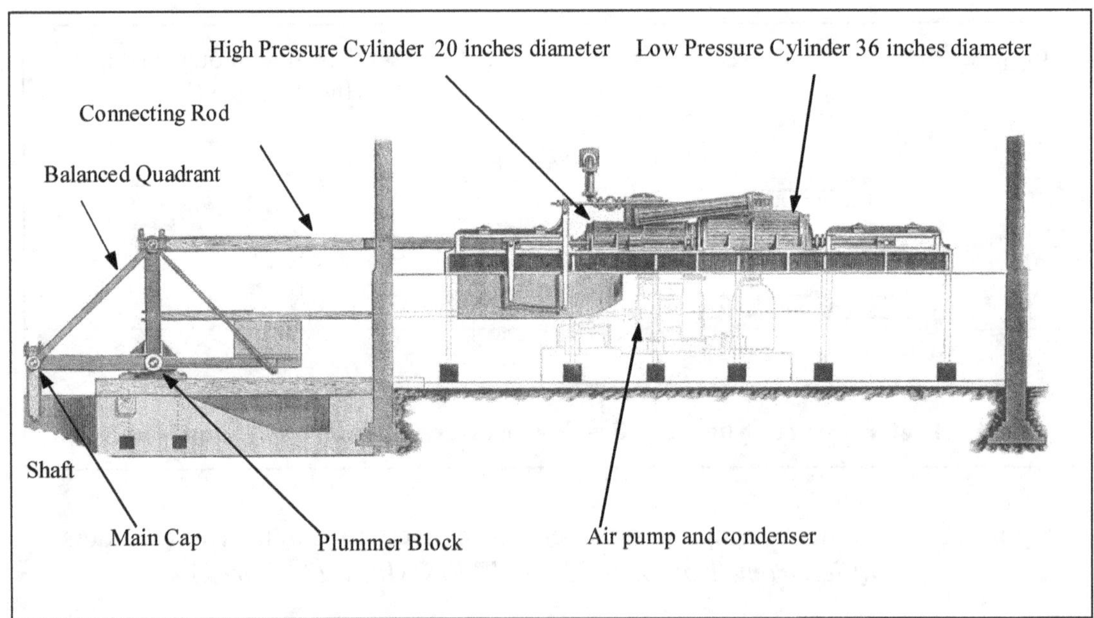

Figure 29. Pumping Arrangements for a Compound Engine at the Solana Mine, Córdoba Province, Andalucía, Spain
(Vernon and Cabrera. 2018. De Re Metallica. v.30. p.83)

Davey also offered an explanation on the merits of using air pumps and condensers. They were usually not part of the engine, but distinct items, and in some cases, such as in underground use they were sometimes worked by a small hydraulic engine actuated by water pressure in the rising main. Other designs had the air pump operated by a tail rod from the main piston. Again, this arrangement for operating the air pump was eventually modified by Davey.

Figure 29 shows a typical pumping engine arrangement at La Solana lead mine, Córdoba, Spain.[16] The engine is mounted on a purpose built block, with the air pump and condenser at the side. A reciprocating horizontal beam (con rod), operated from the crosshead, was connected to a single quadrant to convert horizontal motion to vertical motion. The resulting up-down motion of rods connected to the quadrant operated pumps down the shaft. Orders for Compound Engines were often broken down into four parts: engine, air pump and condenser, quadrants and plummer blocks.

Figure 30. Horizontal Compound Pumping Engine at Cheddars Lane Sewerage Pumping Station, Cambridge. The low-pressure cylinder is in the foreground. The vertical Davey Differential Gear can be seen to the left of the steps.
(Courtesy of Alan Denny, Cambridge)

The Horizontal Differential Compound Pumping engine was sold worldwide. Some of the largest orders were for coalmines in Japan, and a Tasmanian gold mine.

Davey Differential Compound Engines were also manufactured in North America under license to Davey's specification. They were covered by USA patent 167,509 *Improvement in Steam Engines,* filed in 1874, and a license or 'Assignments of Rights' to construct them seems to have been granted to Joseph Moore, San Francisco, California.[17] There was a least one 'Davey Engine', constructed in 1875, for use on a silver mine on the Comstock Lode, Nevada.[18]

Chapter 6 References

[1] Davey, H., 1875, *Improvements in the method of working the slides of steam or water pressure engines and apparatus for that purpose.* UK Patent 2152.

[2] Davey, H., 1876, *Compound pumping engines.* Patent 2814.

[3] HL. 1940. *Hathorn, Davey and Company, Sun Foundry, Leeds.* p..3

[4] Henry Davey Esq., Familiar Faces No. 49. .*Mining Journal,* 5th December 1891, pp. 1366 to 1367

[5] Davey, H. 1874. Differential Expansive Pumping Engine. *Transactions of the North of England Institute of Mining and Mechanical Engineers.* v.23. Newcastle-upon-Tyne. pp.3 to 14 with plates.

[6] Davey, H. 1874. Differential Expansive Pumping Engine. *Transactions of the North of England Institute of Mining and Mechanical Engineers.* v.23. Newcastle-upon-Tyne. pp.3 to 14, with plates.

[7] Davey, H. *Improvement in Steam-Engines* United States of America Patent 167,509, dated 7th September 1875

[8] Hole, J. 2005. The Davey Differential Engine. *Bulletin of the International Stationary Steam Engine Society.* 27.4 - Editors footnote.

[9] Hole, J. 2005. The Davey Differential Engine. *Bulletin of the International Stationary Steam Engine Society.* v.27.4.

[10] Compound Pumping Engine, Chichester Waterworks. *The Engineer,* 8th November 1878 pp.336 and 340

[11] Vernon, R.W. 2017. John Taylor and Sons and their three 'drops of comfort.' *British Mining* .v.103. Northern Mine Research Society. p.39

[17] Davey, H. 1877-78.Direct Acting or Non-Rotative Pumping Engines and Pumps. *Minutes of the Proceedings of the Institution of Civil Engineers* (London). v.53. p.100

[13] Davey, H. 1877-78.Direct Acting or Non-Rotative Pumping Engines and Pumps. *Minutes of the Proceedings of the Institution of Civil Engineers* (London). v.53. p.101

[14] Davey, H. 1873-74. Differential Expansive Pumping Engine. *Transactions of the North of England Institute of Mining and Mechanical Engineers.* v.23. p.3

[15] Davey, H. 1873-74. Differential Expansive Pumping Engine. *Transactions of the North of England Institute of Mining and Mechanical Engineers.* v.23. p.7

[16] Vernon, R.W. and Cabrera, A.M. 2018. La Mina Solana de Belalcázar: Una Borrascosa Operación Minera Escocesa en Córdoba. *De Re Metallica.* Revista de la Sociedad Española para la Defensa del Patrimonio Geológica y Minero. v.30. pp.73 to 89

[17] DLC Box 1

[18] Dickie, G.W. 1914. Men and Machinery of the Comstock -- The Barclay and Davey Engines. *The Engineering and Mining Journal.* v.98:17. 24th October. pp.734 to 738

7. Hathorn, Davis, Campbell and Davey: 1874 to 1877

Henry Davey was made a partner in early 1874, and his name was added to the firm's title. It became known as Hathorn, Davis, Campbell and Davey.[1]

The same year Alfred Davis and Henry Davey were both admitted as members of the Institution of Mechanical Engineers.[2] Davey took this opportunity to promote the Company further, by presenting a paper to the Institution on the 29th October 1874 at their London premises.

The paper entitled, '*On Direct-Acting Pumping Engines and Pumps for High Lifts in Mines.*' was well received.[3] In his vote of thanks, the President of the Institution stated, '*That the very interesting paper which had been read included not only what might be called a novel system of pumping; but also a most ingenious mode of regulating the motion of the slide valve, by coupling up a regularly moving piston governed by a cataract, with one which might be irregularly moving, namely the piston of the engine itself.*' He went on to say, '*He had been struck very much by the ingenuity with which this plan of valve motion had been worked out so completely...*'

The introduction to the paper was very much an elaboration of Davey's previous papers with a summation of the efficiency / inefficiency of various Cornish Beam Engines, followed by an account of Davey's Compound Differential Engine. A detailed description was given of the Compound Pumping engine (27½"+44") installed at the Lambton Colliery, County Durham (Order 2230, dated 6th January 1874, costing £1,700). The engine was apparently used during shaft sinking operations and was worked without a condenser because the water was so muddy!

Davey also discussed pumps and pitwork (the heavy system of rods and quadrants used to actuate the pumps), and referred to the shocks, and thus potential breakages incurred to the pumping system if the load varied, i.e. if the system is not fully charged with water during the steam or suction strokes. Davey argued that this risk was obviated if spear rods, beams and quadrants were entirely dispensed with by placing the engine underground and employing it to force the water directly to the surface. Although this concept of placing either steam or hydraulic engines underground was not new (e.g. Will's founder shaft hydraulic engine, and Ecton deep adit steam engine, Staffordshire), most underground engines were usually located at the top of the pumping system. However, what made Davey's concept revolutionary was the placing of the engine at the bottom of the pumping system and forcing the water to the surface in one lift. At the time of writing the paper, the firm had sold a number of such engines to Davey's specification.

One example of an engine that Davey used to explain this concept was installed at the Navigation Colliery, Mountain Ash, Aberdare, South Wales (Order 2240, dated 15th March 1874).[4] It was a relatively large engine (28"+50", stroke 5' 0" (not jacketed)) and is shown in Figure 31. It is described in the order book as, *'One "Davey's Patent" Compound Differential Expansive Condensing Pumping Engine complete, as per tracings and specification with cylinders.'* The engine drove a pair of pumps, also fixed onto the bedplate, which forced the water 1100ft (335.3m) directly to the surface in one lift.

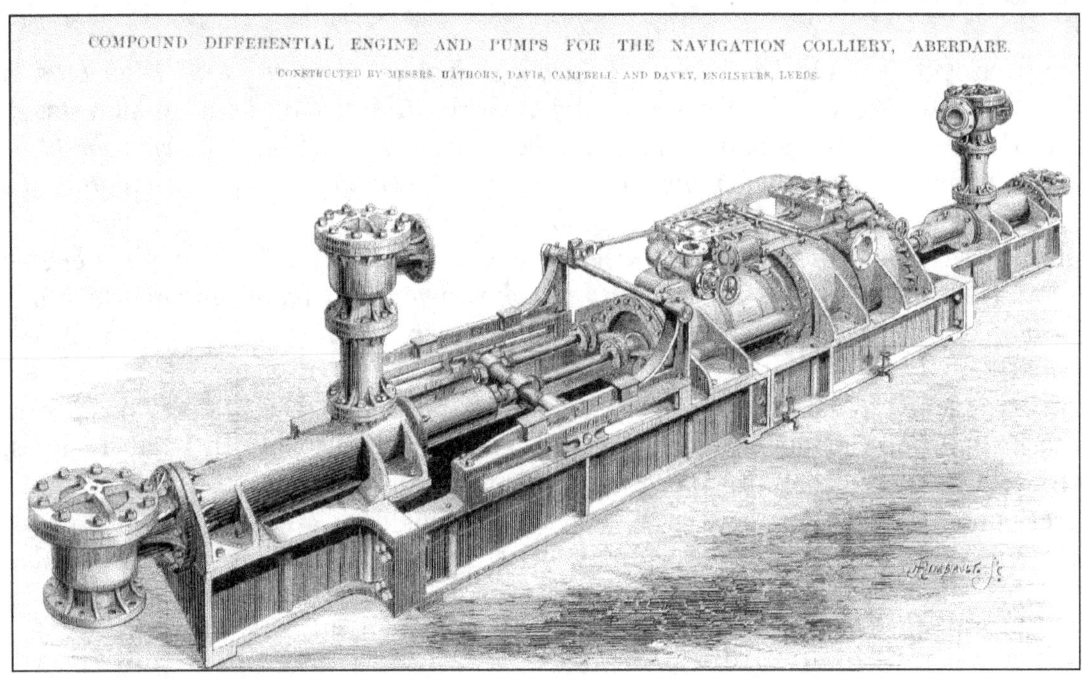

Figure 31. Compound Differential Engine and Pumps, Navigation Colliery, S. Wales. The two single-acting plunger pumps are fixed onto the bedplate at each end of the engine. A tail rod from the low-pressure cylinder piston operated one of the pumps. *(Engineering. 2nd April 1875. p.273 [GG])*

The pumping arrangement at the Navigation Colliery can be seen in Figure 32. Pipes took steam from the surface to the engine, which was located in a chamber at the side of the shaft. This was constructed above the reach of any variations in the level of mine water. The engine operated a pump that drew water from a cistern and forced it to the surface via a rising main. The cistern was fed by a pump, operated by a hydraulic engine (Order 2247, dated 15th March 1874), and was sited in a lower chamber. The full head of water in the rising main worked the hydraulic engine. This was achieved with a smaller diameter pipe that fed water from the rising main to the hydraulic engine.

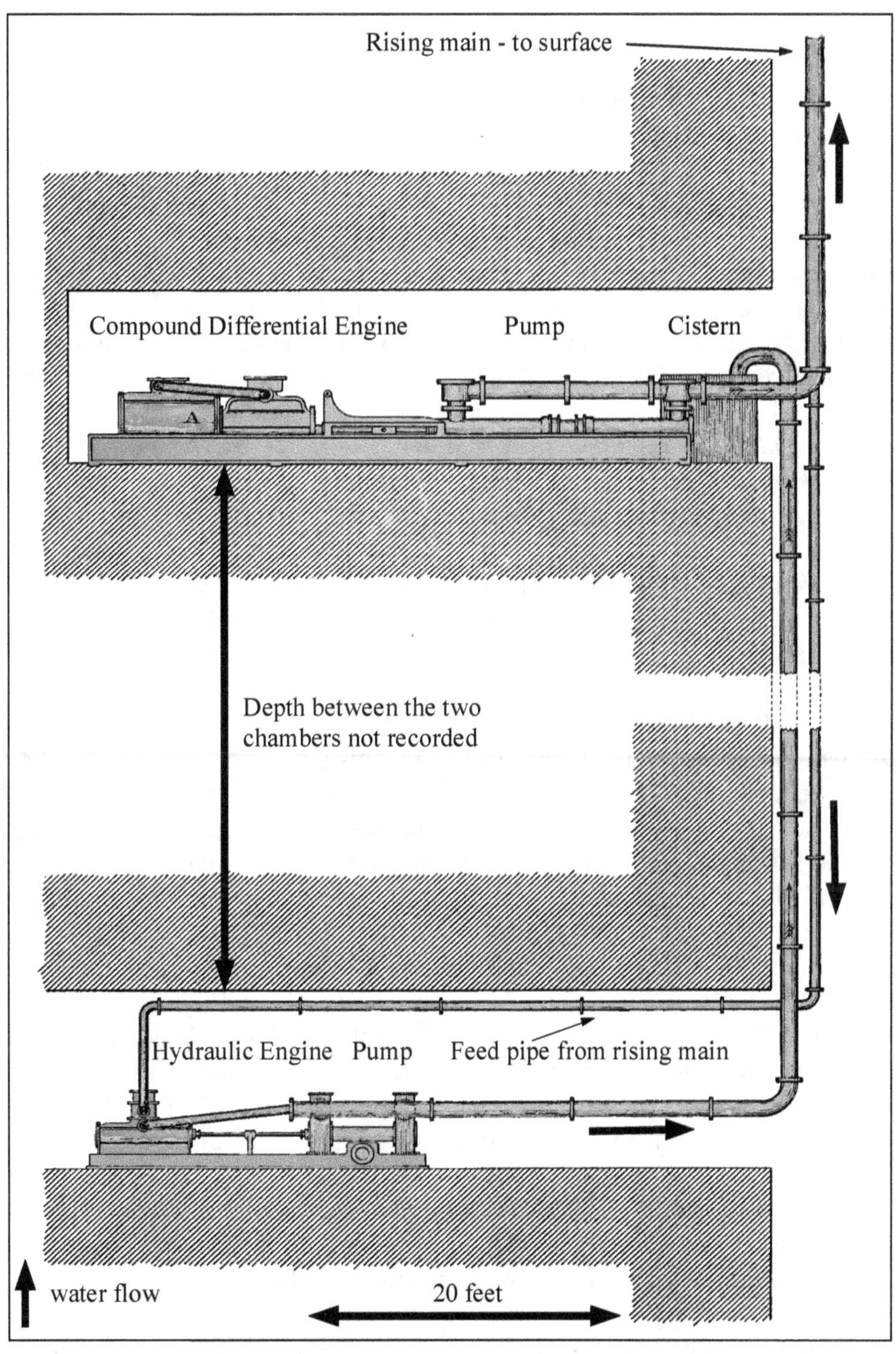

Figure 32. Navigation Colliery. Arrangement of engines and pumps
(Proceedings of the Institution of Mechanical Engineers. 1874. Plate 85)

Pumping under such high pressures meant that there were also other considerations. From his experience Davey specified single or double-acting pumps for such scenarios, fitted with single-beat or mitre valves constructed with rubber discs. The latter could cope with moderate pressures below 50psi.

After delivering his paper, Davey exhibited models of the Compound Differential Pumping Engine together with a working model illustrating the actions of the differential valve gear. There was considerable discussion about the arrangements of the gearing and Davey cited several examples, whereby loading on the engine could change suddenly. At the Lambton sinking, previously referred to, he mentioned that a load of 7 or 8 tons had been placed on the engine on its outward stroke when the bottom clack valve failed. '*In order to save the cylinder covers from being carried away, steam had to be admitted on the opposite side of the piston, reversing the ordinary working of the engine and forming a cushion in front of the piston; and this was entirely accomplished by the automatic action of the differential gear.*'

This was salesmanship at its very best. Davey didn't have to try very hard as the proof was there for all to see, and this was reflected eventually in orders for the differential engine and gearing.

Shortly before Davey presented his paper, the *Prussian Mining and Ironworks Company* had placed one of the largest orders for this type of equipment. The Prussian Company operated the Zollern, Hansa and Erin collieries on the Rhur Coalfield, Germany. The colliery company had been formed by William Mulvany, an Irish capitalist and it was probably Mulvaney's connections with William Coulson, a mining engineer from North-east England that brought Davey's innovations to his attention.[5,6]

Orders 2283 to 2284 were placed with Hathorn, Davis, Campbell and Davey on 7th September 1874 for two Compound Differential Expansive Engines (2 x 35"+60", stroke 6') for the Erin Colliery. They would operate 12½" double-acting ram pumps with a 6' stroke. Each engine was provided with a separate condenser and double-acting air pump (Order 2285). The two hydraulic engines (Orders 2286 and 2287) were specified to pump 500gpm to a height of 250 feet, and were operated by a 910 feet effective head of water. The diameters of the hydraulic engine and pumps were 7½" and 12" respectively, with a 5ft stroke. The orders cost a total of £7,800 (Compound Engines, condensers and pumps £6,200, and the hydraulic engines and pumps £1,600). Figure 33 shows cut-away sections through the engine and pumps.

A more detailed specification of the Erin pumping system is provided in Davey's paper, '*The Underground Pumping Machinery at the Erin Colliery, Westphalia.*' which Davey presented to the Society of Engineers on the 12th June 1876.[7,8] In a similar arrangement to that shown in Figure 32, the main engines were kept out of danger of flooding. The advantage of using hydraulic engines to operate the lower pumps was that they would work under water, and were operated from the main engine room. *'As a further security they could be placed in a water-tight staple, and by such means the hydraulic engines could be under repair, even when the water rose to the main engines 300 feet up the shaft.'* In the following discussion Davey confirmed that the engines had been working constantly for nearly two years, except for short stoppages every to 2 to 3 months, to grind the pump valves that had been damaged by grit in the water.

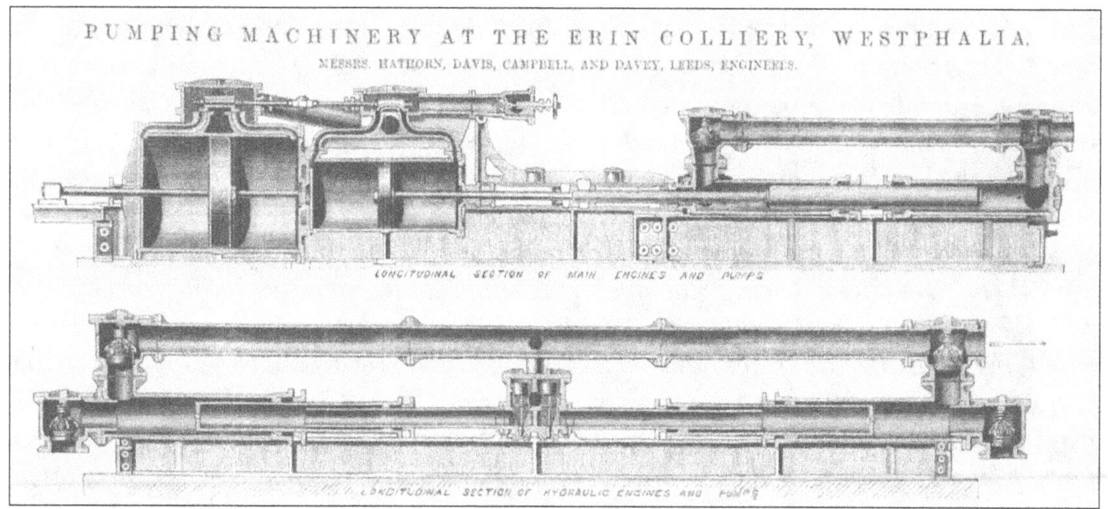

Figure 33. Cut-away sections through the pumping machinery manufactured for the Erin Colliery, Rhur Coalfield, Germany.
(The Engineer. 6th October 1876. p.239 [GG])

On the 1st October 1874, William Mulvany placed a further large order for machinery for the Hansa and Erin Collieries:

Order	Machinery	Cost
2290	1 separate condenser for 72" pumping engine	£340
2291	1 separate condenser for winding engine	£440
2292	2 sets of differential expansive gear complete*	£430
2293	1 hydraulic pumping engine**	£130
2294	2 twin hydraulic hauling engines	£360

* For a 72" engine at Erin Colliery and a large Cornish beam engine at Hansa Colliery.
** The engine at the Hansa Colliery had a 5½" diameter cylinder and a 3' stroke that operated an 8" diameter pump, and was capable of forcing 200gpm a height of 120 feet through 800 feet of piping.

At this time John Hathorn was probably involved with legal and financial matters, considering that he was the principal financier of the operation. Davey was almost certainly involved with drawing up order specifications and making the inevitable day-to-day decisions that such a business entailed. He probably travelled to discuss orders with clients, and he certainly made at least one visit to Germany, It is unclear what roles Davis and Campbell actually played in the running of the company, the former having, it would seem, no engineering background, whilst Davis seemed to have little mention in the affairs of the Company.

During September 1876, The Iron and Steel Institute visited the Sun Foundry, as part of their Leeds meeting.[9] The visit is described in their Journal. *'The works occupy two acres of ground, and give employment to 300 hands. When members visited these works, the proprietors were engaged with engines and machinery for the Croydon Waterworks. The Sun Foundry consists of a foundry, with two 15-ton power cranes. and one 10-ton hand crane, a smiths' shop fitted with a steam hammer, and four machine shops, with two 20-ton travellers, and one of 10-tons, as well as two 6-ton jib cranes. The firm have a boring machine for boring cylinders up to 120 inches diameter by 20 feet stroke. The Croydon engines in course of construction had cylinders 30 inches by 60 inches diameter, and 7 feet stroke, and were capable of lifting 2,500 gallons per minute 200 feet high. Messrs. Hathorn, Davis, Campbell, and Davey had also in hand some hydraulic pumping engines, to be principally employed in draining the "dip" workings of mines, &c. Differential engines have been made by this firm up to 400 horsepower.'*

Evidence suggests that the Company did produce a catalogue of their wares. Although no copies seem to have survived, it is probable that a seven-page listing, in *'Spons Engineers' and Contractors Illustrated Book of Prices of Machines, Tools, Ironwork, and Contractors' Material for 1876'* may have been derived from that catalogue.[10]

Figure 34 show the first page (p.112) of the listing that itemises the many pieces of machinery produced by Hathorn, Davis, Campbell and Davey. Featured machinery included the Compound Differential Engine (p.113), and that is shown in Figure 35. The advert includes a copy of a letter from W. Lishman, of Newbottle Colliery, who concluded, *'Everyone who has seen the engine is much pleased with it.'* The engine worked two quadrants that pumped water via two lifts of pumps.

The following page (p.114) illustrated a Compound Differential Pumping Engine and pumps as applied for underground pumping; Davey's System (Figure 35). The engines could be provided with either surface or injection condensers, and were designed so that pumps of any size could be fitted to them to pump at the same rate as the engine.

The pumps were double-acting, and costly, but were considered to be superior to piston pumps, which in this instance were not affected by sand or grit in the pumped water.

STEAM AND HYDRAULIC MACHINERY.

DAVEY'S PATENT DIFFERENTIAL PUMPING ENGINE.

DAVEY'S DIFFERENTIAL HYDRAULIC ENGINE.

HYDRAULIC RAMS FOR RAISING WATER.

HYDRAULIC HOISTS. HYDRAULIC ORGAN-BLOWERS.

The Differential Steam-Pump.

ROTATIVE STEAM-PUMPS, FOR BOILER FEEDING AND GENERAL PURPOSES.

COMPOUND ROTATIVE STEAM-ENGINES.

THE SEPARATE CONDENSER.

DIFFERENTIAL CORNISH PUMPING ENGINES.

DIFFERENTIAL BLOWING ENGINES.

HAULING ENGINES WORKED BY STEAM, WATER, OR COMPRESSED AIR.

All kinds of Pumping and Winding Machinery, &c., &c., &c.

CATALOGUES ON APPLICATION.

FOR PARTICULARS SEE FOLLOWING PAGES.

HATHORN DAVIS, CAMPBELL, AND DAVEY,

ENGINEERS, SUN FOUNDRY, LEEDS, ENGLAND.

LONDON OFFICE:
4, WESTMINSTER CHAMBERS,
WESTMINSTER, S.W.

Figure 34. A list of machinery manufactured by the firm
(Spons Engineers' and Contractors Illustrated Book of Prices of Machines, Tools, Ironwork, and Contractors' Material for 1876, p.112)

COMPOUND DIFFERENTIAL ENGINE.

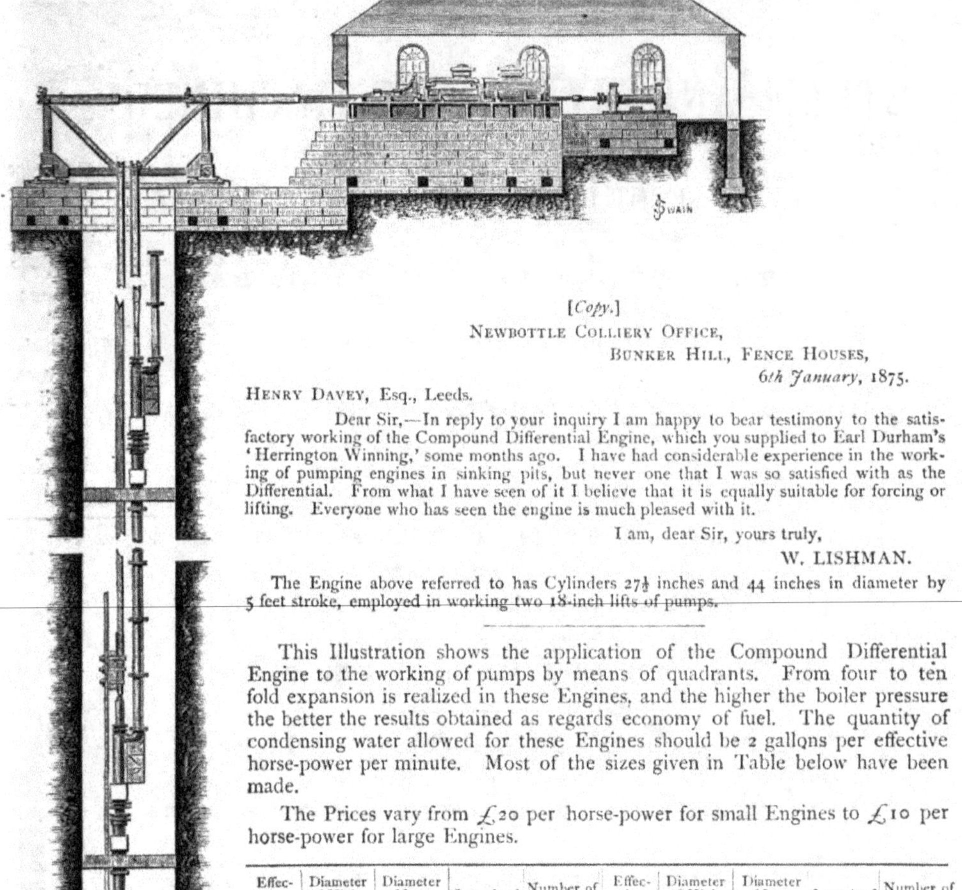

[*Copy.*]

NEWBOTTLE COLLIERY OFFICE,
BUNKER HILL, FENCE HOUSES,
6th January, 1875.

HENRY DAVEY, Esq., Leeds.

Dear Sir,—In reply to your inquiry I am happy to bear testimony to the satisfactory working of the Compound Differential Engine, which you supplied to Earl Durham's 'Herrington Winning,' some months ago. I have had considerable experience in the working of pumping engines in sinking pits, but never one that I was so satisfied with as the Differential. From what I have seen of it I believe that it is equally suitable for forcing or lifting. Everyone who has seen the engine is much pleased with it.

I am, dear Sir, yours truly,

W. LISHMAN.

The Engine above referred to has Cylinders 27½ inches and 44 inches in diameter by 5 feet stroke, employed in working two 18-inch lifts of pumps.

This Illustration shows the application of the Compound Differential Engine to the working of pumps by means of quadrants. From four to ten fold expansion is realized in these Engines, and the higher the boiler pressure the better the results obtained as regards economy of fuel. The quantity of condensing water allowed for these Engines should be 2 gallons per effective horse-power per minute. Most of the sizes given in Table below have been made.

The Prices vary from £20 per horse-power for small Engines to £10 per horse-power for large Engines.

Effective Horse-power.	Diameter of High-pressure Cylinder.	Diameter of Low-pressure Cylinder.	Length of Stroke.		Number of Strokes per Minute.	Effective Horse-power.	Diameter of High-pressure Cylinder.	Diameter of Low-pressure Cylinder.	Length of Stroke.		Number of Strokes per Minute.
	inches.	inches.	ft.	ins.			inches.	inches.	ft.	ins.	
310	34	64	8	0	10	90	20	40	5	0	12
290	34	64	7	6	10	70	18	36	5	0	12
250	35	60	7	0	11	70	16	32	6	0	12
240	35	60	6	0	12	63	16	32	5	0	13
240	30	60	6	0	12	55	15	30	5	0	13
220	30	56	7	0	11	47	15	30	4	0	14
180	30	52	6	0	12	30	12	24	4	0	14
140	30	50	5	0	12	24	12	24	3	0	15
140	25	50	5	0	12	17	10	20	3	0	15
150	24	44	7	0	12	12	8	16	3	0	15
130	24	44	6	0	12	9	8	16	2	6	15
110	24	44	5	0	12	5	6	12	2	6	15

HATHORN DAVIS, CAMPBELL, AND DAVEY,

ENGINEERS, SUN FOUNDRY, LEEDS, ENGLAND. LONDON OFFICE: 4, WESTMINSTER CHAMBERS, WESTMINSTER, S.W.

Figure 35. Advertisement for the Compound Differential Engine
(Spons Engineers' and Contractors Illustrated Book of Prices of Machines, Tools, Ironwork, and Contractors' Material for 1876, p.113)

THE SEPARATE CONDENSER.
(DAVEY'S PATENT.)

This Condenser has been designed to meet the want felt at the present time for a Condenser capable of being easily applied to existing non-condensing engines, to economize fuel and to increase the power. By its application from 30 to 40 per cent. increased power is available, and from 30 to 40 per cent. of fuel is saved.

The Air Pump of the Separate Condenser is worked by means of a small Differential Steam-Engine. In applying the Condenser to an existing engine all that is necessary is to fix it in any convenient place and lead the exhaust and a small steam pipe to it.

For Winding Engines the Separate Condenser is invaluable, because the vacuum is not at all impaired by the repeated stopping of the engine. When this Condenser is applied to engines situated underground in mines, a hydraulic engine is sometimes used to work the Air Pump, the power being obtained from the column or the tubbing.

The advantages claimed for this Condenser are :

1st.—*Economy of Fuel and Increase of Power.* The additional average pressure available on the piston of the engine by the use of the Separate Condenser may be taken as 12 lb. per square inch, consequently, if the engine is not required to do additional work the average steam pressure may be reduced 12 lbs. This, in cases where the average steam pressure is not high, would represent a saving of 40 per cent. of steam, and in the majority of cases the saving would much exceed 30 per cent.

2nd.—The vacuum is maintained unimpaired when the engine is irregular in its working and when it ceases to work, which makes it peculiarly advantageous for winding engines.

3rd.—One Condenser may be made to condense the steam for several independent engines.

4th.—The Condenser being entirely distinct from the engine, it is more accessible, is easier managed, and in the case of requiring repairs it does not prevent the engine being worked as a non-condensing engine.

5th.—A Boiler-feed Pump may be combined with the Separate Condenser, arranged to feed the boilers from the hot well.

In writing for estimates the following particulars of the engine or engines to which the Condenser is to be applied should be stated, viz. :

```
Number of Cylinders      ..   ..   ..   ..   ..
Diameter of Cylinders    ..   ..   ..   ..   ..
Length of Stroke     ..   ..   ..   ..   ..
Greatest number of Strokes per minute    ..
Average pressure of Steam during the Stroke
Type of Condenser, Surface or Injection   ..
Is the Condenser required to increase the
    power of the Engine or to economize fuel?
```

HATHORN DAVIS, CAMPBELL, AND DAVEY,
ENGINEERS, SUN FOUNDRY, LEEDS, ENGLAND. LONDON OFFICE : 4, WESTMINSTER CHAMBERS, WESTMINSTER, S.W.

Figure 36. Advertisement for the Separate Condenser (Davey's patent)
(Spons Engineers' and Contractors Illustrated Book of Prices of Machines, Tools, Ironwork, and Contractors' Material for 1876, p.118)

The Differential Steam Pump was advertised on p.115, while the following page 116 provided details about Davey's Patent Differential Hydraulic Pumping Engine, the type of engine installed in the lower chamber at Erin Colliery.

The penultimate page (p.117) advertised Single Cylinder Differential Pumping Engines and Double Ram Pumps that could be supplied either condensing or non-condensing. Pumps of any size could be put to any of the Engines. The quantity of condensing water required varied with the power developed, but were approximately calculated at about 3 gallons per horsepower per minute.

Lastly (P.118), a separate condenser (Davey's patent) (Figure 36) was described, that had been, '*designed to meet the want felt at the present time for a Condenser capable of being easily applied to existing non-condensing engines, to economize fuel and to increase the power. By its application from 30 to 40 per cent, increased power is available, and from 30 to 40 per cent, of fuel is saved.*

The Air Pump of the separate condenser is worked by means of a small Differential Steam Engine. In applying the Condenser to an existing engine all that is necessary is to fix it in any convenient place and lead the exhaust and a small steam pipe to it.

For Winding Engines, the separate condenser is invaluable, because the vacuum is not at all impaired by the repeated stopping of the engine. When this condenser is applied to engines situated underground in mines, a hydraulic engine is sometimes used to work the Air Pump, the power being obtained from the column or the tubbing.'

Two events occurred in 1875 that had some influence on the direction that the company was going to take in future.

The first occurred on the 1st March 1875 when John Fletcher Hathorn married Charlotte Dick-Lauder at St. Mary's Episcopal Church, Dalkeith, Midlothian. The Dick-Lauder family were also landowners, and owned an estate at Fountainhall, to the south-east of Edinburgh.[11] It is believed that after the marriage Hathorn spent more time attending to the management of his own 3582 acres estate at Whithorn in the historical county of Wigtownshire, now part of Dumfries and Galloway, Scotland.[12]

The second event was the publication of the Public Health Act 1875 that was passed on the 11th August. There had been a previous 1848 Public Health Act that established a Central Board of Health with limited powers but had no financial incentives behind it. It did however provide guidelines to be used by local authorities on ways to improve public health, but unfortunately did not compel councils to implement those recommendations. Consequently, its enactment was sporadic and inconsistent, with schemes like, for example the Sudbury waterworks, being constructed.

The 1875 Public Health Act was a much more robust document, and brought together all previous Acts relating to public health. More importantly, it specified the actions that Councils should take to safeguard the health of their citizens. The act established named local authorities as rural and urban sanitary authorities, replacing the local boards of health. These sanitary authorities would have jurisdiction over the newly created urban and rural sanitary districts.

Until 1875 mining was the dominant industrial sector that placed orders for pumping equipment with Hathorn, Davis, Campbell and Davey, but all that was about to change. The 1875 Public Health Act required local authorities to provide clean drinking water, and establish systems for removing sewage, and certainly the former requirement, would provide business for the Company well into the future. By 1900 the Company issued two catalogues, for waterworks (red cover) and mines (green cover).

The Company received a few small orders for water and sewage pumping engines after the Act became law but by the beginning of 1878, several significant orders had been placed.

Orders 2425 to 2432 - Croydon Waterworks - 22nd January 1876
Prior to the Public Health Act 1875, Croydon had established a water supply from the chalk aquifer beneath the town. Three Beam Engines (2 x 30" and 1 x 60") were employed to pump water from three wells to a reservoir. However, the supply was becoming inadequate and the 60" engine was overworked to meet the demand.[13]

An additional 60 feet deep well was sunk which was supplied with water from a deep borehole and was also connected with the other wells. The new well was to be pumped by a Hathorn, Davis and Co. Horizontal Compound Differential Pumping Engine (30"+60"). One quadrant was employed, and probably for the first time a lever arrangement from it worked the air pump and condenser that was located below the engine plinth. (See Figure 37) The water from the condenser was fed to the public baths.

Working at 12 strokes per minute (7' stroke), a speed easily attained with about 45 lbs steam, supplied by the existing six Cornish boilers, the pumping system (Double-acting piston pump 24" diameter, 7' stroke) was designed to pump 3,200gpm 200 feet high.

The machinery was delivered by 10th November 1876 and cost £5,300.

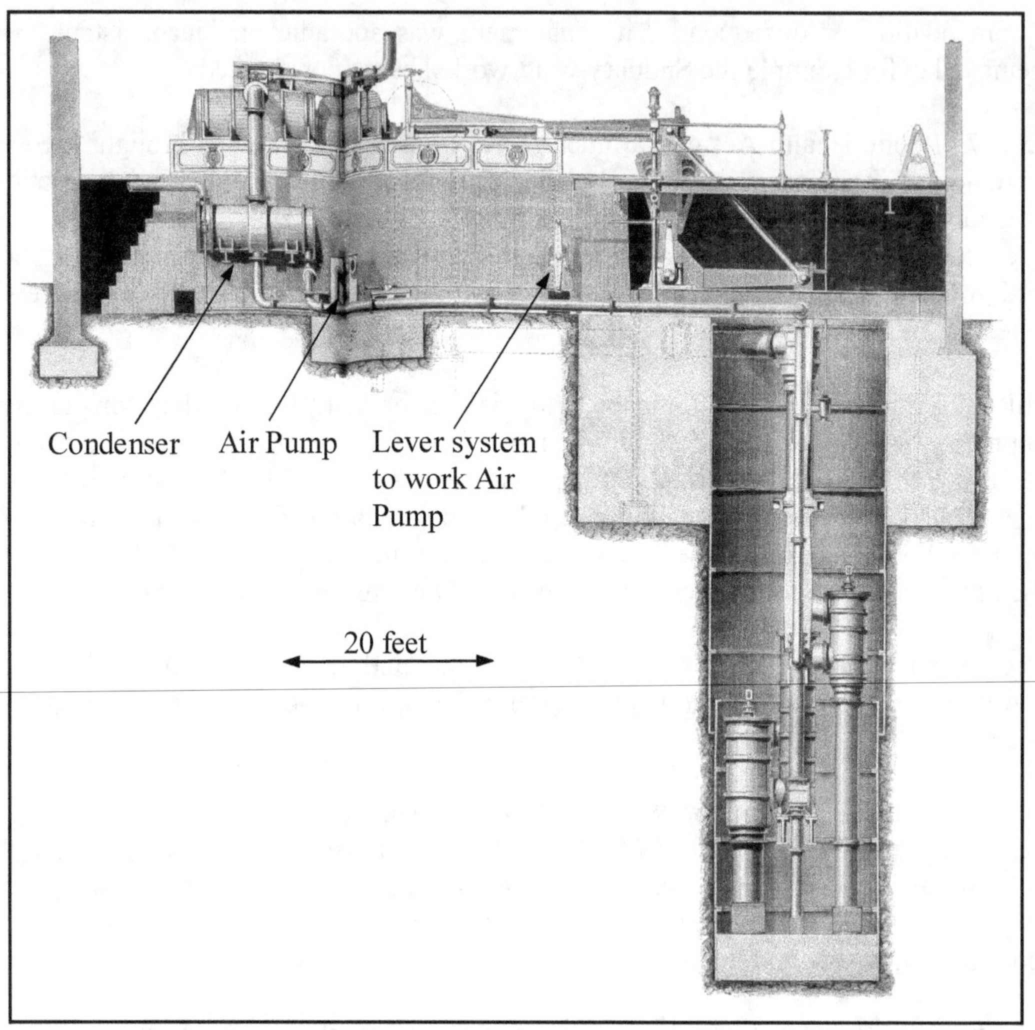

Figure 37. Croydon Waterworks: The Differential Compound Engine used to pump the 60 feet deep well. Note the lever arrangement working directly off the pumping quadrant that worked the air pump. The air pump and condenser are located at the side of the engine-mounting block.
(The Engineer. 9th November 1877. Supplement [GG])

Orders 2492 to 2501 - Melton Mowbray Sewage Works - 20th September 1876
The order was for a pair of Patent High Pressure Differential Condensing Pumping Engines working at 20 strokes per minute. Each engine had one cylinder 14" diameter with a 4' stroke. The order also included for each engine: air pumps, condensers, a pump for supplying mixers, chemical mixers with drive engines, main pumps and quadrants, and two Cornish boilers (14' 6" long and 4' 6" diameter). Total cost was £1,100.

Orders 2550 to 2554 - Saint Helens Corporation Waterworks - 4th May 1877
The Saint Helens Waterworks order was for two Horizontal Compound Differential Engines (14" + 28" cylinders and a 5' stroke) together with air pumps and condensers. Two Lancashire-Galloway boilers (26' long and 7' diameter) were also ordered.

As previously mentioned, it is unclear what role Hugh Fletcher Campbell had in the running of the Company, yet he was taking a share in it, as laid out in an Agreement between the Partners dated 8th May 1876.[14] The Agreement stated:-

'1. That from, and after 8th May 1876, no salary shall be paid to Mr Campbell and to Mr Davey [It is unclear where Campbell was getting an income from, in contrast to Henry Davey who was receiving Royalties from his ever increasing patents, and his private consultancy work.]

2. That from, and after that date, the share of Mr Campbell shall be $^4/_{32}$ nd parts- Colonel Hathorn and Mr Davis giving to him each $^1/_{32}$ nd share of their $^{11}/_{32}$ nd parts.

3. That in ascertaining the profit on Mr Campbell's $^4/_{32}$ nd from 8th May to 8th November 1876 the net profit for the year ending on the latter date shall be halved and of the one half he shall have $^4/_{32}$ nd parts.

4. That Mr Campbell shall accept this addition to his interest in the business as being a final agreement unless through the death of one of the partners or other contingency arise new arrangements may fall to be made.

5. That Mr Campbell shall on or before 1st January 1879 put into his credit with the firm and leave there the sum of £1,000 at least.'

Clearly, Campbell was being treated as reasonably as possible by the other Partners, and was being asked to pay his way, and he had nearly three years in which to do it. However, this was not to be, because on the 10th November 1877, Hugh Fletcher Campbell left the Partnership.

Perhaps Campbell's departure confirmed Hugh Lupton's later statement that John Hathorn had bought the business to provide Hugh Fletcher Campbell with an occupation. Certainly Campbell didn't seem suited for the engineering profession. Campbell filed for bankruptcy in 1890,[15] having tried other occupations in the intervening years.

'In the course of his examination by the Official Receiver, the bankrupt said formerly he was an engineer, and a member of the firm of Hathorn, Davey and Campbell, of the Sun Foundry, Leeds. When he retired from that firm he had nothing to draw,' although

a draft agreement drawn up four months after dissolution suggests that he might have received £1,700.[16] Campbell later became involved with a variety of businesses, including stationary, insurance and a share in the Perran Iron Ore Mines, Cornwall. His appearance in court was a result of his speculations. His debts had amounted to over £14,500.

Certainly Campbell's assertion that he was an engineer, brought to mind a statement in a book written about mining fraud, where one of the fraudsters claimed to be an engineer, based on the fact that he had put coal in an engine's boiler! [17]

After the dissolution a new Partnership was formed, and was titled Hathorn, Davis and Davey.

Chapter 7 References

[1] HL. 1940. *Hathorn, Davey and Company, Sun Foundry, Leeds.* p.1

[2] Members. *Proceedings of the Institution of Mechanical Engineers.* 1874. p.xiv

[3] Davey, H. 1874. On Direct-Acting Pumping Engines and Pumps for High Lifts in Mines. *Proceedings of the Institution of Mechanical Engineers.* pp.258 to 280. Plates 77 to 92.

[4] Compound Differential Engine and Pumps for the Navigation Colliery, Aberdare. *Engineering.* 2nd April 1875. p.273

[5] Dooge, J. 1997. Hibernia im Ruhrgebiet: William Thomas Mulvany and the Industrialisation of the Rhur. *18th-19th Century History*, v.5. Features. Issue 3 (Autumn)

[6] Schmidt-Rutsch, O. 2004. Hibernia, Shamrock, Erin: William Thomas Mulvany and the 'Irish Mines' in the Rhur. *Journal of the Mining Heritage Trust of Ireland.* 4. pp.2 to 10

[7] Davey, H. 1876. The Underground Pumping Machinery at the Erin Colliery, Westphalia.' *The Transactions of the Society of Engineers.* pp.119 to 134, and plates

[8] Pumping machinery at the Erin Colliery, Westphalia. *The Engineer.* 6th October 1876. pp.238 to 240.

[9] Report on visits in connection with the meeting in Leeds. 1876. *Journal of the Iron and Steel Institute* p.xxi

[10] *Spons Engineers' and Contractors Illustrated Book of Prices of Machines, Tools, Ironwork, and Contractors' Material for 1876.* pp. 112 to 118

[11] http://www.thepeerage.com/p37009.

[12] Bateman, J. 1883. *The Great landowners of Great Britain and Ireland.* Harrison (London). p.212

[13] Compound Pumping Engine, Croydon Waterworks. *The Engineer*, November 9th, 1877. p 338 to 339, and illustration in supplement

[14] DLC. Memorandum for Minute of an Agreement dated 8th May 1876.

[15] Leeds Bankrupcy Court, Tuesday. *The Leeds Mercury.* 16th July, 1890

[16] DLC. 1878-06-21 HD Partnership Agreement

[17] Anon. (A Shareholder) 1854. Mining and Miners and Diggers and Priggers. W.Kent and Co., London. (Google Books)

8. Hathorn, Davis, and Davey: 1877 to 1878

The years 1877-78 saw several minor changes in the Company's affairs, but they didn't directly affect the running of the business.

On the 7th January 1878 ownership of the Sun Foundry was re-conveyed from Messrs. Marshall and Telford, to Messrs. Hathorn, Davis, Campbell and Davey, which for legal conformity still included Campbell. This transaction was finalised on the 24th January 1878.[1]

The dissolution of the Partnership was later confirmed in a public notice (Figure 38) in the *Leeds Mercury* (2nd March 1878), and Campbell's share in the business was redistributed to the remaining three Partners.

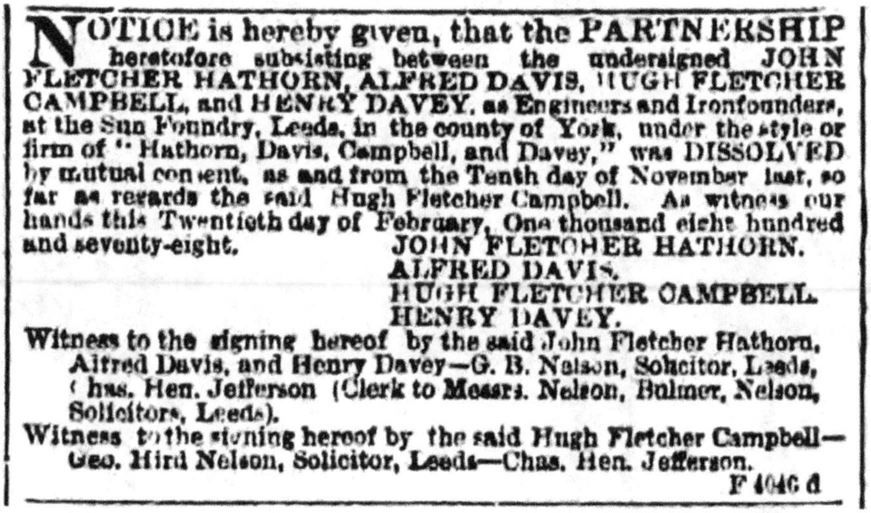

NOTICE is hereby given, that the PARTNERSHIP heretofore subsisting between the undersigned JOHN FLETCHER HATHORN, ALFRED DAVIS, HUGH FLETCHER CAMPBELL, and HENRY DAVEY, as Engineers and Ironfounders, at the Sun Foundry, Leeds, in the county of York, under the style or firm of " Hathorn, Davis, Campbell, and Davey," was DISSOLVED by mutual consent, as and from the Tenth day of November last, so far as regards the said Hugh Fletcher Campbell. As witness our hands this Twentieth day of February, One thousand eight hundred and seventy-eight.　　　　JOHN FLETCHER HATHORN.
ALFRED DAVIS.
HUGH FLETCHER CAMPBELL.
HENRY DAVEY.
Witness to the signing hereof by the said John Fletcher Hathorn, Alfred Davis, and Henry Davey—G. B. Nelson, Solicitor, Leeds, Chas. Hen. Jefferson (Clerk to Messrs. Nelson, Bulmer, Nelson, Solicitors, Leeds).
Witness to the signing hereof by the said Hugh Fletcher Campbell—Geo. Hird Nelson, Solicitor, Leeds—Chas. Hen. Jefferson.
F 4046 d

Figure 38. Public notice confirming that the dissolution of the Partnership
(The Leeds Mercury. 2nd March 1878)

Further orders were being placed with the Company for equipment for water and sewage works. One of significance was:

Orders 2612 to 2623 - Chiswick Main Drainage Company - December 1877
The orders were for a pair of Differential Vertical Compound Condensing Pumping Engines (10"+20", stroke 3') and associated machinery to treat sewage. Both cylinders were jacketed and are described in detail in an article in *Engineering.*[2] Figure 39 shows one of the pair of the engines. Each engine and set of pumps was designed to pump 1.2 million gallons of sewage in 12 hours, 21 feet 6 inches high into the deposit tanks situated 71 feet from the centre of the pump wells.

Figure 39. Chiswick: Differential Vertical Compound Condensing Pumping Engine
(Engineering. 6th June 1879. p.476)

The two engines were designed to work independently of each other. '*To guard against the possibility of any accidents that might stop the system working, the whole pumping plant was ordered in duplicate, with either engine and pumps being large enough to easily carry out maximum pumping requirements.*'

The total order consisted of: two engines (2612), two air pumps and condensers (2613), plunger pumps (2614), edge runner mortar mills and mixers (2615 to 2617) that were located in a room adjacent to the engines. Three 15' by 5' London Cornish boilers (2620) were also supplied as well as piping and sundry equipment. The total cost was £2,310, and included delivery, installation and 'set to work.'

When this order, was placed in 1877 the population of Chiswick, on the west side of London was about 10,000 people, but the new machinery was designed to cope with double that number.

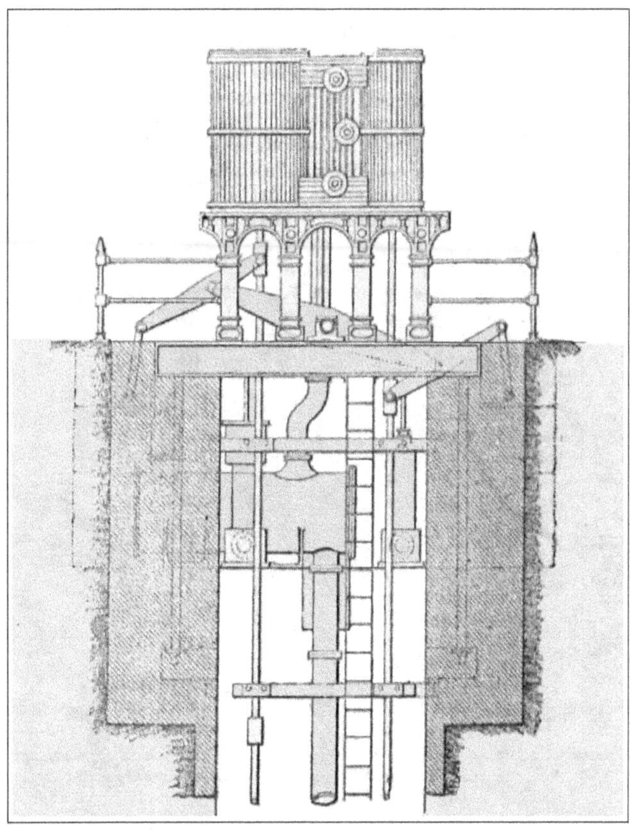

Figure 40. Chiswick: A Vertical Compound in position above a well
(Modified from: Engineering. 6th June 1879. p.478)

The pumping mechanism for the Chiswick order was to a new design by Davey, and is shown in Figure 40.

The high and low-pressure cylinder piston rods extend down beneath the cylinders and operate the plunger pump, and valves, respectively. As the piston rods move up and down they also apply the same motion to side levers on each rod. The movement of the side levers causes a large central beam rock, thereby allowing motion from the low-pressure cylinder piston rod to be transferred to the high-pressure cylinder piston rod, and this assists in working the plunger pump.

A rod attached to an offset cam on the fulcrum of the large beam extends upwards to operate the Differential Gear that regulates the steam input into the high-pressure cylinder. The beam can be seen in Figures 39 and 40 together with the Differential Gear adjacent to the high-pressure cylinder.

To Davey this arrangement was his version of the parallel motion that Watt had devised and patented for the Beam Engine. He refers to it as such, and it is shown in Figure 41. Davey had concluded that, '*the motion was more perfect when the outer ends of the outer beams* [the two short beams] *are guided in horizontal slides, instead of being attached to links, but the link secures sufficient accuracy for practical purposes.*'[3]

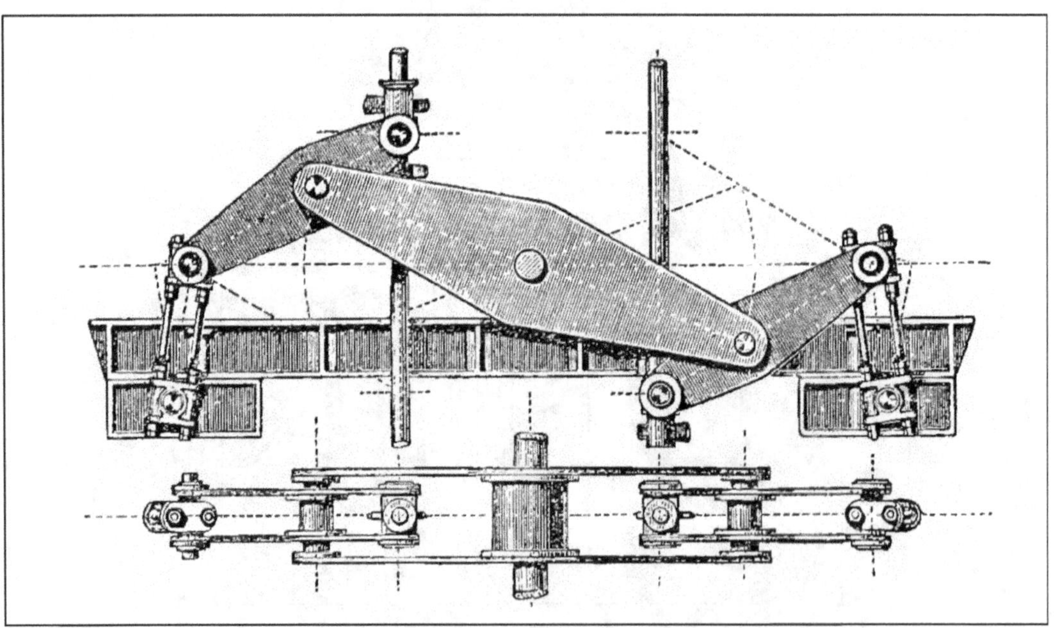

Figure 41. Davey's concept of parallel motion
(*Davey, H. 1905. The Principles, Construction and Application of Pumping Machinery. 2nd edition. p.257*)

The two Chiswick engines were the first Vertical Compound Engines manufactured by the Company. On completion, one of the engines was exhibited at the Paris Exhibition in 1878. The engine was a turning point in engine design, and similar engines were later exported to places as far away as Queensland, Australia. Other designs followed, and one major change was a variation of a concept earlier promoted by Bull, with both cylinders above, and attached at each end of, the pumping beam.

An order for engines specified to be similar to those at Chiswick was placed the following year:

Orders 2635 to 2649 - Lincoln Corporation Sewage Pumping Station. - 1878
The Lincoln Sewage Pumping Station engines were larger (15"+ 30", stroke 5') than those manufactured for Chiswick. The order also included 20" diameter sewage pumps that were also worked directly under each cylinder (Inverted).

The Order Book listing indicates that various revisions were made during the manufacturing period. The final order was for:

2635 to 41	Two compound condensing vertical engines complete, air pumps, condensers, feed pumps and sewage pumps etc.	£3,000
2642/3	Girders, beams, fittings etc.	£200
2645	Four wrought iron girders, crab winch, blocks, chain etc.	£150
2648	Three boilers, mountings and fittings	£1,000
2644/49	Spare set of valves for each pump, spanners etc.	£56
	Extra set of fire bars for each boiler	£30
	Carriage, erection, setting to work, maintenance (6 months)	£720
	Total	£5,156

A selection of other orders during 1878, included a pair of Air Compressing Engines with 33" and 35" steam and air cylinders, respectively for Newstead Colliery, Nottinghamshire (Order 2655 - £1,600); Horizontal Compound Differential Condensing Engines (23"+40") capable of raising 400gpm 720 feet for the Montreal Iron Works, Whitehaven (Order 2676 to 2680 - £2,500); Compound Differential Condensing Pumping Engine (12"+22") capable of raising 30,000gph 151 feet high (Order 2681 to 2687 - £760); Differential Pumping Engine (24") capable of raising 500gpm 300 feet, for Butterknowle Colliery, Darlington (Order 2689 to 2691 - £380).

The Paris Exhibition was held between May and November 1878 and was probably the biggest international exhibition the world had ever seen at that time. Hathorn, Davis and Davey, probably at Henry Davey's instigation, had a stand there, and exhibited various items of machinery.

The catalogue listing for the Paris Exhibition stated: [4]

'Hathorn, Davis & Davey; Engineers; Sun Foundry, Leeds. Patent Compound Differential Pumping Engine, suitable for the supply of towns, &c. (H. Davey, Inventor). One patent Hydraulic Pumping Engine, for pumping water by means of water pressure (H. Davey, Inventor). A Patent Differential Steam Pump for general pumping purposes (H. Davey, Inventor). A working model of Compound Pumping Engine. A working model of Hydraulic Pumping Engine.'

In addition to one of the engines from the Chiswick sewage scheme, other items exhibited included:

Order 2584 One Compound Differential Pumping Engine (10" + 20" and 3' stroke) - no pumps. Dated 24th September 1877

Order 2585 One Hydraulic Dip Pumping Engine (9" cylinder and 18" stroke). Dated 24th September 1877

Order 2652 One Differential Steam Pump (12" Cylinder for engine, 8" Cylinder pump - both with a 20" stroke).

The two models referred to are recorded in the Orders Books. The Compound Differential Engine (Figure 42), together with sinking pumps (Order 2316 - dated 1st February 1875) was originally produced for an Exhibition in Leeds, whilst the Hydraulic Engine (Order 2624 - 10th December 1877) had a 2" cylinder and 6" stroke worked a 2½" air pump. It was made for the South Kensington Museum (Science Museum). Both engines were later listed in 1919 in the *Science Museum Catalogue of the Mechanical Engineering Collection* as items 622 and 636, and stated as having been lent to the Museum in 1888. [5]

Figure 42. The Company's model of the Differential Compound Pumping Engine
(Catalogue of the Mechanical Engineering Collection in the Science Museum, South Kensington. 1919. Part 1. Plate X, No.4)

A correspondent from the *Yorkshire Post and Leeds Intelligencer* visited the Paris Exhibition on the 29th August to write about the English coal, iron and steel industry exhibits, but was very impressed by the Hathorn, Davis and Davey display.[6]

'I must call special attention to the exhibits of Messrs. Hathorn Davis and Davey of your town, which, although not exclusively related to the subject, are of high interest, and have been much remarked. This firm enjoys for its differential pumping engines fame, which is virtually unsurpassed. There is not an engineer, interested the subject, visiting the Exhibition who fails to call on them at their place in the English machine gallery. Davey's patent differential gear has proved so suitable for high lifts, and for the raising of large quantities of water, that naturally it enjoys high reputation. It adapts itself admirably to changed conditions, saves the pumps from heavy shocks and the risk of breakage, and engines combining its advantages are already at work in different parts, representing over 14,000 horse-power. Among the exhibits from the Leeds Sun Foundry, where Messrs Hathorn, Davis, Davey are installed, I must particularly call attention to a vertical compound pumping engine, made for the Chiswick Main Drainage Works, London, which, by the simple arrangement of a beam between the two cylinders, gives not only a constant stream in the main water pipes, but allows the steam to pass in the shortest way from the high to the low pressure cylinder. The differential gear is very neatly arranged, and shows at first sight how effectually it may be regulated, allowing the engine to run at any speed from two to twenty-five strokes simply by opening or shutting a cataract. Similar engines have been constructed for the Croydon, St Helens, and other waterworks. The largest pumping engines made for mines have been built by this firm, and their arrangement is shown in a model of a 500 horse-power engine, with improved plunger lift, made for the South Durham Colliery Company. There are two plungers, each 20in. diameter by 10ft. stroke, employed to force the water in one lift to a height of 700 ft., a feat never before accomplished by any pumping engine. A small working steam pump of about 12-horse power shows a vertical arrangement of the gear; it works, although the water pressure shows the greatest irregularity, most steadily and quite noiselessly, and this at any number of strokes. To show the working of mines when steam can scarcely be used, two hydraulic pumping engines are exhibited. In these the moving power is derived from the main delivery pipe of the large pumping engine. One of them is a model of a hydraulic pumping engine made for a Westphalian Company, and constructed to raise 1000 gallons per minute to a height of 300 feet. The engines have no piston, nor internal packing, likely to get out of order. The packing is all applied in stuffing boxes; the valves are gun-metal water valves, with large areas, from which no throttling need be feared. They work easily, and with a minimum of wear under the heaviest pressure. The engines are under self-control, by means of Davey's differential gear, of which all engineers interested in mine drainage and waterworks have repeatedly

expressed their admiration to Messrs Hathorn, Davis, Davey's able representative in Paris.'

During September 1878, the Paris Exhibition awards were announced. Hathorn, Davis and Davey were awarded a Gold Medal for *'Class 54, machines and apparatus in general.'*[7] This would have been regarded as a significant achievement and it was one that they would prominently and proudly mention in future advertisements (Figure 43).

Figure 43. Two Advertisements announcing the award of a Gold Medal to Hathorn, Davey and Company, Leeds at the Paris Exhibition
(Top: Mining Journal and Railway Gazette 1879, p.171)
(Bottom: Mining Journal and Railway Gazette 1881, p.203)

Whilst the exhibition was in progress, Alfred Davis left the Partnership on the 31st July 1878. There were obviously legal processes to put in motion, and so on the 12th August, a formal Indenture to dissolve the partnership was signed by the three Partners. Hathorn and Davey agreed to pay Alfred Davis the sum of £4,601-16s-11d in full discharge of his share of capital interests, and profits in and arising from, the said Company, and Partnership business, up to the said 31st July last. Alfred Davis would also receive all the furniture in a London property at 31, Duke Street, which the Company used as an office.[8] This was in actual fact the office that Davis used prior to joining the Partnership. By the end of October, the process had been completed and Alfred Davis's share of the Company calculated on his share of the Partnerships fixtures, and stock in trade were re-conveyed to Hathorn and Davey. [9]

With the departure of Davis, the Company would now become known as Hathorn, Davey and Company, as the advertisements in Figure 43 show. It was a name that would endure for over 135 years.

Chapter 8 References

[1] DLC Re-conveyance dated 7th January 1878

[2] The Chiswick Main Drainage Works. *Engineering.* 6th June 1879. pp. 476 to 479

[3] Davey, H. 1905. *The Principles, Construction and Application of Pumping Machinery.* (2nd ed). pp.257 and 259

[4] Paris Universal International Exhibition, 1878. *Official Catalogue of the British Section. Part I.* Her Majesty's Stationary Office.

[5] *Catalogue of the Mechanical Engineering Collection in the Science Museum,* South Kensington. 1919. Part 1. Steam engines and other motors   HMSO. Pp.285,286,290,291 and plate X, No.4

[6] The Paris Exhibition. *Yorkshire Post and Leeds Intelligencer.* 31st August 1878. p.7

[7] The Paris Exhibition, *Yorkshire Post and Leeds Intelligencer.* 12th September 1878. p.5

[8] DLC Indenture dated 12th August 1878.

[9] DLC Conveyance of shares dated 31st October 1878.

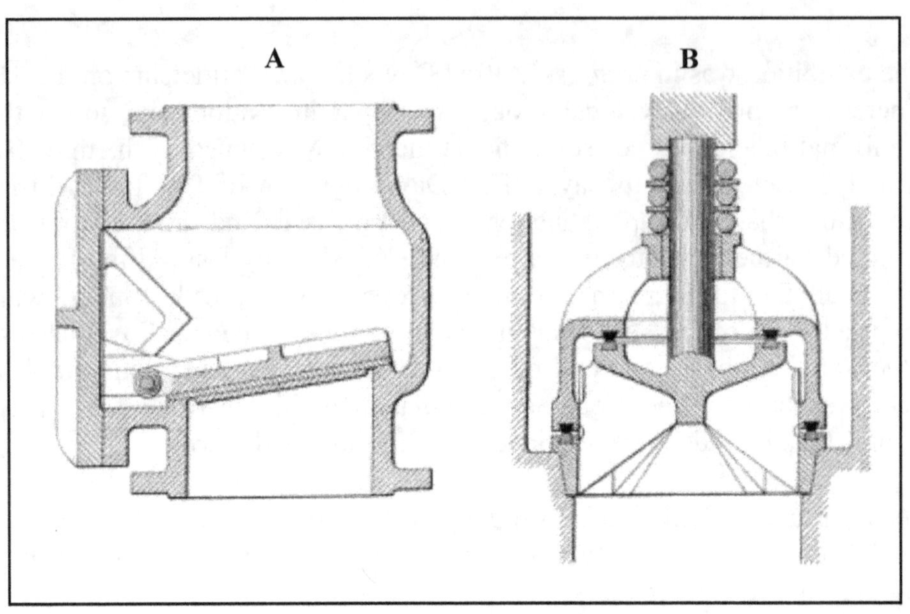

Figures 44A: Clack and Door Valve, and, 44B: Harvey and West Valve.
(Minutes of Proceedings of the Institution of Civil Engineers 1878. Plate 5)

Figure 45. Clack Valve, Beaconsfield Gold Mine, Tasmania
The valve is simply described as a hinged lid with a leather seal.
(Author 2011)

9. Hydraulic Engines and Pumps

Reference has already been made to hydraulic pumping systems. Carrett and Marshall had established hydraulic engines as simple and reliable machines initially for use in collieries, for pumping, and later they produced coalcutting machinery. They also manufactured the hydraulic Joy Organ Blower that worked on the pressure of town water supplies. When Hathorn and Partners took over the Sun Foundry in 1872 much of that trade continued, apart from the production of the hydraulic coal cutter.

It was the simplicity of hydraulic equipment that made it popular. The machinery could be used easily in confined spaces. For a few hundred pounds, a hydraulic engine could be combined with a pump and used to dewater dip headings. Hydraulic systems could be pressurised by a steam engine and pump sited some distance from the shaft top.

Improved pump design was equally important, whether hydraulic or steam powered. This was particularly so for the pumping technology being promoted by Davey. The volume of water (and thus weight), and the heights that the water was being raised in one lift were all factors that had to be considered in pump, valve and pipe work design.

Hydraulic engineering may only have played a small part in Davey's formative years where the emphasis was more towards steam power, and so when he was given the work to reconstruct Milford Docks, it was probably his first significant excursion into hydraulic engineering. In his later position at the Sun Foundry, the production of relatively simple hydraulic organ-blowers would have seemed far removed from his Milford experience, but he soon became involved in larger hydraulic engineering contracts. Those he included in several published papers.

On the 26th March 1878, Henry Davey gave a paper to the Institution of Civil Engineers entitled, *'Direct Acting or Non-Rotative Pumping Engines and Pumps.'* [1] In style, the paper was very similar to earlier papers on the subject, but Davey used it as a medium to present his latest ideas, and elaborated further on how suited hydraulically operated pumps were for underground use.

At collieries the ventilation shaft (upcast) is enclosed at the surface and incorporates a fan to 'pull' the air through the mine. The winding shaft (downcast) is open, and this is important for positioning a conventional pumping engine. A Cornish Beam Engine would be erected at the shaft top. However, the disadvantage with this system is that it clutters the area around the shaft. By using a steam or hydraulic circuit via pipes from a surface installation to operated underground pumps, the implications are that the pipework can be placed in either shaft, and surface machinery can be located away from the immediate shaft top. Hydraulic circuits were usually more economical to use than steam.

Another aspect in favour of this method is that underground water is forced to the surface in one lift, whereas pumping by a Beam Engine, certainly for a deep mine, was done in a series of short lifts that frequently needed constant attention and repairs.

Davey's paper also considered other improvements in underground pumping technology. Previously, pumps had been constructed with pistons, which were usually high maintenance, as they wore out quickly. However, by replacing the piston with a plunger, the life of the machinery was extended, and made the equipment more suitable for 'heavy lifts.'

At the time of presenting the paper to the Institution of Civil Engineers, Davey had designed pumping systems capable of, *'forcing against columns of water up to 1,100 feet in height. This was, without an accident, an immense saving in the cost of maintenance compared with that of ordinary pitwork. Only the pump barrels stand in the shaft; the valve boxes are placed in an archway at the side of the shaft where they can easily be attended to. The old clack and door pieces* [Figure 44A] *so notoriously weak in form, have been discarded, and circular valve boxes have been used instead; and although the total lift of 700 feet* [in Davey's example], *yet a greater margin of strength has been secured at a less cost than by the old system. The pump barrels are provided with relief valves loaded with India-rubber springs to a pressure 30 % above the static column.'*

'The valve in Figure 44B *was selected by the Author* [Davey] *after numerous trials, as the best for heavy pumping. It is a Harvey and West valve, with the bottom beats of gun metal* [An alloy of copper, tin and zinc], *and the top ones of hippopotamus hide* [an animal that spends much of its life in water!]. *The leather beat is preserved from damage by grit by being placed on the top instead of at the bottom.'*

Nevertheless, Davey didn't discard the idea of using clack valves. Figure 45 shows one Hathorn, Davey valve that had been installed in the early 20th Century, at the Beaconsfield Gold Mine, Tasmania.

The pumps that Davey cited in his paper were for two plunger pumps manufactured for the South Durham Colliery (Order 2561, dated 3rd May 1877). In the Order Book, they are described as, *'2-20" plunger lifts each 10 feet stroke, cast of specially tough metal, placed side by side 668 feet from the surface and fixed on a strong girder with supporting columns, as per tracing. Pumps to be specially constructed to ensure great strength, all valve boxes to be circular and to be provided with double beat valves of special make. Each pump barrel to have a relief valve.'* A surface Compound Differential Condensing Pumping Engine (45"+72") actuated double quadrants and two runs of pump rods down the shaft to work the pumps, and was expected to work at 8 strokes, and pump 2000 gallons of water per minute.

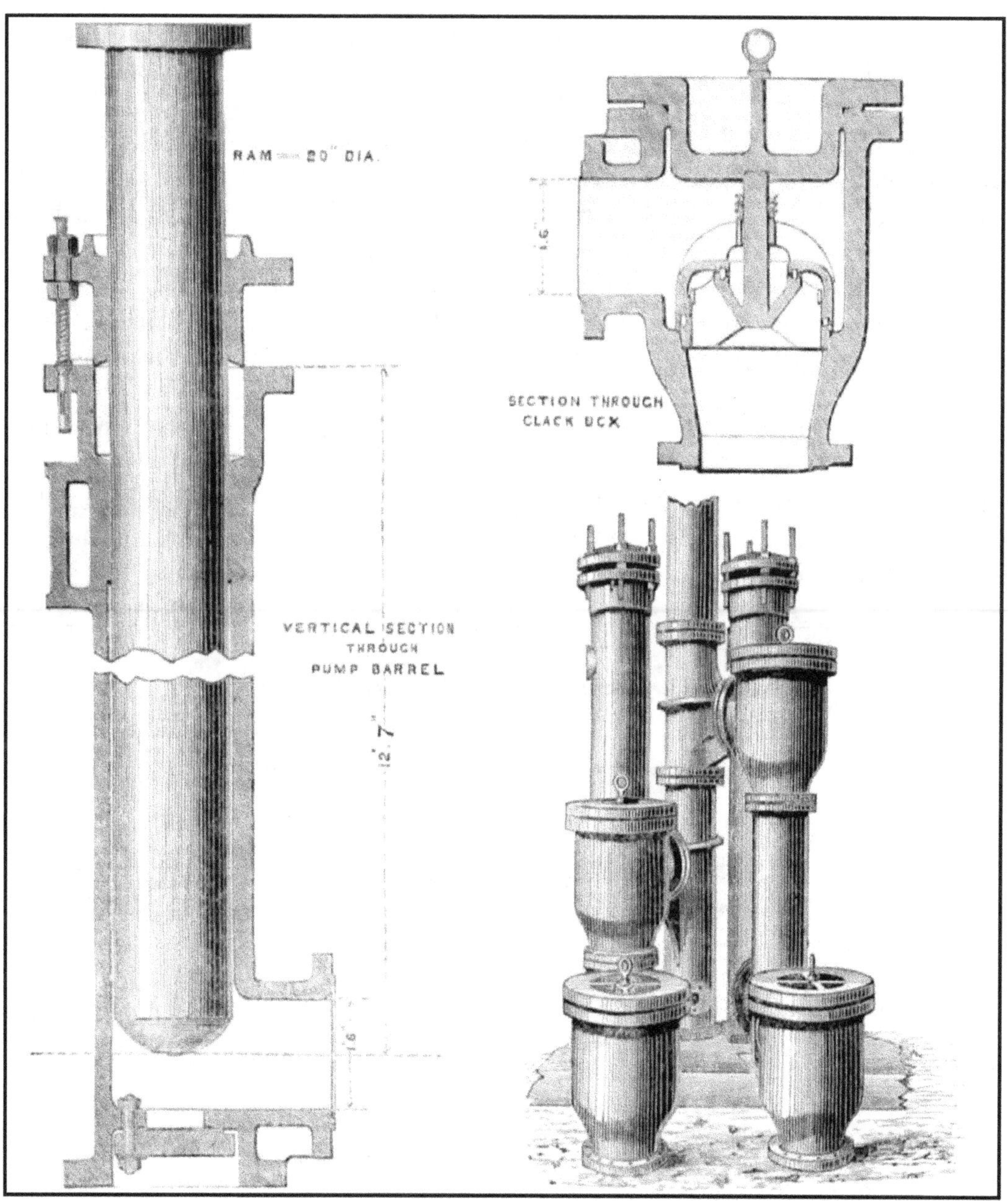

Figure 46. South Durham Colliery: Plunger Pumps with Harvey & West Relief Valves
(The Engineer, January 31st, p.79. [GG])

The rising main is listed as order 2562, and was '*18" diameter, from the pumps to the surface, with bored and turned hydraulic joints and gutta percha rings.*' [A durable rubber like material used for gaskets] Figure 46 is an illustration of the pumps from *The Engineer* (January 31st 1879).[2]

During April 1880, Davey examined the use of hydraulics further when he presented a paper, *'On Water-Pressure Engines for Mining Purposes,'* to the Institution of Mechanical Engineers.[3] The paper looked at a variety of situations where hydraulic engineering was used in underground mining situations. It did not go into detailed discussion on aspects relating to closed hydraulic circuits similar to those installed at the Erin Colliery. Rather, the paper tends to describe hydraulic pumping schemes installed in unconventional situations. Davey did briefly mention a pair of hydraulic engines that the Company had built for the Mansfeld Company's deep salt mine in Germany. Each engine was capable of raising 66gpm of brine 1000 feet high, the hydraulic power being derived from an accumulator constructed on the surface.

The method of draining dip-headings has already been mentioned. In the paper Davey covers this aspect further, and includes an example from Griff Colliery, Nuneaton (Orders 2735 to 2738, dated 28th January 1879, cost £466), where the difficulty of having to draw water from a flooded, very long dip-working (dip was 1 in 6) had to be overcome. The arrangement of the equipment is shown in Figure 47.

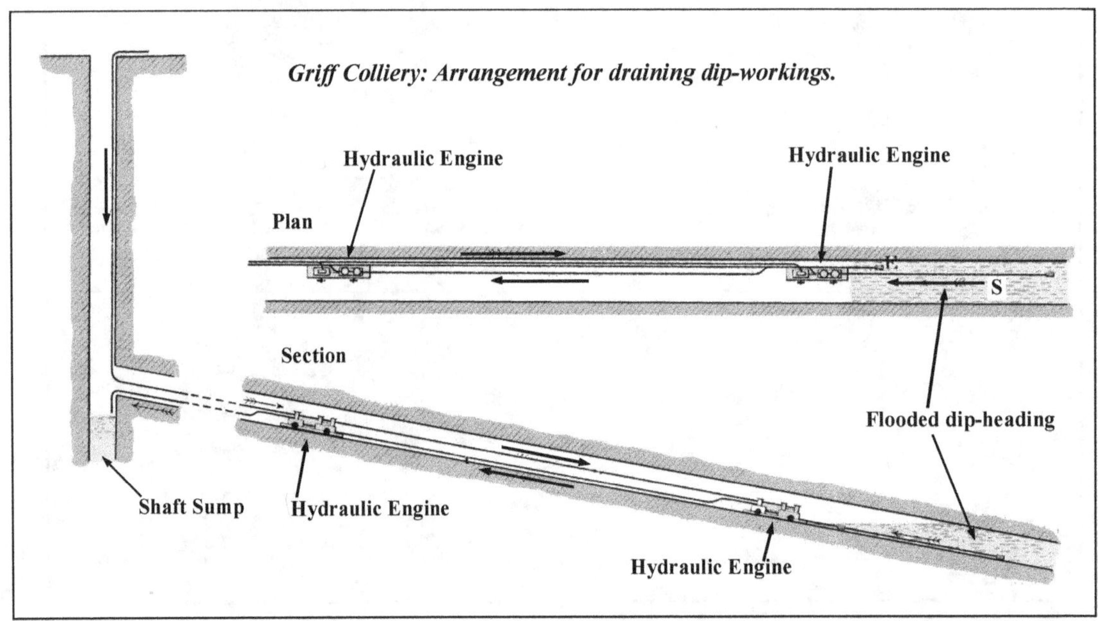

Figure 47. Griff Colliery: Arrangement of the hydraulic engines in a dip-heading
(Based on: Proceedings of the Institution of Mechanical Engineers. 1880. Plates 20)

Two hydraulic engines each capable of raising 150 gpm to a height of 150 feet were positioned in the drift-heading. The head of water that worked the engines was 450 feet. One engine was placed as close to the water edge as possible and a suction pipes added (Figure 47 - S) and water removed until a vertical depth of 25 feet was reached. The second engine was then placed close to the water level and the operation repeated. Using that method, the pumping was uninterrupted, and the heading dewatered. The water pumped from the heading was channelled into the shaft sump for the main pumping engine to pump it to the surface.

Perhaps the most interesting engines Davey refers to are the hydraulic winding and pumping engines ordered by the *A.D. Mining Company, Limited* for their lead mine in Swaledale, North Yorkshire (Orders 2889 to 2893, dated 29th October 1879, cost £820). Equally interesting is the fact that both engines still survive *in situ* underground, albeit with difficult access.

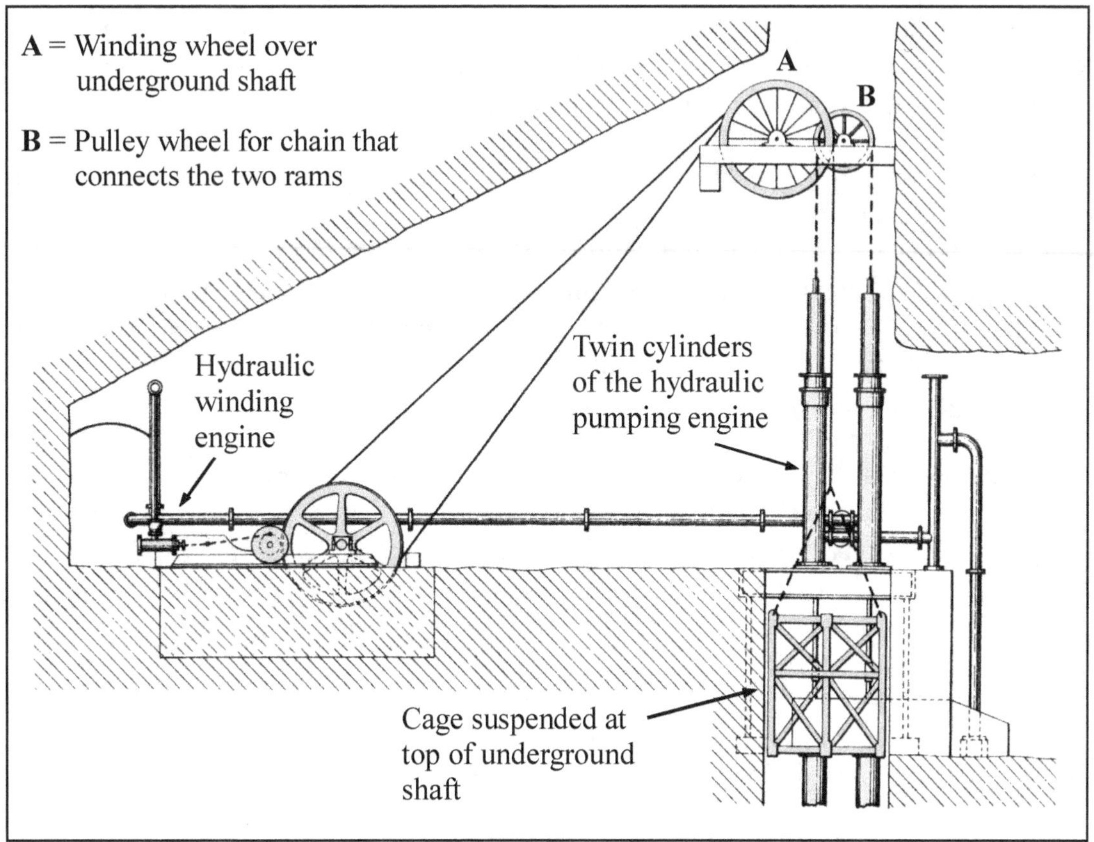

Figure 48. Sir Francis Level: Arrangement of the Pumping and Winding Engines. *(Modified from: Proceedings of the Institution of Mechanical Engineers.1880. Pl. 25)*

Both hydraulic engines were supplied with water from a reservoir constructed some 500 feet above the engines. Pipes led from the reservoir down the hillside for a distance of 1800 feet, and then 240 feet down a vertical shaft to a large underground chamber that housed the engines. Access to the engines is gained via the Sir Francis Adit level, on the west side of Gunnerside Gill, Swaledale. Further up the Gill, the water pipe that supplied the water to the engines can be seen protruding out of a collapsed shaft. It is almost certain that the engines were originally installed by lowering the parts down the shaft, the Sir Francis level bends in some areas, which would make transportation of parts up the level complicated.

The general arrangement of the machinery is shown in the section (Figure 48) taken through the engine chamber and shaft

The pumping engine consists of two vertical hydraulic cylinders each having a power ram 12" diameter by 7' stroke. The rams are connected together by a chain passing over an overhead chain pulley, so that when one ram is on the up stroke, the other ram is descending, and is controlled by valves located on each cylinder. In effect the chain over the pulley acts as a fulcrum for the reciprocating action of the pistons.

A rod fixed to the ram passes through a stuffing box at the base of each cylinder, and is attached to the top of the metal pump rods.

The pumps are 13" diameter. They are simple bucket types, with clack and door pieces and wind bores of the type generally employed for shaft sinking purposes. They were designed to operate at 6½ strokes per minute, and raise 500 gallons from a depth of 360 feet, although the shaft was never more than about 130 feet deep.

The winding engine, '*consists of a pair of double-acting hydraulic cylinders, coupled to a right-angled crank on the drive shaft, which latter is geared to the winding drum by a spur pinion......the gearing has a proportion of 1 to 6 and the winding drum is 6 feet in diameter. The weight to be raised is 2 tons of ore at a time. The cylinders are 5½ inches diameter, 16 inch stroke, and run at 19½ revolutions per minute, giving a speed of 60 feet per minute to the rope.*'

Davey had improved the valve system on the winding engine. Eccentrics operated the valves, and controlled by simple gearing so that the engine could be reversed.

Figure 49. Sir Francis Level. The two vertical cylinders of the hydraulic pumping engine, and valve levers stand at the side of the shaft top.
The top cage is suspended on the chains in the bottom right of the photograph.
(Photograph by the late H.M Parker probably taken in the 1970s)

Various authors have photographed and described the two hydraulic engines in the Sir Francis Level. Probably the first was Lodge (1966)[4], who provided technical information derived from the Hathorn, Davey records, while Crabtree and Forster (1963) state that the overall expenditure for the engines, their installation and construction of housing was about £4,500. They described the engines. *'The winding engine powers a wooden drum 5 ft. 4 in. in diameter and 6ft. 3 in. wide. Two wire rope cables from the drum pass over the shaft headgear to hoist the cages, one cable winding off the drum as the other winds on. Beneath the headgear there is a drop of 12 ft. from the engine room to the sump* [shaft] *surface. Overlooking this drop stands the hydraulic pumping engine which, while quite beyond servicing, is in remarkably good shape considering its age. The cylinders and pipes are all cast iron but at least some of the valves are of brass and still turn quite easily. One valve, which was removed for examination released a strong jet of air, which had been maintained under pressure for 80 years.'* [5] Figure 49 shows the hydraulic pumping engine in *situ*.

In his paper Davey also mentioned a single hydraulic engine sold to the *Hutton Henry Coal Company*, Wingate, Durham (Order 2917). It was used to remove a constant quantity of water that was coming into a shaft that was difficult to pump out at the source. Consequently, the water was conducted by pipe 502 feet down the shaft to a point where the hydraulic engine could pump it to surface. The engine operated on a 260 feet head of water taken from the sump of a shallower shaft, and pumped the water 866 feet to the surface in one lift.

More large orders for hydraulic engines were completed in the following decades, for example at a colliery near Marseilles, France. Orders were also still being taken for the Joy's hydraulic Organ Blower. In 1882 three of Joy's Organ Blowers were installed at the Crystal Palace, London to replace the work done by nine men who previously had manually operated the organ bellows![6]

Chapter 9 References

[1] Davey, H. 1878. Direct Acting or Non-Rotative Pumping Engines and Pumps. *Minutes of Proceedings of the Institution of Civil Engineers.* pp. 98 to 129. Plates 5 to 7.

[2] Plunger Pumps at the South Durham Colliery. *The Engineer*, 31st January 1879. p.79.

[3] Davey, H. 1880. On Water-Pressure Engines for Mining Purposes. *Proceedings of the Institution of Mechanical Engineers.* pp. 245 to 265, and Plates 19 to 29.

[4] Lodge P.D., 1966, 'Hydraulic pumping and winding machinery, Sir Francis Level, Swaledale', *Memoirs of the Northern Cavern and Mine Research Society.* pp. 21-26 and plates 5 to 10.

[5] Crabtree, P.W. and Foster R., 1963, Sir Francis Mine', *Cave Science*, 33. Pp. 1 to 24.

[6] The Crystal Palace Organ. *St. James Gazette*, 29th April 1882. p.11.

10. Hathorn, Davey and Company and a new Partner

With two Partners leaving the Company, it was inevitable that there would be shortcomings in the Company's organisation. It is probable that in the short-term, experienced staff like John Telford who still retained a position, partly on the shop floor and partly administration, dealt with some matters.

It seemed likely that John Hathorn was spending more time in Scotland. Not only did he have an Estate that demanded his attention, he also had a young family; a daughter and son, born 1875 and 1877, respectively. Another reason may have been that he was becoming aware that his health was starting to fail. Henry Davey was obviously the key man in the Partnership, designing the Company's innovative machinery, perhaps making site visits, as well as advertising their expertise in pumping technology with the presentation of papers. It also seems likely that he desired to take on more work in his own right as a consultant.

A new Partner was sought. It had to be someone with initiative, and with an engineering background. How such a position was advertised is not known, but nevertheless, in early-1880 Arthur Herbert Meysey-Thompson was chosen as a new Partner. It seems that he may have been acquainted with, and been introduced by Hugh Fletcher Campbell.[1] Meysey-Thompson did come from a wealthy aristocratic family, but more importantly, he was an engineer.

Arthur Herbert Meysey-Thompson was born in Little Ouseburn, near Harrogate, North Yorkshire on the 5th October 1852. He was educated at Eton and was a former Lieutenant in the Yorkshire Hussars Yeomanry Cavalry.[2] However, as he was not his father's eldest son, he would not inherit the title of Baronet, or the family seat at Kirkby Hall, Little Ouseburn, so he chose a career in engineering.

The 1871 British Census indicates that Thompson (the inclusion of Meysey in the family surname didn't come about until 1874) was a student of engineering at Durham University, from where he probably graduated in 1873. According to Lupton, Meysey-Thompson received his engineering experience initially at *Armstrongs*, the munitions manufacturers, Newcastle upon Tyne, and then at Messrs. *Beyer and Peacock*, engine manufacturers, Manchester.[3]

On the 6th February 1880, Messrs. Hathorn, Davey and Meysey-Thompson drew up a profit sharing Agreement.[4] The current profit sharing consisted of Hathorn $^{10}/_{14}$ and Davey $^{4}/_{14}$. It was presumably proportioned this way because Hathorn had put the greatest amount of capital into the business, and much of Davey's capital was derived from his patents and consultancy work. The patents were later regarded as a capital asset. Under the new Agreement the proposed profit sharing was: Hathorn $^{10}/_{21}$ Davey $^{6}/_{21}$ and Meysey-Thompson $^{5}/_{21}$. The Partnership Agreements seem to have been reviewed and renewed every seven years, so perhaps the denominator was based on multiplying the three partners over the seven years period. However, there was some difference of opinion, as on the correspondence dated 16th March 1880 it is apparent that Meysey-Thompson also wanted a salary to which Hathorn would not agree.[5] Nevertheless, later records do indicate that he did receive a salary.

By the 8th June 1880, the Articles of Partnership had been agreed between the three Partners, and the Company was continued under the style of 'Hathorn, Davey and Co'.[6] Despite Meysey-Thompson's name not being included in the Company's title he was regarded as a very capable person, and presented various papers to promote the Company. Lupton remarks that, '*Mr. Meysey-Thompson did a great deal of travelling for the sale of the company's engines and amongst other countries visited Mexico, where he secured an order for a large engine for the Parral Silver Mine*'.[7]

Seven months later Meysey-Thompson had clearly settled into his managerial role when he chaired a gathering of the Hathorn, Davey and Company workpeople for their annual supper, held at the Star and Garter, Kirkstall, Leeds.[8]

The Institution of Mechanical Engineers held their annual meeting in Leeds in August 1882, and naturally Hathorn, Davey and Company, as well as other important Leeds engineering companies played an important role in the meeting's organisation. After a welcome by the Mayor in the Civic Court of the Town Hall the President of the Institution delivered his annual address, followed by the presentation of various papers.

Meysey-Thompson had the honour of delivering the first paper of the meeting, '*On the History of Engineering in Leeds.*' The paper covered the period from Henry VIII's visits to Yorkshire at the start of the 16th century, and finished with developments in mechanical engineering, that were mainly associated with the coal, iron and textile industries, in the early to middle 19th century. He was clearly referring to some of the achievements of his own Company when he stated, '*The old non-expansive beam pumping engine is rapidly giving way to horizontal compound engines, applied both on the surface and underground; these work with high pressure and large grades of expansion, and consequently with great economy.*'

Henry Davey also gave a paper, '*On Mining Machinery.*' It was a departure from his usual presentations about steam engines. He discussed improvements in boring techniques for mineral exploration, before turning to shaft sinking and steam driven pumping and winding engines. His references to underground pumping gave him the opportunity to mention six hydraulic engines Hathorn, Davey had, and were in the process of constructing for the Mansfeld Salt Mines, Westphalia (Germany). He also gave a lengthy description of pumping engines on the Comstock Lode, Nevada, USA, manufactured there under licence to Davey's design. After briefly covering haulage and ventilation, he mentioned a future with electric power. Concluding his talk Davey stated that, '*From present experience it is perhaps not visionary to suppose that the time is not far distant when tall chimneys will cease to pour blackness over our towns and country; and when coal, still continuing to be the chief source of power, will be completely consumed in the shape of gas, or burnt direct under pressure. Modern experience points to the hot air or gas engine as the prime mover, and to electricity as the transmitter of power, in the future.*'

There was an Appendix to Davey's talk that was given by Messrs. Mulvany who outlined the water problems they had experienced at their German collieries, particularly in the shafts, and how the many problems had been overcome. There was also lengthy discussion on the paper, and on shaft sinking methods used in Westphalia.

The following two days were devoted to visiting various engineering establishments, one of which was the Sun Foundry. The visit was described. '*These works were established by Messrs. Carrett, Marshall and Co., about thirty years ago, for the manufacture of pumping and hydraulic machinery. For the last ten years they have been in the possession of the present firm, and have turned out a large number of powerful pumping and hydraulic engines, both for water supply and for mining purposes, many of them on Mr. Davey's compound differential system; and also mining machinery of all kinds. There were in hand several large pumping engines, and a centrifugal pump for raising 100 tons of water per minute, which is to be used for draining purposes. A small steam motor with flashing boiler was on view; also an engine-recorder for pumping engines, which registers on a sheet of paper from day to day a complete chart of the working of the engine, and shows the quantity of water pumped during every hour of each day.*' The small steam motor mentioned, was Davey's Domestic Engine, which will be described in a later chapter.

A further report of the visit in the *Engineer* described the construction of the standard quadrants, and other components, used in the pumping installations.[9] '*These are built of timber, plated strongly and tied; and for cutting out the holes in the wrought iron side plates, a machine, specially made by Messrs. Joshua Buckton and Co., is employed, the cutter head being annular in form, and thus cutting out circular plates of the metal, or, when the machine is used for the heavy crossheads of large pumping engines, large cylindrical blocks. The cutter head is supplied with a constant stream of water, by which the cuttings are driven back through holes on the back of the cutter head. The bed plates or girders of the direct-acting pumping engines are often of great length, and too great a length for safe transport; it is therefore necessary to send them away in two pieces. They are, however, cast in one length with separating core plates, which leave only a small thickness of metal continuous on the outer face of the girder. When cold these are broken in the ordinary way, and upon bolting the two parts together again in place the joint is perfect and cannot very well be seen.'*

Other machines were also described, *'Amongst the machine tools in these works is a boring machine by Buckton and Co., chiefly used for boring blowing cylinders, which it will take up to 80in. in diameter and 10ft.stroke. The machine is of the simplest kind, consisting simply of a boring bar, driving shaft with a long bearing in a heavy headstock, and having one large worm wheel on the shaft driven direct by a worm with large rubbing surface. '*

'Among the machinery in course of construction is a fine tandem compound engine with 18in. and 34in. cylinders, of 3ft. stroke, and to run at sixty-five revolutions per minute, or a piston speed of 390ft. per minute. This engine is to be connected by gearing to a large centrifugal pump, which is intended to raise 100 tons of water per minute 16ft. high, the pump being for a Cambridgeshire fen. The pump has a vertical shaft, and will be placed so that the water is forced the 16ft.'

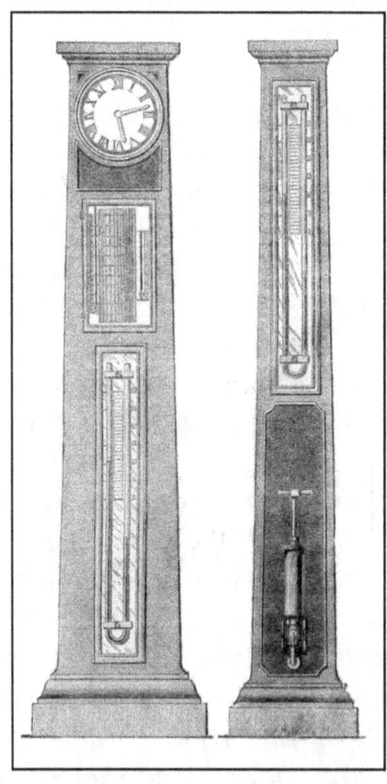

Figure 50. Henry Davey's meter for recording the performance of steam engines
(The Engineer, July 28th 1882. p.59 [GG])

The Engineers' Reporter was particularly impressed with a meter for recording the performance of steam engines designed by Henry Davey. *'Mention should be made also of Mr. Davey's new pumping engine recorder for Cornish and other direct-acting pumping engines. This comprises a well gauge, and stroke recorder, or a counter, or a well gauge alone, as shown by the engravings in the preceding column (Figure 50). The well gauge consists of an ordinary U water column pressure gauge, attached to an empty pipe, except of air which dips into the well. The pressure upon the air, of course, varies with the height of the water in the well, and the gauge is more trustworthy than the usual float and wire or cord over pulleys, especially when the wire has to pass round several corners. The recorder consists of a barrel, upon which is placed a sheet of paper to show the quantity of water pumped through in twenty-four hours. The barrel is worked by a clock in the upper part of an ornamental case, and the pencil which makes the record on the paper is actuated by gearing which receives its motion directly from the piston rod. The performance of the engine is thus recorded, and short or full strokes or stops registered. The exact quantity of water pumped is thus recorded, and not simply the number of strokes, whether short or long. A small pump is attached to the well gauge pipe, so that by a stroke of this the air in the pipe may be renewed to make up for any loss.'*

The engine for a Cambridgeshire fen referred to in the report was the first of several engines supplied by Hathorn Davey to drain the low lying fen areas of Cambridgeshire, and adjacent counties. The drainage of Burnt Fen that comprised of 33,000 acres of land 12 feet below the level of the adjacent river was carried out using two steam engines, under the control of the Burnt Fen Drainage Commissioners. However, there was some anxiety that should one of the existing pumping engines fail, the surviving engine would not cope with the additional load, and so the decision was reached to install a further engine. Consequently, they put out tenders to three companies, and Hathorn, Davey's quotation for £2,690 was accepted.[10,11]

The Burnt Fen engine and centrifugal pump (Orders 3510 to 3519, dated 17th April 1882) are shown in Figure 51. The engine was a Compound Condensing Engine (18"+30", stroke 3') and was provided with a variable expansion valve working on the back of the high-pressure valve. The three Lancashire boilers were 20' by 7'.

The centrifugal pump, (6' 6" diameter) was geared to the engine by means of a steel bevel gearing and was virtually noiseless. It was the first of this type to be made by Hathorn, Davey. The pump was specified to raise 100 tons of water minute 13 feet high, with an assumed maximum discharge of 121 tons per minute 9 feet high. On testing, the pump was capable of exceeding those quantities by 50%. It was 64% efficient under normal working conditions. *'This is an excellent result, but Messrs. Hathorn, Davey and Co. proposed to make a further trial, and hoped to reach 70%.'* [12]

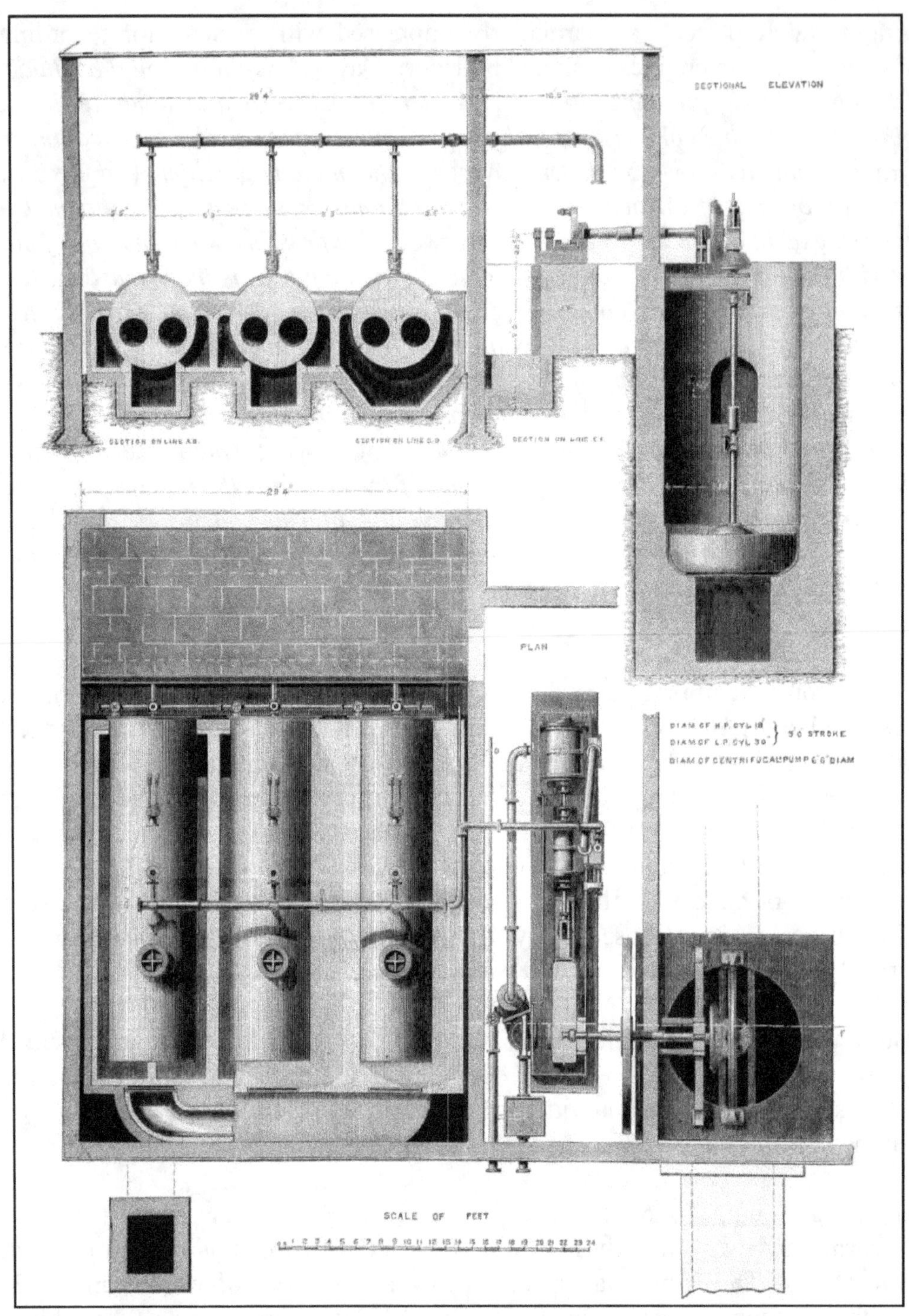

Figure 51. Burnt Fen: Plan and section of the Pumping Station
(The Engineer, 22nd February 1884. p.148 [GG])

During negotiations with Hathorn, Davey the Commissioners initially changed their minds and decided they would like the pump to discharge at the lower rate of 75 tons per minute. This was possibly as a cost cutting exercise. It was during the installation of the engine that Hathorn, Davey suggested that the speed of the engine could be increased to 100 tons per minute, by the addition of extra gear wheels. Although initially declined, it seems likely that eventually this additional work was accepted at a cost of £70. Although later, there was a long argument over the cost of a set of spanners![13]

It is apparent from the listings in Order Book No. 2 (February 1879 to January 1883), that British collieries, still dominated the orders received. Usually the orders were for one-off items, or spares and these would often include valves, delivery pipes or even a Differential Gear. H. Stobart and Company of Etherley, was a long standing customer and was just one example from the north-east coalfield. Elsewhere, orders came from Altham Colliery, Accrington, Lancashire, as well as Garforth and Woolley Collieries, Yorkshire, and a few from the Midlands coalfields. There was also an order from further away that included the Yate Colliery, near Bristol (Orders 3232 to 3235, dated 18th February 1881) for a Compound Differential Condensing Pumping Engine (22"+35", stroke 4') with 6" double-acting ram pumps.

Sometimes there was a penalty clause written into the order stipulating that the client required the engine by a specific date. If that date wasn't met, then it usually meant that the cost of the order could be reduced by a percentage. The penalty percentage could increase significantly if the order was late by months. Hence in the Order Books, the delivery date is written in red ink over the order page. Occasionally, and for various reasons, the bargain to deliver on time if not fulfilled could result in Court proceedings. This is demonstrated by a writ delivered by the *Woolley Colliery Company* solicitors to Hathorn, Davey on the 14th December 1882. An example of a procedure that may not have been uncommon![14]

The *Woolley Colliery Company* claimed that, *'On or about 6th May 1881, the Plaintiffs (Woolley) gave the defendants (Hathorn, Davey) a verbal order for a Compound Differential Condensing Pumping Engine on certain terms as to price, delivery and payments, and the Defendants by letter dated 9th May 1881, accepted such order and agreed to supply and deliver such Engine to the Plaintiffs on the same terms, which terms they embodied in their said letter.'*

The orders under dispute were 3294 to 3298, dated 13th May 1881 for an engine (35"+60") with air pumps, condenser, quadrants, plummer blocks and con rod. It was later confirmed that the engine should be capable of lifting 1500gpm 390 feet high at a speed of 8 strokes per minute. According to a letter dated 9th May 1881, Hathorn,

Davey had agreed to supply the engine for £2,264, and this would include the services of a skilled Erector. The engine was to be delivered to Wheatley Wood Colliery in 16 weeks from the date of the order. A penalty would be incurred of £10 per week for non-delivery.

The terms of payment were that one quarter of the cost would be paid to Hathorn, Davey on delivery of the bulk of the work at the collieries, a further quarter four weeks afterwards, and the remainder paid four months after delivery.

The Plaintiffs stated that the engine was required to prevent damage to Wheatley Wood Colliery by flooding.

However, Hathorn, Davey had apparently breached their agreement by only delivering the engine 25 weeks after the agreed delivery time of 16 weeks. The Plaintiffs were entitled to compensation of £250.

In addition, as a consequence of the engine not being delivered on time, the colliery had become flooded, damaging roof and roadway. The mine had sustained damage amounting to £750. The Plaintiffs also claimed interest on the £250 until it was paid, and the potential for other claims to be made, as the case may require.

In their defence, Hathorn, Davey responded on the 5th January 1883 refuting the Plaintiff's claims. They did not admit to receiving a verbal order from the Plaintiffs, but did admit to receiving a letter from them on the 11th May [which links with the order entry dated 13th May]. Hathorn, Davey also denied that the Plaintiff had told them the purpose of the engine, and that the engine had been completed and fitted together on the premises (Sun Foundry) at the commencement of September 1881, and was at that point ready for delivery to the Plaintiffs.

On or around the 9th September 1881, all engine parts had been delivered to the Plaintiff for the purpose of commencing the erection of the engine. Hathorn, Davey had within the meaning of the contract with the Woolley Colliery Company delivered within the period of 18 weeks from the date of the order. They considered it normal practice within the trade for completion to mean the erection of the engine as being ready for delivery. The Defendants however, did admit that they had defaulted by being overdue in the delivery by two weeks, and had brought to the Court £20, and stated that it was sufficient to satisfy the Plaintiff's claim, and denied that they were liable for damage to the colliery.

'By way of Set-Off and Counter-Claim,' Hathorn, Davey responded by stating that the Plaintiffs had not accepted the various parts of the engine when they were notified that

it was ready for delivery. In addition, they had been deprived of the services of the skilled Erector sent by them, under the contract, to superintend the erection of the said engine for an unreasonable time and had incurred a loss. Hathorn, Davey also sued the Plaintiffs for money payable to them in respect of workmen, machinery, tools, and materials supplied.

Hathorn, Davey's claim amounted to £15-8s for the Plaintiffs not accepting the various parts of the engine within a reasonable time, and £81-19s-11d for workmen, machinery, tools and materials supplied.

The outcome of the Court case is not recorded, but probably it was settled out of Court to everyone's mutual satisfaction.

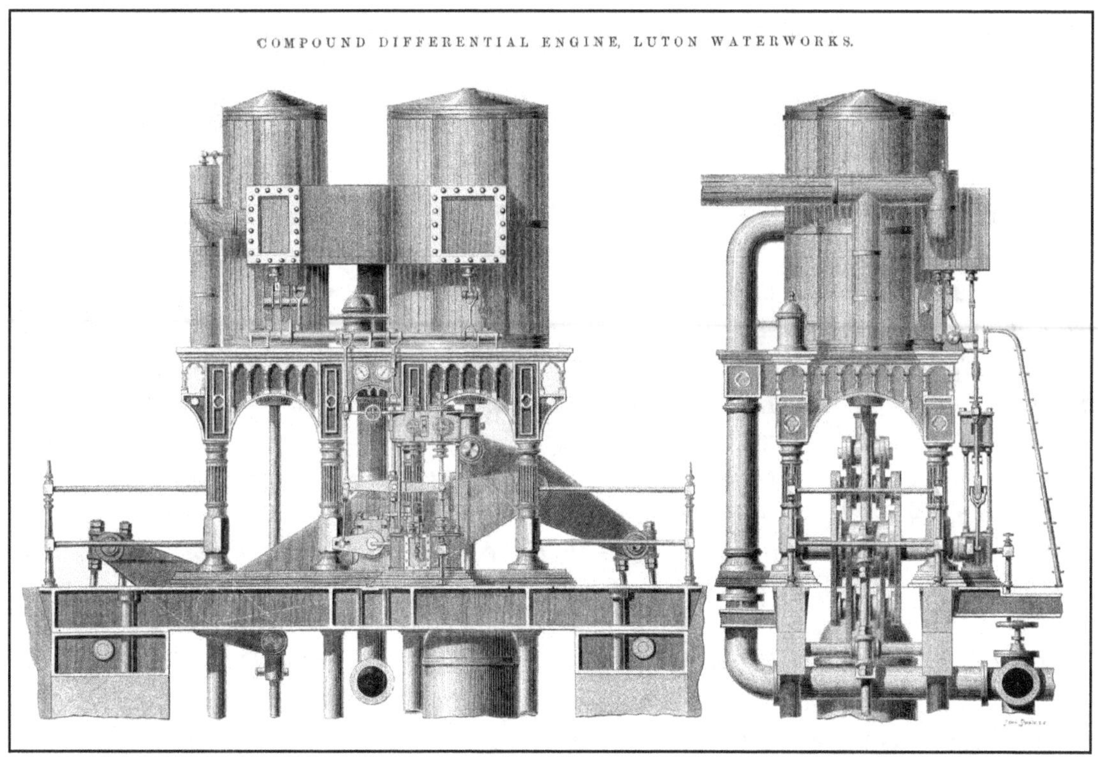

Figure 52. Luton Waterworks: Compound Differential Condensing Pumping Engine
(The Engineer. 1st December 1882, p. 412 [GG])

Orders from water companies, and sanitation boards were also increasing at this time. *The Staffordshire Potteries Waterworks Company* bought several engines and this was the start of a long relationship that Staffordshire water companies had with Hathorn, Davey. Other water companies and sanitation boards bought equipment, e.g.,

Gainsborough Corporation; the waterworks at Yeadon at Leeds; Chester; Ventnor, Isle of Wight; Burton on Trent; and the Luton Waterworks Company (See Figure 52).

The Luton order (Orders 3250 to 3266, dated 6th March 1881) was for a Compound Differential Condensing Pumping Engine (28"+56", stroke 6') to work two 19" plunger pumps, with a lift of 225 feet.[15] Steam pressure was 50lbs. The engine can be compared with the Chiswick engine shown in Figure 39 to see how the design was modified, particularly Davey's parallel motion.

Iron mining companies were also purchasing the standard Horizontal Compound Engines, with pumping quadrants. *The Yarlside Mining Company* for example, purchased two Compound Differential Pumping Engines. Orders 2897 to 2904, dated 13th November 1879 and orders 3584 to 3590, dated 19th September 1882 were for engines with 34"+60" and 38"+66" cylinders, respectively. Both engines had 10', double strokes, and each worked two 20" pumps. The latter engine pumped from a depth of 550 feet in two lifts.

The Yarlside iron mines are located in the Carboniferous Limestone of the Furness district of Lancashire, and they were notoriously wet mines. One description in 1898 examined both the mines drainage and pumping arrangements that included the still working Hathorn, Davey engines.[16] *'The large quantity of water that has now to be dealt with being unforeseen in the earlier development of the mine, engines and pumping plant of comparatively small size only were provided; but, as necessity has arisen, one pumping-engine and plant after another has been erected until, at the present time, there are 6 and sometimes 7 engines delivering water to the surface.' 'The quantity of water pumped varies from a maximum of 6860 gallons per minute in winter to a minimum of 4,940 gallons in summer. The average quantity per minute pumped during 1898 was 5662 gallons, which is equal to over 8,000,000 gallons in 24 hours; and the maximum quantity in winter is 9,878,400 gallons in 24 hours.'*

A comparison was made between the Cornish Beam Engine and Davey's Compound Engines. Considering that the Hathorn, Davey engines had been working for 15 years, the conclusion is quite complimentary. *'For the Cornish engines, the lengths of lift vary from 240 to 360 feet, but it is found that the shorter lifts are much better in working and are not so troublesome as regards keeping the joints tight and maintaining the glands in good order. With the Hathorn, Davey engine, there is no trouble with blown-out joints in the pumps, although the lifts are over 378 feet in vertical height: they work without shock, and the glands will last for 2 years without repacking.'*

The 1898 description also provides an explanation as to why many foundations for Hathorn Davey Compound Engines have survived. *'It has been found that it is*

This is certainly the case at Yarlside where the foundations of a substantial engine bed still exist.[17] Figure 53 is a satellite photograph that shows the elongation concrete engine bed for the 1882 Horizontal Compound Engine (38"+66") at the No. 11 pit. The lower walled area on the north side of the engine bed was probably the location of the air pump and condenser. The mine site is located about 1 kilometre east of the ruins of Furness Abbey.

Figure 53. Satellite Photograph of Yarlside Mine No.11 Pit showing the substantial foundation of the Hathorn, Davey Compound engine-bed.
(Google Earth)

Another interesting contract in this period was to supply the two Compound (20"+35") Pumping Engines (Orders 2962 and 3009 - February and April 1880) to drain water from pilot heading sumps during the construction of the Mersey Railway Tunnel between Liverpool and Birkenhead. They were probably the first engines the Company supplied for a major civil engineering project.

However, the international market for engines was also starting to expand. By the early 1880s the British Empire was reaching all corners of the world. It seems that entrepreneurialism was obligatory, and a whole range of London registered companies were being established to build railways, waterworks, establish factories and sink mines in foreign countries. Consequently, numerous new business opportunities were starting to open up for Hathorn, Davey.

Chapter 10 References

[1] SNA GD 455/76

[2] Fashionable Yorkshire Wedding. *The Leeds Mercury*. 2nd June 1896

[3] HL. 1940. *Hathorn, Davey and Company, Sun Foundry, Leeds*. p..2

[4] DLC Partnership Agreement 6th February 1880

[5] DLC Correspondence 16th March 1880

[6] DLC Articles of Partnership 8th June 1880

[7] HL. 1940. *Hathorn, Davey and Company, Sun Foundry, Leeds*. p.4.

[8] Local and District. *The Leeds Times*. 12th February 1881. p.3

[9] Visits to the Provinces. Messrs Hathorn Davey, and C's Works, Leeds. *The Engineer*, 28th July 1882. p.59.

[10] Beckett, J. 1984. *The Urgent Hour: The Drainage of the Burnt Fen District*. (2nd Edition). Ely Local History Publications Board. Pp. 27 to 30.

[11] Compound Engine and Centrifugal Pump. *The Engineer*, 22nd February 1884. pp. 145 and 148

[12] ibid. *The Engineer*, 22nd February 1884. p.145.

[13] ibid. Becket, p.30.

[14] DLC Woolley Colliery Company 1883

[15] Details of Pumping Engine, Luton Water Works. *The Engineer*. 1st December 1882. pp. 412 and 416.

[16] Davison, J. 1898-99. Description of the Pumping Plant at the Stank and Yarlside mines in the Furness District of North Lancashire. *Transactions of the North of England Institute of Mining and Mechanical Engineers*. Vol.48. pp. 148 to 156.

[17] Sandbach, P. 2009. Stank and Yarlside mines. *CAT The Newsletter of the Cumbria Amenity Trust Mining History Society*. 95: May 2009. pp. 26 to 29

11. Hathorn, Davey and Company: The International Market

The 1880s can be regarded as a period of major expansion for the Hathorn, Davey Company. In Europe, the Company had previously supplied businesses in Germany, but now orders were starting to come from France and Spain. With the spread of British influence worldwide, Hathorn, Davey were also receiving orders from companies working in China, Chile and Japan, the latter country undergoing an industrial revolution. Previously, small orders from further away had come from countries in the British Empire. It seems that the only buffer to this influence was the United States of America, who were evolving their own brand of business practice and it was their companies that were sometimes in competition with British companies for contracts.

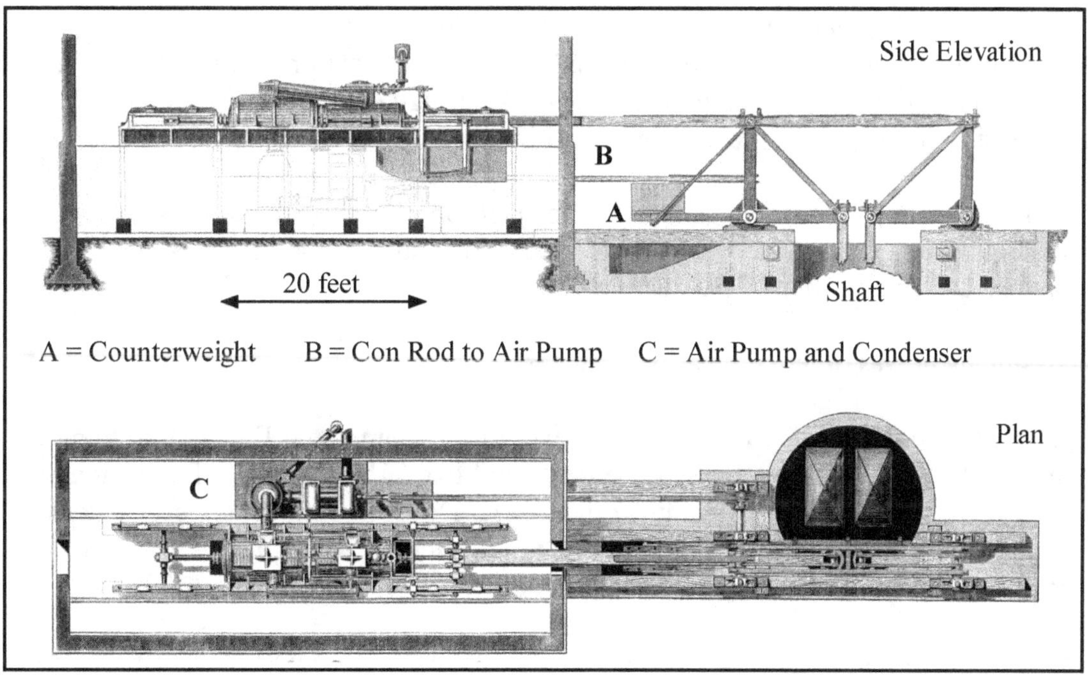

Figure 54. Kai ping, China: Side Elevation and Plan of the Pumping Engine
(Modified from: The Engineer. 1st April 1881. p.237 [GG])

The *Chinese Engineering and Mining Company* was formed in 1878 to develop and work coalmines on the Kai ping Coalfield, located about 100kms east of Beijing, China. By February the following year they had placed very large orders with Hathorn, Davey. Initially, the mining company proposed to sink and equip two shafts, served by both winding and pumping engines.[1]

The *Chinese Engineering and Mining Company* seems to have commissioned Hathorn, Davey to manufacture the majority of pumping, winding, and ventilation equipment, including spare parts, and also a reserve engine to operate the ventilation fan.

During August 1879, it was reported that, *'Operations were suspended in the winter, pending the arrival of men and machinery from England, but three men and a portion of the machinery have now arrived.'* [2]

The proposed layout of the colliery surface was very typical of the period. Two shafts were to be sunk, an upcast and a downcast. Buildings were constructed around a quadrangle with the pumping engine house located centrally between the shafts. Figure 54 shows the pumping engine. Note that the pumping equipment is offset from the shaft centre, so that the shaft can also accommodate a double cage system. The pump rods were counter balanced with a counterweight. The engine also had a tail rod from the low-pressure cylinder attached to a crosshead. This suggests that the engine could have been used to work pumps in the second shaft if required.

The two winding engine houses were located on the two opposite sides of the quadrangle obviously to face each shaft. The boiler houses and chimneys were to be constructed on the far side of the quadrangle.

The equipment was ordered in various batches:

Orders 2752 to 2772 dated February 1879 were for a Compound Differential Pumping Engine (30"+50") with boilers, two quadrants and two 21" Bucket Lift Pumps together with a Double Cylinder Winding Engine; A Double Cylinder Non-Condensing Winding Engine (Figure 55 and 56) with three winding ropes and a cage and two boilers. The total cost was £2,530, and the order was completed in June 1879.

Orders 2842 to 2849 dated 2nd August 1879 were for a Double Cylinder (8") Semi-Portable Winding Engine and boiler with wire ropes, steam pipes etc, and three wrought iron kibbles. The total cost was £791 plus £108 for an extra set of steam pipes.

Orders 2942 to 2958 dated 27th January 1880 were for miscellaneous items that included quadrants, pipes, pumping rods and spears, suction pipes. The cost was £1,688 for delivery to Liverpool, or £1,708 to London.

Orders 2960, 2961 and 2987 were for two Vertical Donkey Pumps, clack valves with seats, and eight pairs of water heaters, respectively.

Orders 3025 to 3026 were for two cast iron windbores and eight wire ropes.

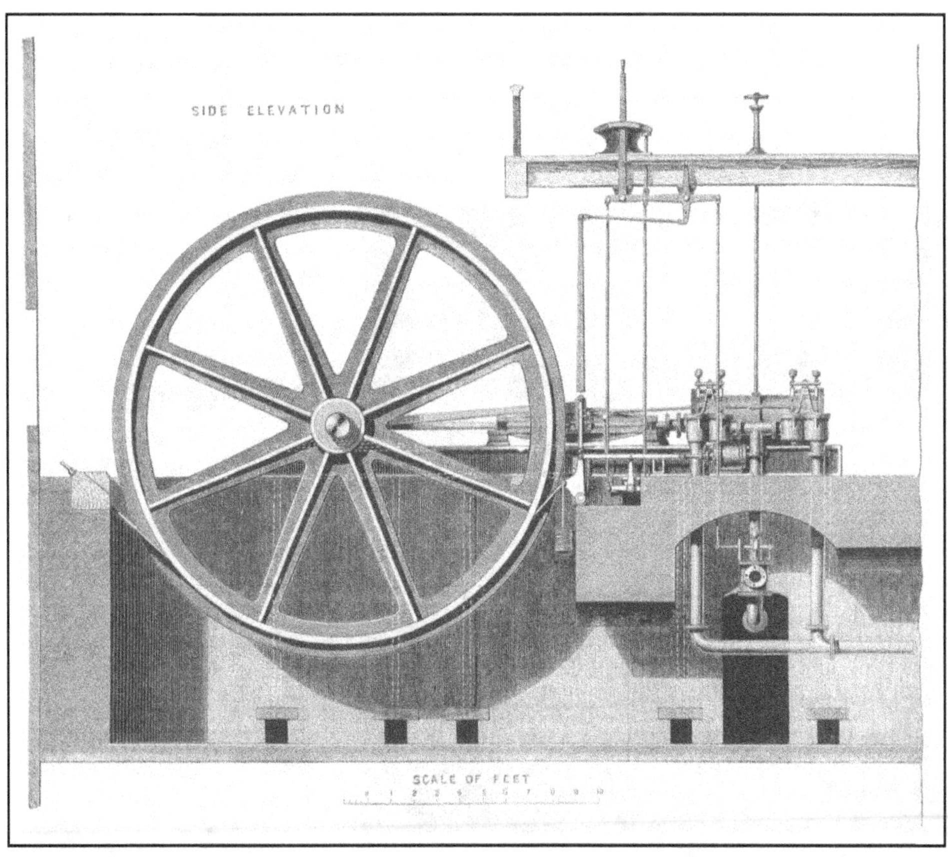

Figure 55. Kai ping, China: Side Elevation of the Double Cylinder Winding Engine
(The Engineer. 1st April 1881. p.240 [GG])

Orders 3031 to 3034 (£1,012) and 3039 to 3046 (£955 - 19th May 1880) were for mixed items including more water heaters, guide and winding ropes (850 feet), pipes, kibbles valves, and pipes. Order 3091 was for pipes.

Orders 3127 to 3136 were for a Compound Differential Pumping Engine (30"+50") complete with air pumps and condenser, two quadrants, pump rods, 20" plunger pumps and spare parts. The order was dated 19th October 1880 and cost £1,635.

Orders 3139 to 3142 were for a Guibal Fan (30' x 10') with a full velocity of 100,000 cubic feet of air per minute, worked by a Non-Condensing Compound Engine (16"+32" with 2' stroke). One high-pressure engine (24" with 2' stroke) was to be used as a reserve in case the Guibal Fan engine broke down. Both engines were to be tested under steam at the works [Sun Foundry] at a cost of £20.

Orders 3220 and 3268 were for drain cocks, spur wheel and drum end.

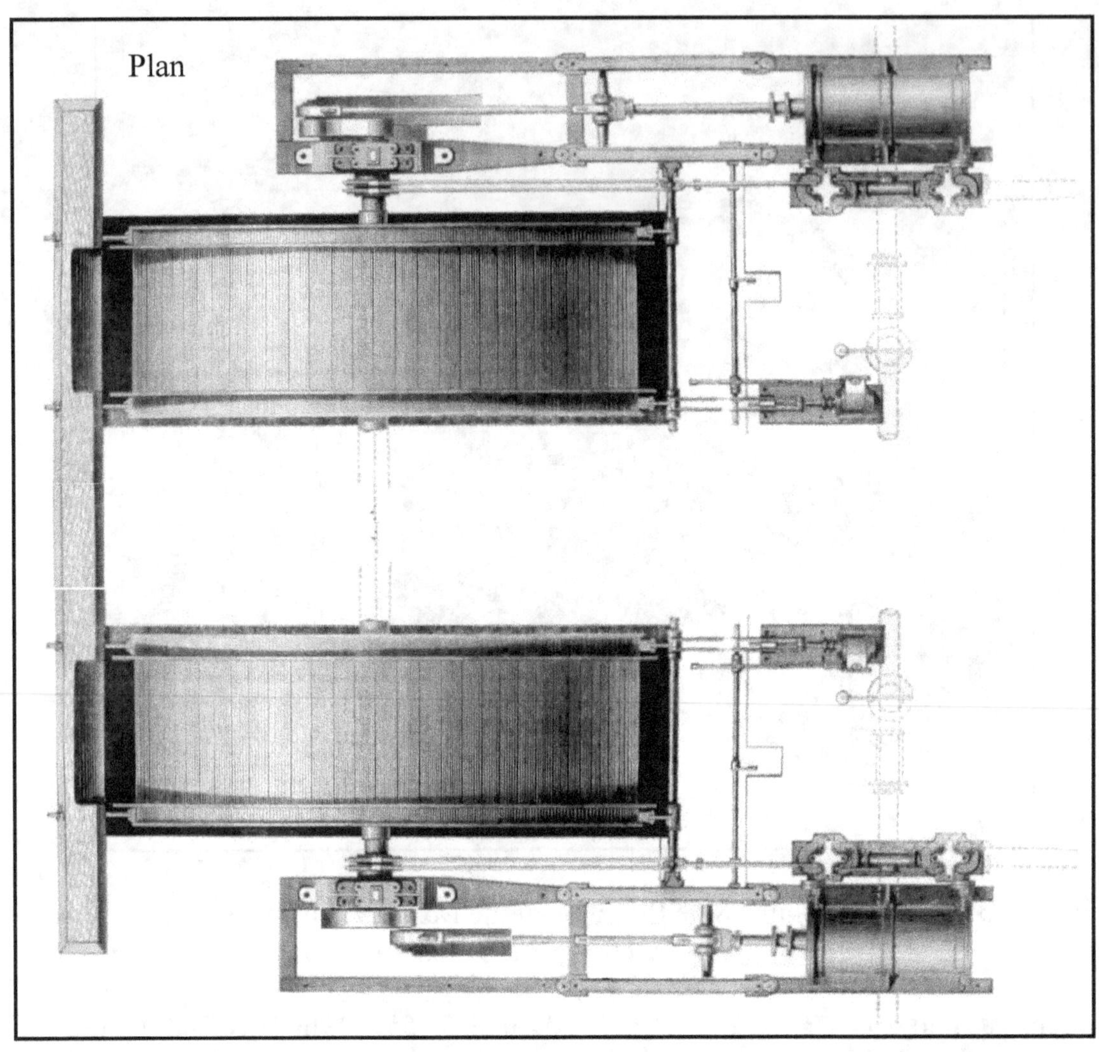

Figure 56. Kai ping, China: Plan of the Double Cylinder Winding Engine
(Modified from: The Engineer. 1st April 1881. p.240 [GG])

Hathorn, Davey manufactured very few winding engines. The winding engine with the 18' diameter drums was probably the largest winding engine manufactured by the Company. It was designed to wind 800 tons of coal per day from a depth of 500 feet.

Elsewhere in China, a group of merchants formed the *Shanghai Waterworks Company, Limited* in November 1880 with the purpose of providing water to the Shanghai International Settlement in the Yangpu district of the city.[3] The waterworks were the first to be built in China, and at the time, were the largest in eastern Asia. The building to house the machinery was designed by the Company's engineer J.W. Hart, in the style of an English castle (Figure 57).

Two engines, three Cornish boilers, and associated equipment was ordered from Hathorn, Davey on 24th December 1880 (Orders 3175 to 3185) followed by several small orders for spare parts and valves (Orders 3241 to 3243). Order 3175 was for a Compound Differential Engine (18"+34", stroke 5') with air pump and condenser, and order 3176 for an *'ordinary rotative type with steam cylinders (18"+34" with 3' stroke) placed on separate bed plates, and the cranks were to be placed at the angle of 90° to each other with the flywheel between them.'* The cost was £1,650 for the engines and £1,012-10s for the boilers.

Figure 57. Shanghai Waterworks. The engine and boiler house
(https://www.virtualshanghai.net/Asset/Source/dbImage_ID-2037_No-1.jpg)

Japan started to develop as an industrial nation during the mid 1800s. Until that time, Japan had been an isolated country, with only a Dutch trading colony at Nagasaki, but it was perhaps due to the appearance of the United States Navy in Edo Bay in 1853, that the centuries old policy of seclusion was lifted. [4]

In 1863 five Japanese students (the Choshu Five) made a clandestine voyage to Britain, against the strict policy that forbade people to leave the country. On their return to Japan, they promoted industrialization as a strategy. Known as the Meiji Period, or Restoration, industrialisation took hold very quickly, and in just 50 years traditional Japan was transformed into an industrial society. Some of the earliest industrial remains relate to iron manufacture in an attempt to produce cannons, and naturally this led to the development of the Japanese coal industry.

On the 15th October 1883, the first of a number of 'Davey's' Compound Differential Engines was ordered for the Japanese coal industry. Orders 3902 to 3904 were for a Underground Horizontal Compound Differential Condensing Pumping Engine (12"+24", stroke 4') with 9½"ram pumps, as well as an air pump and condenser, and capable of raising 300gpm a height of 240 feet. The total cost of the order was £997 and it was to be completed in 12 to 14 weeks and, *securely packed for a long voyage.*

The order had been placed by the *China and Japan Trading Company* for installation at the Miike Mine, Chikugo, on the southern island of Kyūshū. This was to be the first of many pumping engines to be sent to the Miike coalfield that will be described in a later chapter.

In addition to the Asian market, orders also came from South and Central America. An order placed by *Charles Lambert and Company* for the Brillador and San Carlos Copper Mine, Coquimbo, Chile on the 27th May 1880, was the first from South and Central America. It was also uniquely interesting as it was for the only man-engine ever manufactured by Hathorn Davey.

Man engines were a means of transporting miners to and from the surface in situations where mine workings were very deep. The man-engine consisted of a single line of counterbalanced rods that were always kept in tension. All the rods were guided on rollers. The stroke in this instance was about 12 feet, at five strokes a minute. At the shaft top the horizontal reciprocating motion of the flat rods was converted by a quadrant to produce a reciprocating motion in vertical rods in the shaft. The vertical rods formed the man-engine.

A miner going to the surface, for example, would step onto a platform attached to the rods, rise 12 feet and step off the moving platform onto a fixed platform. Constantly repeating this process, the miner would be raised to the surface. Thus a miner could ascend or descend the shaft at a rate of 60 feet per minute.[5]

The order contained three engines. Numbers 3051 to 3059 were for a Compound Differential Condensing Pumping Engine (26"+46", stroke 8') with the air pump worked by the tail rod. The shaft installation included a 10" rising mains, 90 feet long, for a 9" bucket lift pump, and 984 feet of 7" rising main for two sets of 8½"ram pumps, and a pitch pine con rod between the engine and the quadrants.

The combined capstan / man engine (Order 3065) had two cylinders (both 11", stroke 20"). At the time of installation, the shaft was 1000 feet deep, but it was proposed to sink the shaft down to a depth of 2,200 feet. The top 423 feet of the shaft was vertical, but below that depth it was inclined at about 60°. On the crankshaft of the capstan/man engine there was a pinion that geared into a spur wheel to operate the capstan drum (6' x 5' 3") that had been adapted for lifting a load of 5 tons. A further pinion could also be geared into a 14' spur wheel that worked the man engine (See Figure 58).

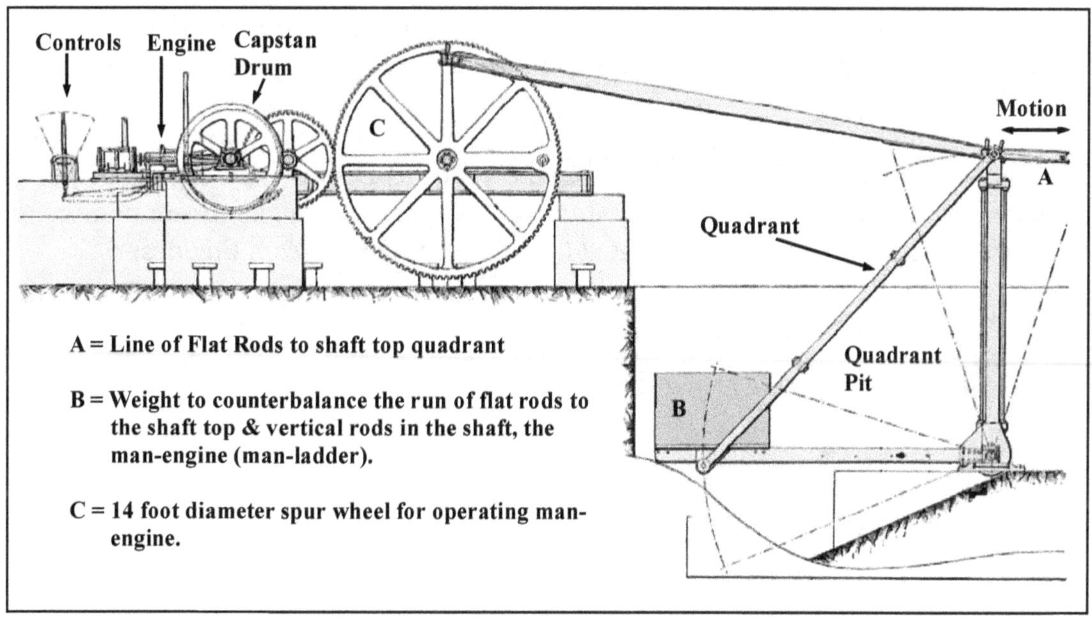

Figure 58. Brillador and San Carlos Copper Mine, Chile
The arrangement of the combined capstan and man-engine
(Modified from: The Engineer, 18th March 1881, Supplement [GG])

The third engine (11" cylinders with a 20" stroke (Order 3067) was for working two winding drums. The total order contained quadrants for the man-engine as well as various guide rollers for supporting the rod runs on both the surface, and down the shaft.

There were also several other orders from South and Central America in the first half of the 1880s.

Orders 3729 to 3734 (February 1883) were for miscellaneous items that included bucket pumps, iron rods and couplings and a quadrant, and appear to have been required to couple to an existing engine. The *Ouro Preto Mining Company* operating in Brazil had placed the orders. It was a company managed by *John Taylor and Sons* of London. This seems to have been the first order placed with Hathorn, Davey by *John Taylor and Sons*. Eventually, Taylors would also order many pumping engines for their gold mining operations in India, and Australia.

During May and August 1883, the *El Bole Mining Negotiations Company*, Yacatecas, Mexico ordered a significant amount of machinery in two batches. Orders 3764 to 3775 (dated 22nd May 1883) were for a standard pumping system with a Compound Differential Engine, quadrants etc. Orders 3860 to 3873 (August 1883) were for numerous miscellaneous items that included one new pump for a windmill. It is possible that Meysey-Thompson made a trip to Mexico at this time in connection with this contract.

The European market was also expanded in the 1880s as Davey's innovations gained a reputation for efficiency and reliability. In addition to engines, a significant number of orders were taken for Davey's Differential Gear, to be fitted to Beam Engines.

Chapter 11 References

[1] Coal Mining in China - Chinese Engineering and Mining Company. *The Engineer*. 1st April 1881. pp.204, 237, 240 and 242.

[2] Coal in China. *Aberdeen Evening Express*, 23rd August 1879. p.2, c..4.

[3] The Shanghai Waterworks Company, Limited - Abridged Prospectus. *London Daily News*, 13th November 1880, p.1. c.4., and https://en.wikipedia.org/wiki/Yangshupu_Waterworks.

[4] Kato, K. 2019. *Sites of Japan's Meiji Industrial Revolution*. Department of Industrial Heritage, Cabinet Secretariat, Government of Japan. 1st and 2nd editions.

[5] Pumping Machinery and Man Engine. *The Engineer*, 18th March 1881. p 281 and supplement.

12. Hathorn, Davey and Company: The European Market

By the 1880s, the Horizontal Compound Differential Engine was a *'market-leader'* with a variety of applications. It's sales had always been robust in the British coalfields, and to a lesser extent the metal mining industry, where the success of a mining company was more dependent on the fluctuations of metal prices and the richness of the mineral deposit. However, the more sustainable iron mining companies of Lancashire, Cumbria and North Yorkshire did show an interest in it.

Whereas Hathorn, Davey engines were being exported to China, Japan and later, India, they only received minor attention in European countries. This may have been due to local competition, and perhaps the dominance of Beam Engine technology. British companies tended to 'export' the technology to their Continental operations.

Spain, and to a lesser extent Portugal, had been slow to embrace steam pumping technology, despite several attempts at the start of the 19th century to introduce Boulton and Watt engines to indigenous mining concerns. Such engines were purchased for the Buarcos coal mine at Figuiera da Foz, Portugal,[1] and the Almadén mercury mine, Spain.[2] The former engine was never actually erected, whilst the engine at Almadén operated for several decades. In Spain, other Boulton and Watt engines were also purchased to operate in saw mills and ironworks.[3]

In the 1840s, Spain, our nearest relatively under-developed country with vast mineral reserves, provided a ready pocket for British investment. This was helped by a relaxation of Spanish import tariffs, and also to changes in British legislation for the formation of companies. Mining Companies were set up, and Beam Engines were introduced primarily to the lead mining district of Linares, Andalucía. Even today, the area is typified by the remains of a large numbers of Cornish type Beam Engine houses, the greatest concentration of such buildings outside of Cornwall and Devon.[4]

Had Davey's pumping technology been available in the 1840s / 50s then almost certainly it would have been purchased. However, even when available, it had become established practice at many of the Linares mines, to simply use the resources at hand and move the Beam Engines from shaft to shaft, even in the 20th century.

It was left to a Scottish mining company to introduce the first Hathorn, Davey engine into Spain in 1883. *The Belalcázar Silver-Lead Company, Limited* was registered on the 22nd February 1883 in Glasgow, Scotland, with the purpose of working mines in Spain.[5] They appear to have worked only one mine, the Solana silver-lead mine near Belalcázar, northern Córdoba Province, Andalucía. During April and August 1883 the Company ordered two Compound Pumping Engines from Hathorn, Davey.[6]

Figure 59. Solana Mine: the Hathorn, Davey engine and boiler house
(Author 2017)

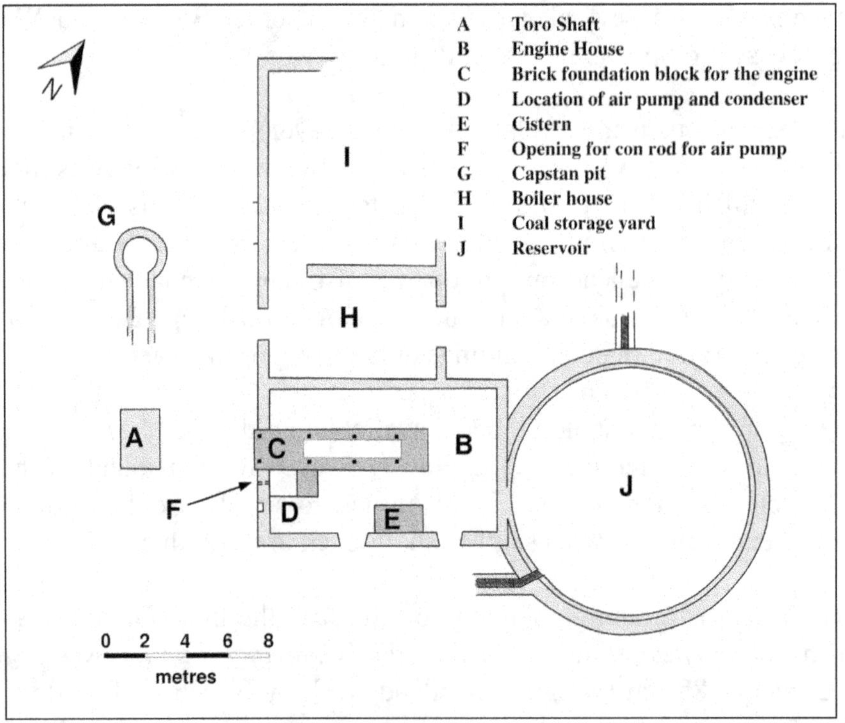

Figure 60. Solana Mine: plan of the engine house and associated structures.
(Vernon and Cabrera, 2018. De Re Metallica v.30. p.87)

Orders 3752 to 3757, dated 11th April 1883, were for one Compound Differential Condensing Engine (16"+35", stroke 6') lagged with polished mahogany with brass bands, together with one air pump (worked from quadrant), condenser of ample capacity, one balanced quadrant, plus plummer blocks, connecting rod, and main caps for the top of the shaft spears. The order cost £785 and it was delivered to Hull for shipment on 18th June 1883.

However, on the 23rd August 1883 the *Belalcázar Silver Lead Company* ordered a second engine (Orders: 3878 to 3884). A note in the Order Book stated, '*This engine is to replace 3752, and following numbers, which have been lost at sea.*' The replacement engine was slightly larger (20"+36", stroke 6'), but the rest of the machinery remained the same.

The Company was reformed several times, as the *Solana Mining Company, Limited* in 1889[7], and then *Toro Mining Company, Limited* in 1892.[8] However, this was a short-lived affair and a resolution was passed to wind up the company in February 1894. The Hathorn, Davey engine was then sold on to a Spanish mining company.

Figure 61. Solana Mine: Brick built mounting block for the Hathorn, Davey engine
(Author 2017)

The ruins of the Hathorn, Davey engine house still dominate the Solana mine site (Figure 59). It is one of the best surviving examples of this type of structure, and still contains a substantial brick built mounting block for the engine (Figure 61). Figure 60 is a plan of the engine house and associated structures.

There were also several other orders for Compound Differential Engines from Spanish owned lead mining companies.

The Société Anonyme Miniere de la Province de Murcia ordered a complete pumping system on the 2nd August 1881 that included a Compound Differential Condensing Pumping Engine (18"+34") (Orders 3338 to 3347) On the 21st November 1882 the Company purchased a second and larger engine (24"+48") and pumping machinery (Orders 3648 to 3681).

However, not all purchases of the Compound Engine were fulfilled. On 14th February 1884, José Bover, the Inspector of Mines for Almería Province, Spain, had ordered a pumping system for La Solana del Fondón, Almería. The orders (3980 to 3995) included a compound engine (12"+20") and a complete set of pumping machinery, as well as a Cornish boiler (14'+5') and 1,200 firebricks on which to support it. The total cost was £1,195. A second order (4060 to 4076, dated 12th June 1884) is also listed for the La Solana del Fondón Mine, for a larger engine (20"+36") that probably superseded the order placed on the 14th February 1884.

Apparently, Bover had placed the order on behalf of the Marchioness Torrealta, probably the landowner, via *Chives and Fraser*, British merchants established in Almería.[9] However, on the 27th January 1887, Hathorn Davey wrote to *Chives and Fraser* asking why the engine had never been collected. *'The work was completed nearly three years ago and sent to the station.' 'The railway company after repeated application for us to take it away, which we duly passed on to Mr. Bover.'* This was to no avail. So since that time they were stored in their original packing at the Sun Foundry. Hathorn, Davey had clearly intended to charge for storage, and had expected to come to some sort of settlement with the Marchioness. Again no progress in dispatching the machinery was made. Hathorn, Davey thought they could use the equipment, and asked for permission to exhibit the machinery at an exhibition at Newcastle. Again, nothing was done. Eventually Hathorn, Davey, via their Solicitors (Nelson and Son, and Lupton, Leeds) wrote to the Marchioness in May 1887 stating that if the machinery were not removed by 11th June 1887, it would be sold.

The author has visited the mines in the Fondón area on several occasions, and has found no evidence for a Hathorn, Davey engine ever been erected in the area. So it is presumed that Hathorn, Davey did resell the engine. Who it was sold to is not recorded.

Elsewhere in Spain, the British *Rio Tinto Company* placed an order for a Compound Differential Condensing Pumping Engine (12"+20") with a 6½" double ram pump (Orders 3891 to 3898, dated 29th September 1883).

During the early 1880s, there was one significant order from France. The Société Anonyme de Charbonages, Bouche de Rhone, ordered a hydraulic pumping system, and other machinery for their Lhuillier coalmine, Fuveau near Marseilles. Davey was later to describe the hydraulic system in detail.

Orders 3711, 3776 to 3779A, 3798 (12th July 1883), 3812, 3888, 4148 included a rotative compound engine, horizontal compound condensing pumping engine (30"+50", and 3' stroke at 23 strokes per minute), condensers and air pumps, a horizontal accumulator, and 3 single acting ram pumps.

The engine was a departure from Davey's standard horizontal compound in that it was a cross-compound engine. Figure 62 shows one half of the engine and illustrates the high-pressure cylinder, forcing pump and the combined air pump / condenser. The accumulator is located at the side of the engine.[10]

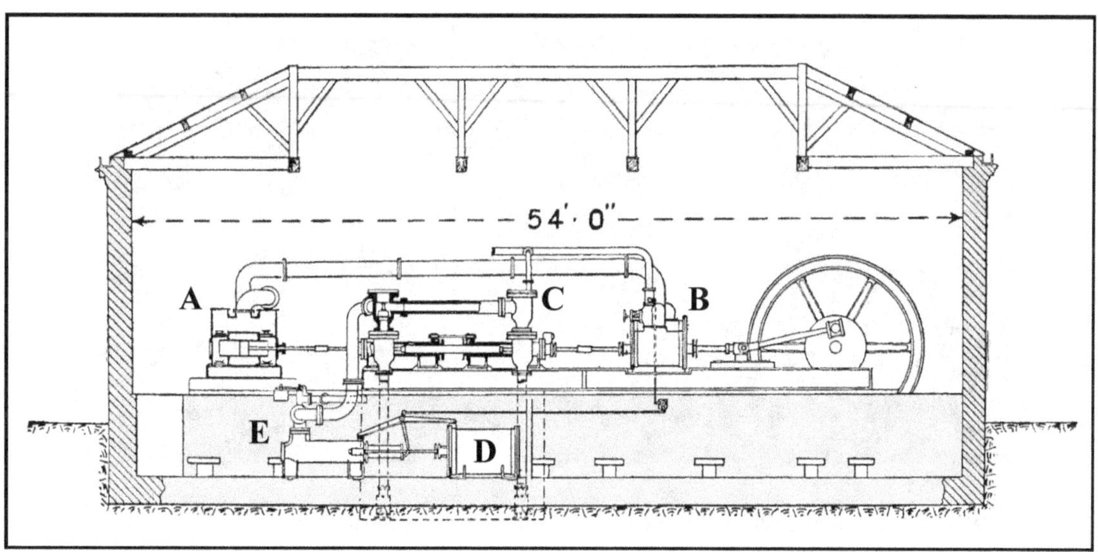

Figure 62. Lhuillier Colliery: The Surface Pumping Installation.
A= Combined air pump / condenser; B = High-pressure cylinder; C = Hydraulic circuit forcing pump; D= Accumulator engine; E= Accumulator piston.
(Davey 1905, Principles, Construction and Application of Pumping Machinery, p.180)

Figure 63 is an illustration of the Hathorn, Davey surface installation at Lhuillier. The high-pressure cylinder is to the left of the person in the Figure. At the far end are the top chambers for the surface forcing pumps. The accumulator is located below floor level. Otherwise, this was a typical closed hydraulic system that pumped water up to an adit from a depth of 1400 feet, where the lower hydraulic pump chamber was located. To function, correctly, '*All that is required is a constant pressure in the steam pipes, and that is secured in this case by the employment of a horizontal steam accumulator, or regulator.*' This consisted of, '*A plain steam cylinder having a piston, the rod of which is attached to a plunger working in a hydraulic cylinder in communication with the power pipes.*' '*The front end of the steam piston is in communication with the atmosphere; the back end is connected by means of a pipe to the main steam pipe, so that the boiler pressure is maintained at the back of the steam piston, and the relative areas of the steam piston and hydraulic plunger are such as to maintain the required pressure to the power pipes.*'

Figure 63. LHuillier Colliery: Illustration of the Hathorn Davey Engine
(Biver, 1885, p.83)

During a trial of the system, the work done by the pump in raising the water was 124 horsepower, whilst the steam engine operated at 229 horsepower, thus giving a total efficiency of 54½ %, which also included a small amount of work done by the engine to lift condensed water to a cooling pond. The cost of the machinery and pipes was £5,430. The total cost that included the underground excavation of a pump lodge buildings etc., was £8,630.

Davey was clearly very pleased with the efficiency of the installation at Lhuillier coalmine, as he wrote several pages that offered details of its mechanical efficiency, and cost effectiveness. His calculations were based on a description of the engine that was published in France. [11, 12]

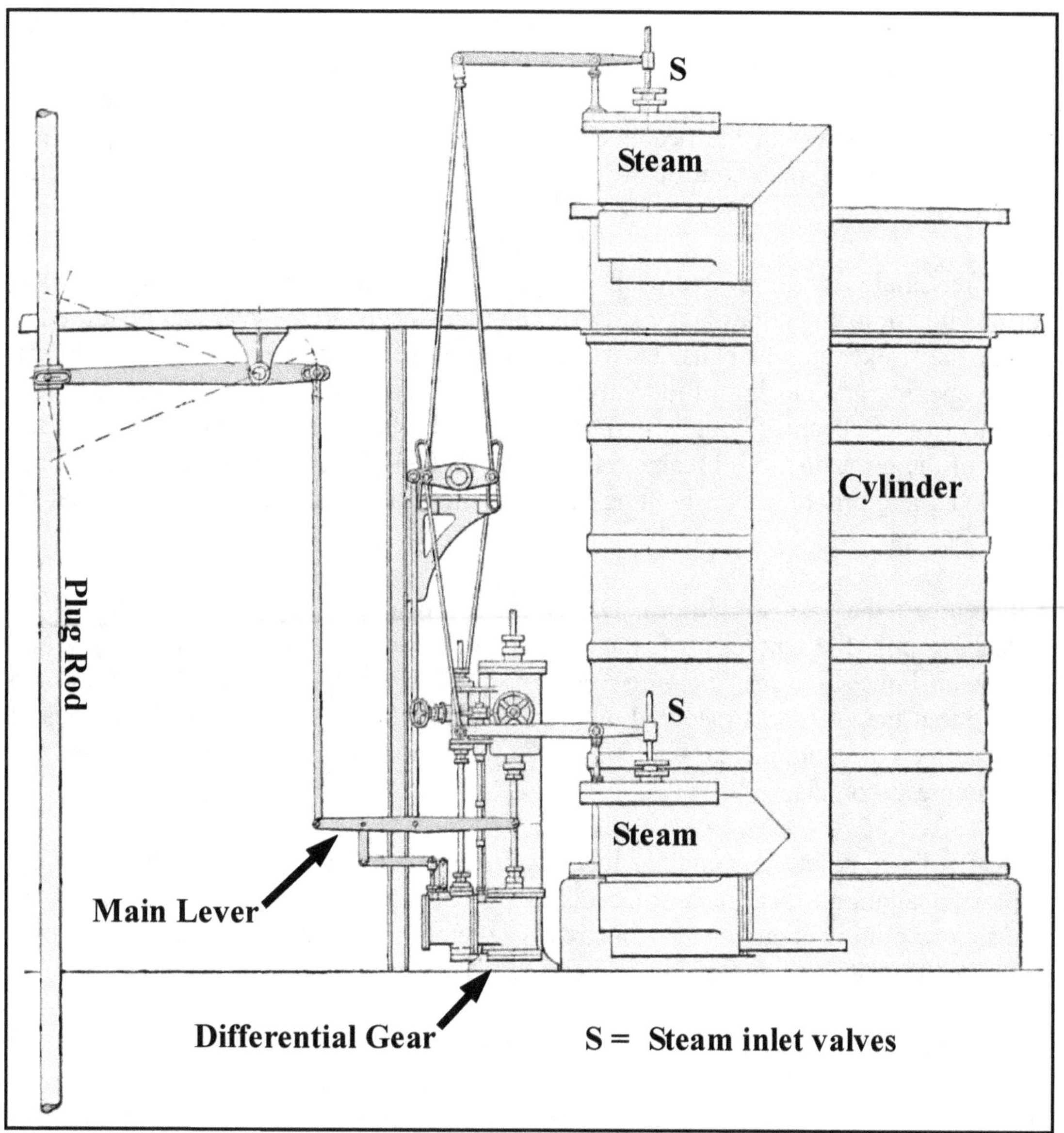

Figure 64. Davey's Differential Valve Gear fitted to a Cornish Beam Engine
(Engineering, 26th February, 1875. p.168 [GG])

It was a well-known fact amongst Cornish engineers that the sudden removal of a Beam Engine's load could serious damage to the engine. Consequently, the Differential Gear became popular for fitting to the Cornish Beam Engines. During the 1880s, many Differential Gears were sold to Germany, and elsewhere.

Figure 64 illustrates how the differential gear was fitted to a Beam Engine. Under normal operations, steam is distributed alternatively to either side of the engines double-acting piston by valve S. The plug rod is connected to the beam overhead, and moves up and down with the beam. Under normal operating conditions the plug rod activates the valves that control steam going into, and exhaust going out of, the cylinder.[13]

The Differential Gear connected to the main lever acts like a second engine and takes over this function, and controls the valves. The two wheels on the gear can be turned to regulate the stroke, and the pause between strokes. In effect the movement of the beam, and the opening and closing of the valves by the Differential Gear is in phase with the movement of the beam. Should the movement of the beam get out of phase with the motion of the gear, due to a breakage for example, then the gear introduces steam into the cylinder and provides a cushioning effect on the piston, and stops the engine before damage occurs.

At the end of the 1880s, Hathorn, Davey produced a special leaflet advertising the application and the success of the Differential Gear. It was entitled, *'Davey's Differential Valve Gear for Pumping Engines, with Latest Improvements.'* In addition to advising that the Gear was patented, and was suitable for use with Cornish engines, it described the Differential Gear and concluded with a list of pleased purchasers, and ten approving testimonials.

The leaflet listed ninety-one engines that have been fitted with the gear. Fifty-two were sold for installation on engines at German Collieries, and two others were bought by Spanish lead mining companies, at Linares. The remaining listed gears were for a range of British sites that included mainly mines and waterworks.

One testimonial from George Hewitt of Coalpit Heath Collieries, Bristol, dated 8th February 1882 stated, *'Gentlemen, - I have much pleasure in bearing testimony to the efficiency of the Differential Gear which you attached to our 84½ in. Cornish Pumping Engine, and shall be pleased to show it to anyone wishing to see it in work.'*

Five of the other testimonials were from H. Wipperman, who seemed to be one of the main agents for purchasing the gear for German Colllieries. One detailed testimonial was forwarded to Hathorn, Davey by Wipperman dated 11th July 1881. It was from

Edwards Elbers, the Representative of the Mining Committee of a Colliery at Westhausen. *'The Differential Gear which you put to the direct acting Pumping Engine at Westhausen Colliery, and which was started on the 23rd of April last, has worked since then most satisfactorily. As we can work the Engine very much faster than with the old Gear, the efficiency of the engine has risen fully by one-third. The danger of the balance-weights striking through, in case the spears should break, has been reduced to a great extent, if not altogether removed. The working of the Gear itself is so certain and quiet, that with a little care it is almost impossible to get out of order.'*

Davey's name will always be associated with the Differential Gear, but he will also always be remembered for the Davey Domestic Safety Motor, which Hugh Lupton was later to describe as, *'Another good line.'* [14]

Chapter 12 References

[1] Brandão, J.M., Vernon, R.W., Callapez, P.M. and Soares Pinto, J.M. 2019. *The attempted introduction of pumping technology at the Buarcos coal mine: an ambition of the superintendent, and travelled mineralogist, Bonifácio de Andrada (Portugal, early 19th century).* Conference: Santos Rocha, Arqueologia e Territorios da Figueira da Foz, Portugal, Figueira da Foz, Portugal

[2] Sobrino, A.H. and Aparico, J.F. 2005. *La Bomba de Fuego en Almadén.* Fundacion Almadén, Spain

[3] Boulton and Watt Collection, Birmingham Central Library

[4] Vernon, R.W. 2017. John Taylor and Sons and their Three 'Drops of Comfort. ' *British Mining 103.* Northern Mine Research Society.

[5] SNA, BT2/1217 - Belalcázar Silver-Lead Company, Limited

[6] Vernon, R.W. and Cabrera, A.M. 2018. La mina Solana de Belalcázar: Una Borrascosa operacion minera escocesa en Córdoba. *De Re Metallica* v.30. SEDPGYM, Madrid, Spain. pp.73 to 90.

[7] SNA BT2/1880 - Solana Mining Company, Limited

[8] SNA BT2/2305 - Toro Mining Company, Limited

[9] DLC. La Solana del Fondón.

[10] Davey, H. 1905. *The Principles, Construction and Application of Pumping Machinery (Steam and Water Pressure).* 2nd Edition. pp. 178 to 186.

[11] Ibid. Davey, 1905. pp. 183 to 186.

[12] Biver, E. 1885 Installation D'une machine D'epuisement. *Le Génie civil: revue générale des industries françaises et étrangères France.* 8:6. 12th December. pp. 81 to 86.

[13] Davey's Differential Valve Gear for Cornish Engines. *The Engineer.* 26th February, 1875. pp. 168 to 169

[14] HL. 1940. *Hathorn, Davey and Company, Sun Foundry, Leeds.* p.4

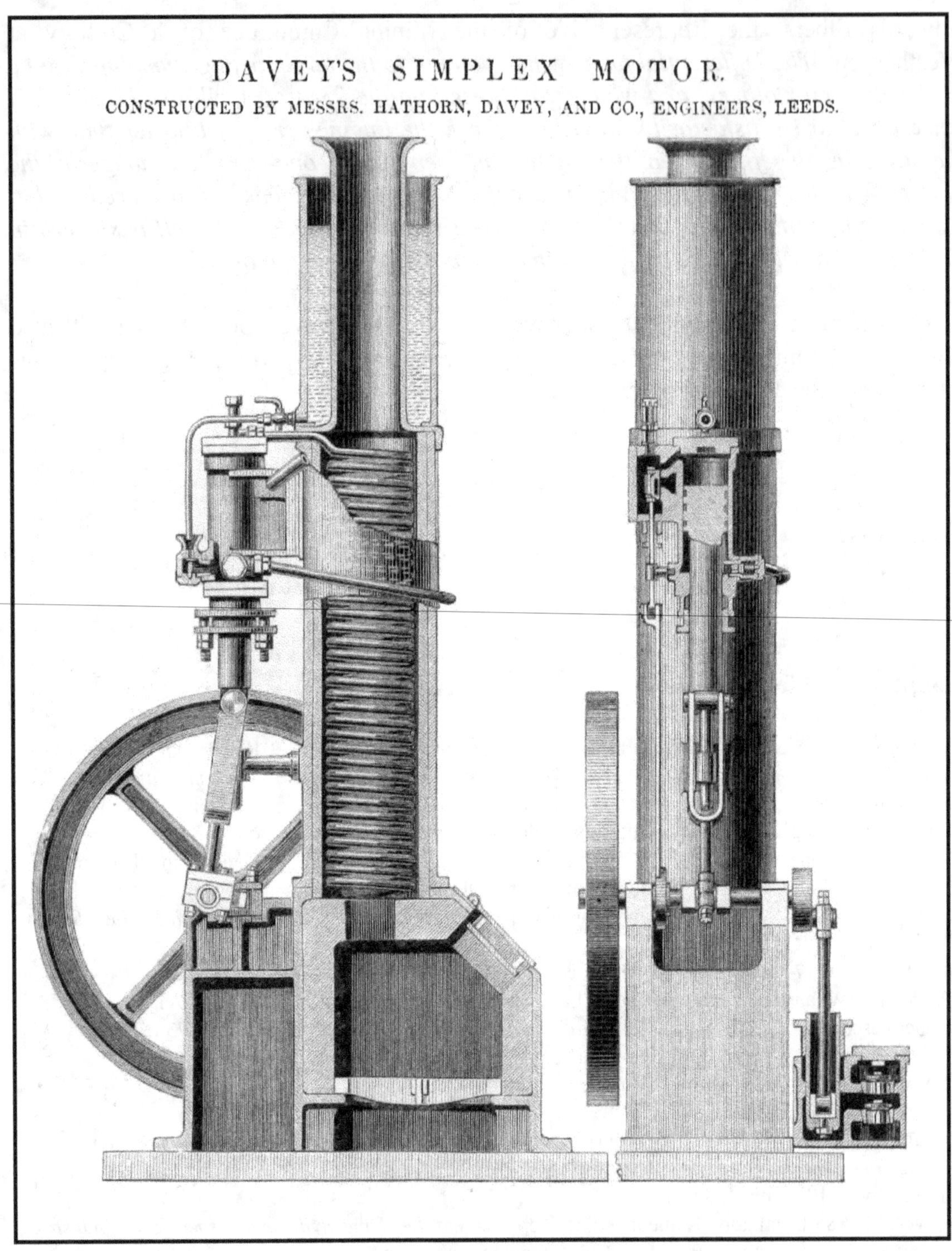

Figure 65. Davey's Simplex Motor.
(Engineering, 24th September 1880, p.251 [GG])

Chapter 13. Davey's Domestic Safety Motor

Davey's Domestic Safety Motor was indeed a *'good line'* for the Company. It was a simple low-pressure 'off the shelf' steam engine that could safely be used, without the risk of a boiler explosion. It was initially designed for use in the home, hence 'domestic.' As the Motor evolved in both design and power many other applications were found for it that included use on farms, and in small factories.

The engine seems to have made its first public appearance in August 1879 at the Yorkshire Show, where Hathorn, Davey had a stand. '*Messrs. Hathorn, Davey and Co., of Leeds, besides the pumping apparatus they have at Stand 27, show a water meter capable of measuring 500 gallons per hour, and two natty little domestic steam motors, one of 1½-horse and the other 1-horsepower, for small machinery and pumping water. The "simplex motor" is claimed to be the simplest, safest and cheapest form of steam engine yet invented.*' [1]

A year later an article in *Engineering,* described the engine in detail.[2] *'This little engine is exceedingly simple and direct in its construction, and it is probable that it may take a not unimportant place among the small power motors in the improvement of which so much has been done of late years. Mr Davey's machine is in reality a steam engine, in so far that it works entirely by steam, but as a steam engine it has the special feature that it has no boiler, in the sense at least of any vessel containing a considerable quantity of water. A reference to our figure (Figure 65) will show that it has a single cylinder only, made with a very large piston so that the area above the piston is much greater than that below. The space above the piston is, in fact, the real working cylinder, while the space below is only a compressing pump. The steam distribution is effected by a slide valve, while the pump chamber has connected to it two small single-beat valves, one opening inwards and the other opening outwards into a coil which lies within the furnace, this coil taking the place of the boiler. It is enclosed in a cast-iron casing lined with firebrick, and the fire is placed below it, as shown. The way in which the engine works is as follows: On its up-stroke the piston draws a quantity of air into the cylinder below the piston, and along with this air is a small quantity of water is always taken in. The last comes about by the help of the little cup above the suction valve, into which a fine stream of water is constantly running. On moving downwards the mixture of air and water is first compressed up to a point determined by the working pressure of the engine, and then pushed through the delivery valve into the coil, when the little puff of water is at once flashed into steam. There is no valve between the delivery valve of the pump and the slide valve, but perfectly free communication, and each time a new portion of water is introduced into the coil, a corresponding portion of steam passes away to the steam cylinder. Here it works exactly in the usual way.' ' The air does not appear to play any appreciable part in driving the piston; its chief purpose is to insure that the water, when sent into the coil, is really blown in as a spray, and not allowed to drop or run in.'*

One of the first engines of this type (3½"cylinder, stroke 4") was used for experimental work at the Engineering Laboratory of University College [Oxford]. Davey also made some modifications. Firstly, he employed a pump at the top of the cylinder to replace the pump action of the space under the piston. This meant that secondly, he could make the cylinder double acting and increase its size. *'This little motor is very substantially made, it takes up little space, is easily started, and has no explosible boiler.'*

Up to this point, there had been very little interest in Hathorn, Davey's products from the American market, but after the Simplex engine appeared in *Engineering*, the same illustrated article was published the following month in *The Scientific American*, and undoubtedly generated some American interest in this unusual motor.[3]

The 'Motor' was in effect a small condensing engine that worked at atmospheric pressure. Davey filed his first patent for the Domestic Motor in 1879. Patent 844, dated 3rd March 1879 was titled *'Motive Power Engines and Generation'* and provided a detailed specification of the Simplex Motor. During 1884 and 1885, Davey took out other patents (Patent 15135, dated 23rd February 1884) on the motor to cover the improvements he made to increase the engine's power. He described it as a Low Pressure Steam Motor. Eventually the motor would have a boiler and engine that were cast in one piece. The water feed into the system would be regulated by a float working in a chamber that also doubled as a water level gauge when observed through a glass panel. It is probable that the exact numbers of sales of Domestic Motor are not recorded in the Order Books as they were often ordered and produced in batches. Orders 4270 and 4272 dated 18 November 1885 for example, show that some eighteen motors were manufactured, that varied from 0.5 to 1.5 horsepower.

To protect against infringements of his patent in America, Davey applied for an American patent. Described as a *'Low Pressure Steam Motor'* a patent was filed for the motor on the 7th July 1884, and granted as patent 310,387 on the 6th January 1885. Some of the drawings in the American patent are shown in Figure 66, and Davey describes his invention as follows:

'A is the fire-box within the boiler B, which is an upright cast-iron vessel, projecting forward at the bottom to give space for the firebox A and to carry the bearings for the crank shaft C, on which at one side of the boiler is the fly-wheel D. The upper part of the boiler also projects forward, accommodating within it the cylinder E. The piston-rod extends down through the cylinder - bottom, and is linked by the connecting-rod F to the crank on the shaft C. From the firebox A, a flue tube, G, extends up through the

body of the boiler and its steam-space, and in order to increase the heating-power water-tubes H may be fixed across the flue. K is the condenser, which is a vertical tube placed at the side of the boiler, to which it is fixed. At the top it receives the exhaust steam from the cylinder by a branch, L, which may have a cock or valve, L, as shown, to regulate the speed of the engine by more or less throttling the exhaust. A pipe, M, from a cistern a little higher than the water-level in the boiler, supplies an injection branch, I, opening into the condenser, and a feed branch, N, opening into a vertical tube, U, which communicates with the steam and water spaces of the boiler. The feed branch has an opening fitted with a conical plug, V, which is attached to a float, W. The water in the tube U maintains the same level as that in the boiler, but is quiescent, and the float W, by raising or lowering the plug V, regulates the feed so as to maintain the water at its proper level. P is a small air pump worked by a crank pin on the fly-wheel shaft. This pump draws air and water from the condenser and discharges the water into a well, from which there is a lateral outlet, Q.'

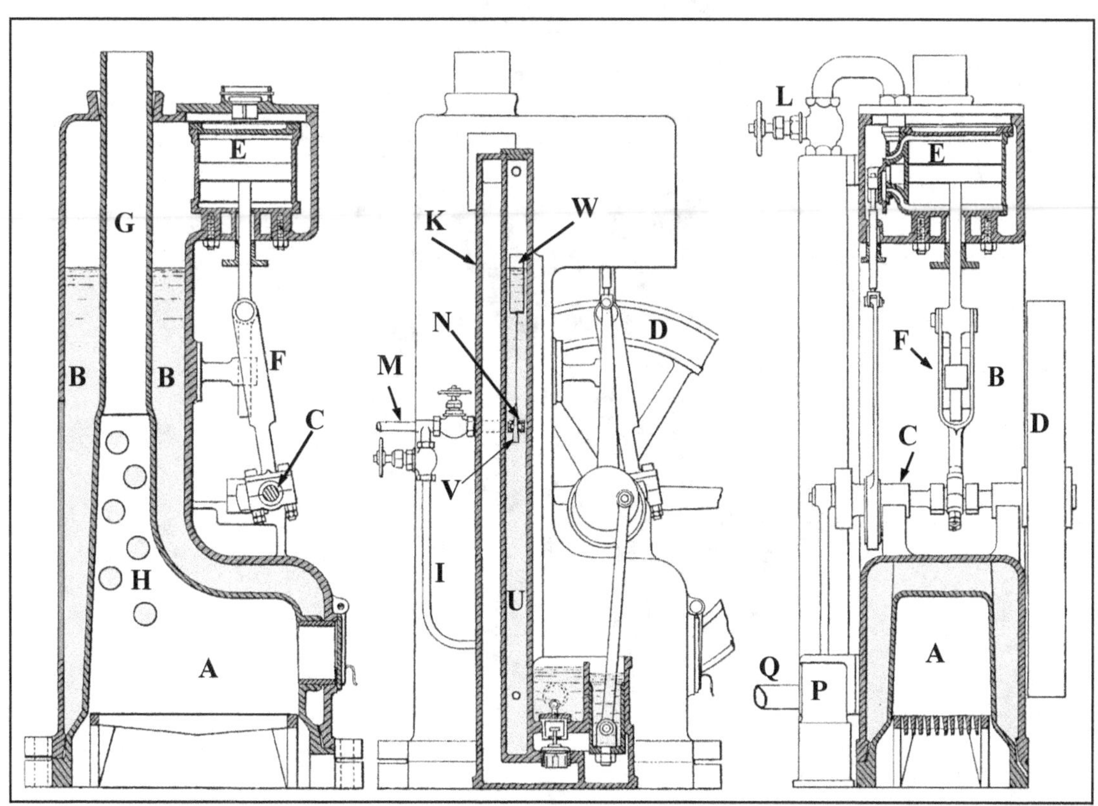

Figure 66. Davey's Low Pressure Steam Motor
(United States of America Patent No. 310,387 dated 6th January 1885)

The sale of Hathorn, Davey engines in the USA had always been elusive. A previous attempt to get a foothold in the market with the Compound Differential Pumping Engine, by granting a manufacturing license to Joseph Moore, Engineer, San Francisco, California, had met with only limited success.[4]

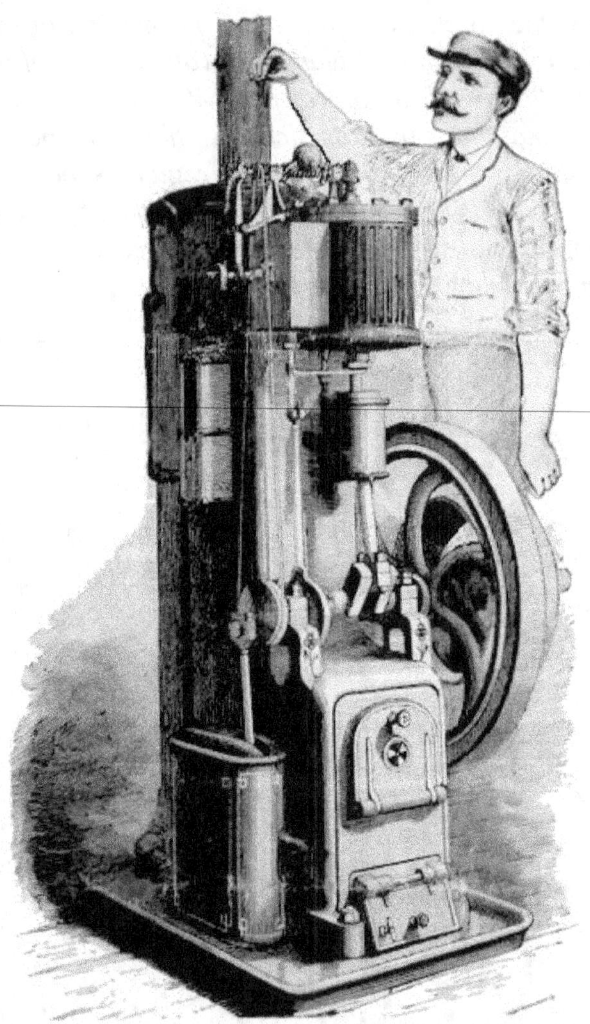

Figure 67. Willard's Advertisement for Davey's Low Pressure Steam Motor
(Source: Unknown)

The Domestic Motor however, was a different matter, in theory at least. It appealed to a wider market and had the potential to become a popular item of equipment, so consequently it promoted a further attempt by Hathorn, Davey to sell their products in the USA.

A suitable American engineering company was sought and eventually a draft Agreement was prepared in December 1884 between Hathorn, Davey and Charles Powell Willard of Chicago that would grant Willard exclusive rights to manufacture Davey's 'Domestic Motor' in the USA.[5] *Charles P. Willard and Company* were already manufacturers of portable, stationary and marine steam engines as well as boilers, steam launches, steam yachts and tug-boats, and apparently were capable of producing 600 engines a year.[6]

Willards was a successful company in it's own right and later added aviation engines to its products. It is not known how many 'Domestic Motors' were produced, but certainly the Motor did feature in the Company's catalogues (See Figure 67).

During 1885 the Domestic Motor was exhibited in London and Paris. At the International Inventions Exhibition (Order 4154 - A to F) held in London for which the Society of Arts awarded Hathorn, Davey a gold medal for the invention.[7]

Also some interest in the Domestic Motor was shown in France when it was exhibited at the Paris Exposition. It was also described in detail in the *Revue Industrielle* 1885. The exhibited engine for all intents and purpose was similar to that shown in Figure 67. The article included a table that detailed four trials carried out on the engines performance, over 108, 240, 221 and 600 minutes. In the case of the latter time period, the engine completed 75,428 revolutions, averaging 125.7 rpm. Fuel (coke) and water consumption amounted to 53 and 300kgs, respectively.[8]

Probably encouraged by the interest shown in the Domestic Motor at the Exposition Auguste Albaret, who ran an Agricultural Engineering business at Liancourt-Rantigny, Oise, about 30kms north of Paris decided he would like a license to manufacture the Motor. Albaret had trained as an engineer. He had also spent some of his years working on railway projects in Spain and Portugal before specialising in agricultural machinery, for which he had been awarded the Legion of Honour.

The draft Agreement dated 1885 between Hathorn, Davey and Albaret is lengthy. It confirmed that Davey had taken out various patents for the Domestic Motor. In addition he had also been granted Letter Patents in France dated 31st May 1884 for improvements to the Motor. In essence, Hathorn, Davey wanted a payment of £100 from Albaret for the right to manufacture the motor under license. This was in addition to an annual Royalty.[9] Perhaps the terms were sufficient to dampen Albaret's enthusiasm, and no further correspondence seems to have occurred.

However, William Preece, famous for his work in telecommunications, was not that impressed with the Domestic Motor. In a lecture about Domestic Electric Lighting, given in 1886, he stated, *'A simple effective safe domestic motor has yet to be introduced. Hathorn, Davey, and Co., of Leeds, exhibited at the Inventions Exhibition the best that has yet been made, but it has the disadvantage of requiring a large supply of water to condense the steam. It is a vacuum-engine working at atmospheric pressure, and is, therefore, free from explosion.'*

Henry Davey responded to this doubt cast on the Motor's economy, by citing how it was employed at his home. *'Mr. Preece in speaking of the Davey motor at the Inventor's Exhibition, raised an objection to the quantity of water used for condensing. It is true that it is not always convenient to provide the water necessary, but when it can be obtained form the town's mains, the cost does not preclude its use, and where it can be obtained without cost, as it can in many country places, the water difficulty vanishes. I would have given some facts on this subject if there had been an opportunity. I have had the motor running for two years lighting my house. On the average, there are in use seventeen 16 candle-power lamps. The Corporation of Leeds charge £2 per annum for water, which is obtained direct from the mains. the consumption of coke is one-hundredweight per day, and the fuel is only added once in twenty-four hours. The fire burns continuously, regulated by means of an automatic damper. The only attention the motor receives is that given, one hour a day, by a labourer. I do not keep a man servant. After 5 o'clock the man goes to his home, and the motor is left to itself for the remainder of the day. At 4 o'clock the next day the man comes to fill up the boiler with coke, by that time about two-thirds empty; he wipes down the engine, oils it and at 5 o'clock leaves. At bed time one of the household turns one tap which shuts off the motor, and another tap which stops the engine gradually, by letting in air to destroy the vacuum. So perfectly does the automatic damper keep the steam at exactly at atmospheric pressure that I am enabled to run the installation without a governor, and the lights have been perfectly steady. Of course I have to keep all on, but I think it is a wonderful result; in two years I have only lost one lamp.. The coke costs me about £4, and the water £2 per annum. A heating coil in the house is supplied from the motor. From careful experiments I find that the consumption of steam is 30 lbs., and the consumption of water 45 gallons per independent horse-power per hour. I think such an economical result (obtained from an engine so safe that it may be placed in charge of the housemaid, and so simple that it may be started and worked by a boy or labourer) is worthy of serious attention. As to the cost of installation, I agree with Mr. Preece's figures, but I think his estimate of the cost of working is very high indeed. One word as to the character of the fittings. I agree with Mr. Preece in what he said. One of the great charms of the electric light is that it lends itself to beautiful simplicity. My fittings cost me a few shillings. A Japanese umbrella, costing 2s. 3d., covers the "plaster vegetable" in the centre of the dining-room. From the centre, one bare lamp hangs; around it, at some distance from the umbrella, hangs three other lamps, under flat glass mirrors, six inches in diameter. Another lamp on the sideboard complete the lighting of the room. The light is beautifully diffused, and there is no glare. The lights are out of the line of sight, and are not noticed, if attention is not directed to them, - then the extreme simplicity is admired.'*

The Domestic Motor was exhibited at many English Agricultural Shows. In 1889 it was exhibited at the Royal Windsor Show where one correspondent summarised the motor's good points.[10] *'This motor, which we illustrate, has many points to recommend it to the attention of those who require an engine of this kind. In working it is practically noiseless, and its safety is assured inasmuch as no explosion is possible with it in use. It is constructed on extremely simple lines, no renewable parts or special lubricant being necessary. It is claimed also to have great economy over other small motors in practice. It has no exhaust and there is no escape of burnt oil or noxious odours. It requires no engineer, the fire only requiring occasional attention, while it is asserted that the risk of fire is no greater than that in the case of an ordinary house stove. It is certainly compact and economical in space, and requires, the makers inform us, less fuel than any other motor of equal power. No special packing is required, nor is any special care necessary to prevent damage to the engine or generator. The motor when combined with pumps has been found of especial service in connection with water supply, and for numerous other purposes, which will readily suggest themselves, the "Davey" motor seems admirably adapted.'*

Hathorn, Davey lent a Domestic Motor to the Science Museum, London in 1888. It was constructed to Davey's 1884 patent, which by this time had become the standard design. The horsepower of the motor is not stated, but it did consume 6 pounds of coke per hour.[11] Apparently, the Science Museum still has the model.[12]

By the end of the decade, the Domestic Motor had undergone a few more design changes. It had become a larger machine to cope with more demanding tasks. In addition it was awarded a silver medal by the Royal Agricultural Society. Such was the Motor's success that by 1891, Hathorn Davey had produced a twenty-page sales catalogue to publicise the motor's versatility.[13]

The term 'Domestic' was dropped, and replaced by a new title, 'The Davey Safety Motor.' The front page of the catalogue proclaimed that it could be used at Reservoirs, Estates, Villages, Water Towers, and Country Houses.

A table on page 9 of the catalogue (Figure 68) indicates that for pumping from shallow wells (about 30 feet deep) the motor came in five sizes, up to 5 horsepower. The cost of the 5 horsepower Motor with pumps was £253, and the engine and pumps weighed four tons! It required a minimum floor space of 8 feet 3 inches by 5 feet 0 inches.

No.	Approximate Indicated Horse-power.	Price with Pumps.	Price without Pumps.	Diameter of Suction and Delivery Pipes.	No. of gallons pumped per hour approximate.	Gross weight of Motors and Pumps for Shipment including Pipework to top of Well.	Weight of Motors without Pumps.
	H.-p.	£	£	In.		Cwts.	Cwts.
1		64	58	1 1/4	250	18	16 3/4
2	1 1/2	115	82	1 3/4	800	25	18
3	2 1/2	140	96	2	1,300	35	25
4	3 1/2	185	130	2 1/2	1,900	60	40
5	5	253	165	3 1/2	2,700	80	49

These Motors and Pumps are suitable for Wells 30 ft. deep, and will deliver to the height of 60 ft. above the well. For greater depths and higher delivery, special prices will be quoted.

The Pumps will draw from a depth of from 20 to 26 ft. from the floor on which the Motor stands.

Figure 68. Table taken from the catalogue showing the horsepower,
sizes, weight and cost of the Davey Safety Motor
(Davey Motor Catalogue p.9 - Courtesy of Jon Knowles)

Figure 69 is an example of one of the larger version of the Davey Safety Motor suitable for working two or three throw pumps. It would have been used to supply a small village or country house with water. Smaller versions of the engine could be manufactured with the condenser placed in a 30 inches square galvanized iron tank. This method allowed the engine to run for two or three hours without changing the water. Different pump configurations in wells are also described in the catalogue, as well as various other uses for the motor.

The final sections of the catalogue listed some of the purchasers. They varied from Railway Companies, Government Departments, and landowners, for example, the Duke of Westminster. There was also a page of Testimonials. Messrs. T. Hardcastle and Son, Firwood Works, Bolton, stated, *'The 1 H.P. Motor which you made for us works satisfactorily. It drives a chaff cutter which cuts up the food for eleven horses, and also steams the chaff.'* (November 1886)

The Davey Safety Motor was also exported. Order 4370 and A, dated August 1876 was for one 10 horsepower motor, probably one of the largest built, for the General Post Office, Wellington, New Zealand. Another engine was supplied to the Government Printing Office, Rangoon, Burma.

Figure 69. A larger version of the Davey Safety Motor
(Davey Motor Catalogue p.4 - Courtesy of Jon Knowles)

From 1879 Hathorn, Davey manufactured the Simplex / Domestic / Safety / Davey Motors in sizes up to about 10 horsepower. Hugh Lupton must have had fond memories of the Motor when he remarked that the motor, *'was fitted with a glass panel and prompted one customer, when writing for a replacement, to describe it as, the window of the engine,'* an amusing anecdote that had stuck in his mind for some 50 years! [14]

Chapter 13 References

[1] The Yorkshire Show: Machinery and Implements. *Leeds Mercury*, 7th August 1879. p.8

[2] Davey's Simplex Motor. *Engineering*. 24th September 1880, p.251

[3] Davey's Simplex Motor. *The Scientific American*. 30th October 1880. p.278.

[4] DLC Draft Agreement between Davey and Moore dated 21st May and letter dated 13th December 1875

[5] DLC Draft Agreement between Hathorn, Davey and Willard December 1884.

[6] Cope G.W, 1890, 'Charles P. Willard and Co. *The Iron and Steel Interests of Chicago*, Chicago, pp. 77 and 78

[7] *International Inventions Exhibition, London* 1885. Awards of the International Juries. Executive Council of the International Inventions Exhibition, and the Council of the Society of Arts.

[8] Moteur Domestique. *Revue Industrielle* 10, 5th March 1885. pp.93 to 94.

[9] DLC Patent of Safety Motor assigned to French Firm. 1885

[10] Machinery Exhibits at the "Royal" Show at Windsor'. *Mining Journal*, 20 July 1889, p. 819,

[11] *Catalogue of the Mechanical Engineering Collection in the Science Museum*, South Kensington. 1919. Part 1. Steam engines and other motors   HMSO. p.51

[12] https://collection.sciencemuseumgroup.org.uk/objects/co468557/model-of-a-hawthorn-davey-safety-domestic-engine-a-model-safety-domestic-steam-engine

[13] I am grateful to Jon Knowles for bringing the Davey Safety Motor Catalogue to my attention.

[14] HL. 1940. *Hathorn, Davey and Company, Sun Foundry, Leeds*. p.5

14. Hathorn, Davey: The late 1880s and more Partnership changes

There were more changes in the Hathorn, Davey Partnership in the 1880s when during the latter half of the decade John Hathorn died, and a new partner was appointed.

Another event also took place at this time. Henry Davey got married sometime during 1886 / 87. The exact date of his marriage has not been found, but he married Elizabeth Barbenson le Ber, who was born on the Island of Alderney, Channel Islands. Very little is known about Elizabeth. Her marriage to Henry may have been her second. It is also possible that she was an old friend of Henry's.[1]

It is speculation, but living in Leeds may not have been too appealing for Elizabeth. Perhaps to partly appease her, and to fulfil his own ambitions, about 1887 Davey indicated that he would like to play only an advisory role in the management of Hathorn, Davey. He proposed to establish himself in London as an engineering consultant. Their only son, Norman Henry, was born on the 5th May 1888 at Malden, southwest London, where the Davey's had set up their new home.

Although the Hathorn, Davey Partnership was formed for all intents and purposes on the 8th November 1879, it didn't become legally binding until the 8th June 1880, with all three partners signing the Agreement. Under the Agreement's terms, the Partnership would last a period of eight years (originally this was for a period of seven years), and all Partners would benefit from any patents and inventions attributed to an individual Partner, until the Partnership ended.[2] Hence, leaving the Partnership before it ended might cause complications! Nevertheless, it was John Fletcher Hathorn, with perhaps the least to lose financially, but with failing health, who made the first moves to leave, and to try to retrieve any money that he had put into, or was owed from, the business.

It is probable that John Hathorn's health had started to deteriorate during 1883 or early 1884. In a letter to his Solicitors dated 2nd December 1884, he expressed his wish to reduce his share in Hathorn, Davey and Company, *'as he was still and invalid.'* It is not known what his health condition was, but prior to writing the letter, he had undergone six months treatment in London, and he considered that it had done him some good. It seemed that if possible, he actually would have liked to leave the Partnership completely because while he was against taking on a fourth partner, he would not object to someone replacing him. He thought that he had £19,000 to £20,000 in the firm, but he would accept £12,000 initially, and the balance could be paid over a period of three years. His wish was conveyed to the Partnership's solicitors on 25th December 1884.[3]

A report with the Company accounts dated 31st July 1884 indicate that taking on a new Partner as a replacement, may well have been Colonel Hathorn's preferred option. It would mean that a new Partnership agreement would have to be drawn up, but that he could retire with a minimum share of the capital he had put into the firm.

Another suggested option was that a Bond could be taken over for the freehold of the Sun Foundry (valued at £16,865-4s-6d) together with the plant and machinery (valued at £16,082-0s-0d), but to accept this, Hathorn would have to be sure that no heavy losses were being made by the future business. It was added that if the present Partnership continued to the termination date (November 1887), then it was likely that if no arrangements had been made, then realising the assets of the Company would probably be done at a loss.

The report concluded with an assessment of the Company's turnover. '*It is also right to observe that with a turnover of £40,000 per annum, it is not likely that the profits for each year can amount to more than £1,600 or £1,700 after payment on the interest on the Partners Capital, Royalties to Mr. Davey and other charges against the profit in the present state of trade not more than 12% can be realised on the turnover.*' [4]

Davey must have considered that another solution to disbanding the Partnership was to form a new Company with Limited Liability. On the 8th June 1885, Hathorn responded. '*Dear Davey, Before turning Hathorn, Davey and Company into a Limited Company I would like to know that there is reasonable hope that the Company could be formed, and also how and when I should get my capital out. If you will let me know this and how you propose to start it I will give you an answer as to whether I approve it.*' [5]

Davey replied promptly on the following day. '*I have no idea how the problem is to be solved but as you are both very anxious to go out of the business, I thought it was no harm to suggest it. However, I suppose I must bear everything for two years longer.*' [6]

It would seem from Davey's response that Meysey-Thompson was also considering his position in the Partnership. If the situation couldn't be resolved, then Davey was quite happy to wait until the Partnership terminated naturally in November 1887. Davey was quite adamant that if a Limited Company were to be formed then it would not be possible unless another partner was found. '*As far as Colonel Hathorn is concerned, he very much desires to get out as in his present state of health he must be rid of all worry.*' (10th June 1885)

By the 13th June 1885, Davey was getting weary with the exchange of views between himself and Hathorn's solicitors. '*I have no proposals to make. I am tired of making proposals. If Colonel Hathorn wants his money let him find someone to buy the place, I am ready to sell.'*

Eventually, a new partner was found, which resolved the problem and for the time being, postponed any thoughts of forming a Limited Liability Company.

The new partner was Hugh Lupton. The Luptons were a prominent Yorkshire family, with their fame extending back to the reign of Henry VIII.[7] By the 19th century they owned woollen mills, and were part of a group of Liberal minded Leeds merchants and industrialists. Hugh Lupton was the youngest of five brothers. He was born in 1861, attended Rugby School, and then went to University College, Oxford to read Modern History. He graduated in 1881, and eventually took up an apprenticeship with Hathorn, Davey and Company. An Apprenticeship Indenture indicates that this was as late as 18th February 1886.[8] Having gained sufficient experience in practical engineering and business skills, he became a Partner in 1887, the same year that Davey left Leeds.

On the 16th June 1886, Hathorn had requested a copy of the Partnership Agreement because the Partnership was now near the end (November 1887),[9] so perhaps he was having second thoughts, or the new arrangement would give him the ability to leave without complications? Considering Colonel Hathorn's stance on wishing to leave the Partnership, it is unclear why his name should appear as one of four partners on a draft of a new Partnership Agreement dated 1887.[10]

The Agreement was open to renewed negotiation as illustrated by one point raised in April 1887. Davey, it seems, was a strong negotiator, especially as he held most of the cards, in the form of patents and engineering skills. Under the new terms of the Agreement he had negotiated receiving half of the profits, and a quarter of any incurred losses. He had taken the view that this was not an important division of profit and loss, as he had agreed with the majority of points raised by the other Partners. The Company Solicitor felt that it was impossible to press him further on this matter. They were only waiting for Colonel Hathorn's approval.[11]

Both Lupton and Meysey-Thompson agreed that under the circumstances they were in a safe position because under the old Agreement, the patentee [Davey] took 5% on the invoice price of orders, irrespective of profit! On the 27th April 1887 Colonel Hathorn's Solicitors took another view. *'I think Mr. Lupton and Mr. Thompson yield far too easily to Mr. Davey in all the points raised in the fear that he will leave the concern, altogether which he really cannot do as he has only had a share in his patents in the past.'* [12]

Nevertheless, an Agreement was reached.[13] It stated that the Partnership would last for seven years from 1st August 1887, with a capital of £40,000 allocated among the partners as follows:

J.F. Hathorn	£16,000
H. Davey	£5,000
A.H.M. Thompson	£9,000
H. Lupton	£10,000

This was an indication of the size of the Lupton's investment in the Partnership. No part of the capital could be drawn out during the Partnership without the consent in writing from all the co-partners. Each partner, before the division of profit would be entitled to interest on his capital at the rate of 5% per annum.

Several sections of the Agreement concerned the various patents under Henry Davey's name that were regarded as business assets. The Company was to have sole rights to use them. Very generously, Davey would receive half of the profits to cover Royalties and other claims in connection with the patents, but pencil notes suggest that this may have been amended.

It is apparent that when Henry Davey moved to London he was going to set up a London Office for the business. *'A London office shall immediately be opened by Mr. Davey in his own name and he shall have the exclusive charge thereof, but if desired by the other Partners, the name of the firm shall appear on the office address. He shall act as a consulting engineer and also endeavour through his connections to get orders for the business and shall devote his whole time to the work of the Company. The whole expenses in connection with the London Office shall be paid by him and the whole profits arising there from the consulting fees or otherwise shall be accounted for by the said Henry Davey to the firm.'*

Elsewhere, the Agreement stipulated that the resident partners in Leeds would be Meysey Thompson and Hugh Lupton. John Hathorn would devote as much time as he felt fit. None of the Partners would engage in other business without the consent of the co-partners. Henry Davey was required to visit the Sun Foundry only when the business required, in the interest of the Company.

Other clauses relate to Partners lending money, or Company assets, and taking on apprentices without consulting the co-Partners. The active Partners would also draw salaries before any profits were divided. The salaries were as follows:

Henry Davey £300 / Meysey-Thompson £200 / Hugh Lupton £100

The final clauses, dealt with bankruptcy, expulsion, or death of any of the Partners. If a Partner were to die for example, their share of the capital would be repaid to their Executers with interest.

The final copy of the Agreement has not been discovered but it is likely that much of the un-amended sections were implemented, with items that related to Davey and his patents being modified.

There seems to have been some reluctance by John Hathorn to sign the final Agreement because on the 19th August 1887, Lupton wrote to the Mr. Nelson, the Company Solicitor, *'I am very glad that Col. Hathorn has decided to sign the Partnership agreement and much obliged to you for writing to tell me.'*[14]

The following year, all this was to change again with the death of John Fletcher Hathorn on the 16th November 1888. He was 49 years old.[15]

ASSETS

I. Freehold		
1. Cottages acc.	£4033-00-00	
2. Land acc.	£3532-16-00	
3. Buildings acc.	£8937-00-00	
		£16502-16-00
II. Plant and machinery		£13437-00-00
III. Stock		
1.Materials	£2426-00-00	
2. Model Wood	£64-00-00	
3. Stationary and printing	£30-00-00	
4. Drawings	£300-00-00	
		£2820-00-00
IV. Work in progress		£10300-00-00
V. Book Debts.		
1. As per Appendix II *	£12706-12-04	
2. London Office fees owing	£371-10-00	
		£12706-12-04
VI Models		
1. Present value of ?	£2000-00-00	
2. Sinking fund acc.	£150-00-00	
		£2150-00-00
VII. Patents		£600-00-00
VIII. Office furniture (London)		£40-00-00
IX. Cash in hand		
1. As per Firms books	£35-12-11	
2. As per petty cash book	£73-11-07	
		£109-04-06
Total		£58665-12-10

LIABILITIES

I. Capital		
1. Lieut. Col. Hathorn	£16814-14-01	
2. Mr Davey	£7846-07-05	
3. Mr A.H. M. Thompson	£9814-03-01	
4. Mr Hugh Lupton	£10030-13-07	
		£44505-18-02
II. Loans		
1. On Freehold No.2 Cottages acc.	£2000-00-00	
2. From Bankers per cashbook	£2069-15-10	£4069-15-10
III. Trade Creditors		
1. As per Appendix III.*	£6134-12-02	
2. Doubtful debts	£700-00-00	£6834-12-02
IV. Amount Received on account		
of contract from Sir Andrew Fairbavin		£450-00-00
V. Suspense acc.		£750-00-00
Balance - being Profit		£2055-06-08

* Appendices II and III not seen.

acc = account.

Figure 70. Hathorn, Davey: Balance Sheet dated 31st July 1888
(National Archives of Scotland, Edinburgh. GD 455/75)

Hathorn's Executors included Charlotte his wife, Hugh Fletcher Campbell, who was now an accountant, residing in Leeds, and Henry John Fletcher Campbell, his older brother. Obviously they needed a valuation of his holdings in Hathorn, Davey, and a copy of the Hathorn, Davey Balance Sheet and Profit / Loss Accounts dated 31st July 1888 relating to the valuation survives in the National Archives of Scotland, and is summarised in Figure 70.

The Profit and Loss Account for the year ending 31st July 1888 is very detailed. Items under loss included: Stock that included purchase (£23,160-11s-10d), Wages (£10,940-4s-4d), Coal and Coke (£327-2s-1d), Foundry Expenses (£814-6s), and Horse Account (£124-18s-6d). It also included the allowances for Messrs. Davey, Thompson and Lupton that were £300, £200, and £100, respectively.

The prime source of profit was from sales that amounted to £36,648-15s-2d. Work in progress was valued at £10,300 and stock-in-hand was £2426. Other items of income included apprentice fees (£150) and London Office fees (£474-5s-3d).

The amount of capital that Colonel Hathorn had invested in Hathorn, Davey and Company was eventually calculated to be £16,871-7s-9d.

In the same year that Hathorn died, Hathorn, Davey was listed in a publication about England's Great Manufacturing Centres with reference to, 'Business Men and Mercantile Interest Wealth and Growth.'[16] The listing was very complimentary towards the firm. After providing a brief introduction to the Company's history, the article continues, *'With business-like promptitude, the latter firm at once proceeded to put down new machinery and appliances suitable for the construction of engines of all kinds, and of the largest size. The now well-known "differential gear" of Mr Davey was speedily adopted throughout the works for pumping engines of the largest sizes, with a special view to the requirements of collieries and waterworks.'*

The article also mentions Davey's Domestic Motor which, *'proves a formidable rival to the petroleum and gas engines, both as a first cost and economy of working. It has of late been largely adopted by the authorities of Trinity House for lighthouse purposes, air-compressing, &c. The motor was adopted by the Royal Agricultural Society for driving their dairy machinery at the recent Preston Show, and also received the highest award made by them for engines.'*

The article also refers specifically to some of the pumping machinery the business produced. *'It may be said, with every feeling of safety, that from the works of Messrs. Hathorn, Davey and Co. there has probably been within the last few years the largest pumping engines turned out which have ever been constructed. Among a host of others may be specially mentioned the engines made for the South Staffordshire Mines Drainage, which are conjunctly capable of raising the enormous quantity of seven and a half millions gallons per day to a height of 450 feet.'*

The South Staffordshire Mines Drainage Commission actually bought two pumping engines, and other equipment, from Hathorn, Davey for a mines drainage scheme that was both unique and original in its concept.

Chapter 14 References

[1] Henry Davey may have first met Elizabeth in the 1860s when she boarded at the Hope Corner School, Rowbarton, Taunton (1861 Census). There is also some evidence from a later Channel Island Census that Elizabeth may have been married there.

[2] SNA. GD455 / 77

[3] ibid. SNA GD455 / 77.

[4] ibid. SNA GD455 / 77

[5] ibid. SNA GD455 / 77

[6] ibid. SNA GD455 / 77

[7] https://en.wikipedia.org/wiki/Lupton_family

[8] Hugh Lupton -Apprenticeship Indenture 18th February 1886. Courtesy of Sulzer, Leeds

[9] DLC Correspondence from John Hathorn 16th June 1886

[10] DLC Memorandum of Agreement for the future conduct of the business of Hathorn, Davey & Co. Sun Foundry, Leeds. 1887

[11] SNA GD455 / 75

[12] ibid. SNA GD455 / 75

[13] DLC Partnership Agreement that includes Lupton. 1st August 1887

[14] DLC Letter: Hathorn agreeing to Lupton as partner. 19th August 1887

[15] DLC John Hathorn's death. 16th November 1888

[16] Leeds and Bradford, Dewsbury, Batley Keighley, &c. England's Great Manufacturing Centres with reference to, *Business Men and Mercantile Interest Wealth and Growth.* 1888. Copy held at WYJS, Leeds. p.171.

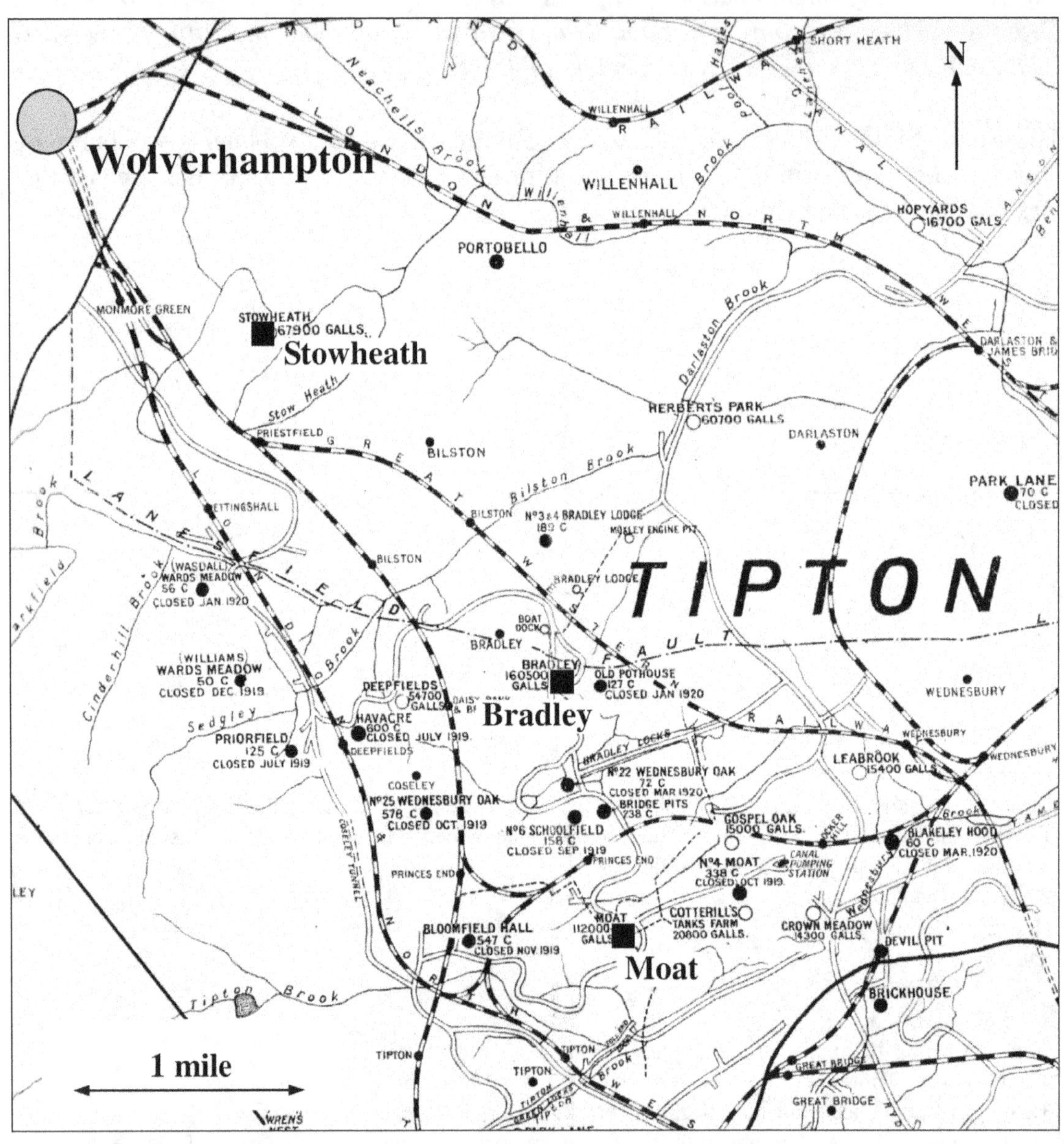

Figure 71. Plan showing the locations of the Bradley, Moat and Stowheath engines. The Lanesfield Fault originally formed the boundary between the Bilston and Tipton Drainage Areas before it was realised water migrated across the fault zone.
(South Staffordshire Mines Drainage. 1920. Report. Based on Plate 6)

15. Hathorn, Davey and the South Staffordshire Mines Drainage Commission

By the mid-1850s, the 'Black Country' to the west of Birmingham was an area of heavy industrial activity. The Thick Coal was the principal source of fuel and was intensely mined throughout the region until water became a major impediment to those operations. Many mines employed steam-pumping technology, but some engines were becoming overwhelmed by the quantities of water they had to deal with. It was remarked that by 1854, the Tipton area which occupied a north-central area of the coalfield, was waterlogged apart from the shallow workings. Only a united effort by the mine owners could resolve the situation.

In 1870, the Old Hill Mines Drainage Company Limited was formed for the purpose of pumping dry the area to the south of Tipton, achieved in part by erecting large pumping plant, and acquiring the leases of other pumping engines.[1]

However, this scheme was inadequate and in 1873, the South Staffordshire Mines Drainage Act was passed to deal with dewatering the majority of mines in the Black Country. To achieve this, up to forty Commissioners were appointed, and they were given the power to levy a rate for that purpose, and also to levy a separate drainage rate on the mineral raised. The Commissioners had the power to appoint three Arbitrators who would assess the rates for each mine owner, based on how their mining operations were affected by water. This included such aspects as the transmission of water from adjacent mines. In 1874, the Arbitrators divided the area into five districts that were sub-divided into Pounds. The boundaries were mainly fixed by the geology, which included, faulting, anticlines and coal outcrop. The largest district was Tipton covering 33.35 square miles. [2]

The scheme was supposed to be totally self-financing, and the rates raised by the assessment of levies would be used to pay salaries and wages, pay interest on any borrowed money, and execute drainage work. Under such a system it was inevitable that arguments would break out between mine-owners on the fixing of levies, and ultimately with the closure of mines, income into the scheme diminished.[3]

One major problem with the Tipton district was that the Lanesfield Fault, the barrier that divided it from the Bilston district, was ineffective, so in 1881, part of Bilston was merged with Tipton. The following year, a Sub-Committee was formed to examine the situation in detail. The cost of dewatering the Tipton district was costing more than the mines drainage rate. *'At that time, 16 or 17 pumping engines were being worked by the Commissioners, all of them old and of an uneconomical type.'*[4]

On the 6th June 1883 a recommendation to remedy the situation was put to the Commissioners. It suggested that new Hathorn, Davey pumping engines should be installed at Bradley and Moat at a cost of £9,333 and £8,873, respectively, plus a further outlay of £2,446 for pumps (bottom lifts). In addition, the Stowheath Beam Engine would be fitted with one of Davey's Differential Gears at a cost of £150.[5]

All three sites are shown on Figure 71, together with the many surrounding collieries that were pumping water in the Bilston - Tipton drainage area.

By August 1885, all three engines were in operation, and by 1886 the number of adjacent working engines had been reduced to six. Together with underground work to divert water to the engines, the mines drainage problem in the Tipton District was resolved.[6]

A comment in The *Engineer* suggests that the Moat and Bradley engines once erected would probably be the most powerful pumping engines in England.[7]

The Bradley Engine

The Bradley engine, a Horizontal Compound Condensing Pumping Engine, known locally as 'The Drainer', together with its associated pumping equipment, is listed as orders 3786 to 3797, July 1883 (52"+90", stroke 10'). It worked two 27" plunger pumps that could force water to a height of 500 feet. The engine had a surface condenser that was cooled by the pumped mine water. It was capable of raising four millions gallons of water in 24 hours, from a depth of 400 feet.[8]

Probably due to the engine's size, Davey reverted back to an older design where both cylinders were connected to the crosshead by three piston rods, one from the high-pressure piston and two from the low-pressure piston, the latter passing outside the high-pressure cylinder, supported in long brass bushed bearings.[9]

'The effect of expansion and contraction in such long cylinders is very great, and if not allowed for would produce very heavy and dangerous stresses. To fix the two ends of such cylinders rigidly to a bed-plate would lead to a breakage. A long bed-plate is necessary to distribute the thrust of the engine on the foundation. Mr. Davey has provided for expansion and contraction, and at the same time has distributed the stresses uniformly over the foundations by an abutment between the two cylinders, rigidly fixed to two strong girders which form the bed-plate of the engine, and runs its entire length. To this abutment, which is a massive casting, the two cylinders are bolted. The ends of the cylinders away from the abutment are supported on cross girders attached to the engine bed, but the cylinders are not fixed to the cross girders,

they simply rest on planed surface between guides, to preserve a central position. The cylinders are thus free to expand and contract, without in any way straining the beds or being strained themselves.'

'Another feature is the construction of the pistons. These are made very deep to provide a large area of supporting surface. The supporting surface of the piston is such that the weight of the piston imposes a pressure of less than 8psi on the underside of the cylinder.'

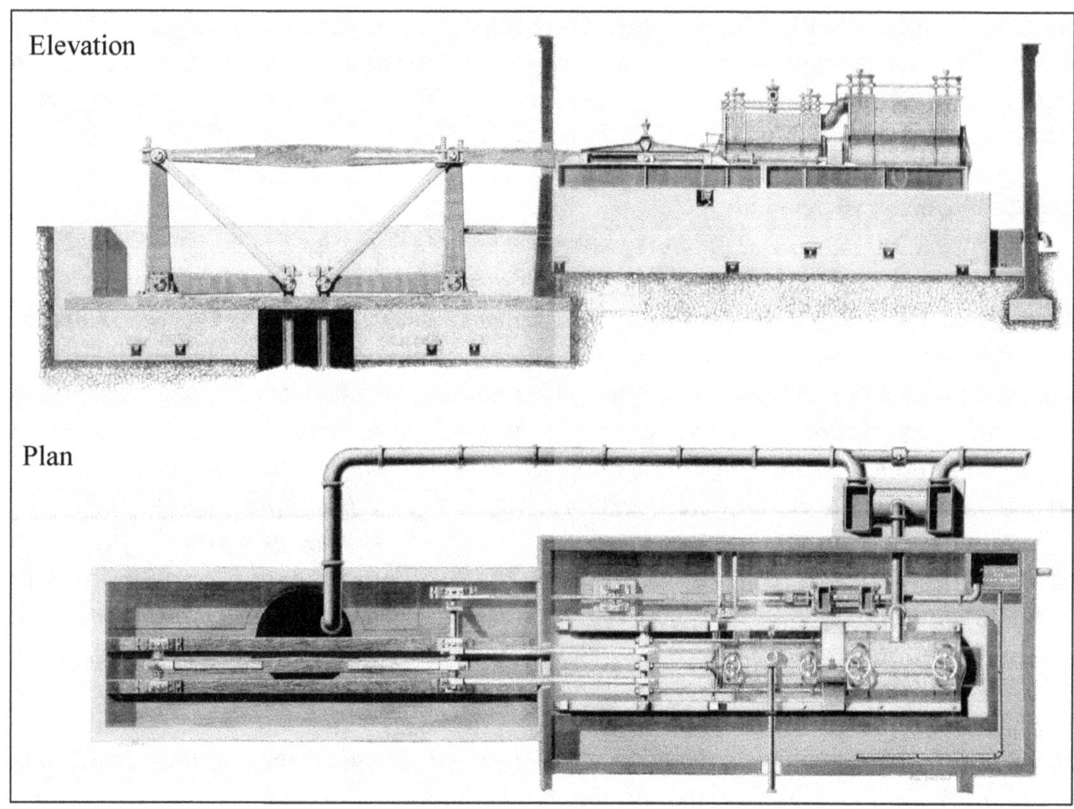

Figure 72. The Bradley Engine: It is a typical Compound Engine pumping arrangement for working a double set of pumps in a shaft. The plan shows that the rising main (emerging from the shaft) and is connected to the condenser.
(Modified from The Engineer, January 23rd 1891. Supplement.)

Davey also provided the engine with the latest version of the Differential Gear. *'The effect of the old gear was to throttle the exhaust as well as the steam when the engine raced or the gear was not properly adjusted. In the new form of gear the exhaust is always free, and the effect of racing is to cut off steam both in the large and small cylinder, and to shut altogether the exhaust from the high-pressure cylinder.'*

Provision was also made for shutting off either pump from the rising column in the case of an accident. *'The pump rods when at work balance each other, so that the engine has precisely the same work on both strokes, but when one pump is out of work there is the full load on the other. By means of the gear the engine can be adjusted to work under such circumstances. It was worked for some time under that condition when a pipe connection to one of the pumps failed.'*

Given the engine's specification, it is no wonder that Davey took exception when a report on the South Staffordshire Mines Drainage Commission in *The Engineer* (1888), stated, ' *The Commission has had one serious misfortune during the past year in the breakdown and temporary stoppage of the Bradley engine, a new and improved addition to their plant, of which great things were expected. Instead of economising the funds of the Commission, this engine has caused an additional outlay of two thousand pounds.'* [10] Davey responded: [11]

'Sir,- In your notice last week of the pumping operations at Bradley you have fallen into an error in stating that the engine had broken down. The engine has never broken down, has not been repaired, and is as perfect as it was the day it was first started. You have made a further mistake in stating that the engine has been the cause of the additional expenditure. You are aware that such statements are likely to cause considerable damage to the reputation of the designer and constructor of the engine if they are not contradicted. I hope, therefore, you will give the same prominence to this correction that you did to the statements in question. I will now give you the facts from which the false report is likely to have arisen. The Bradley engine ....works two 27in. by 10ft stroke plunger pumps, by means of two levers, to a depth of 400ft. from the surface. The water in the "pound" above the pumps is kept back and allowed to come on to the pumps only as fast as the pumps take it by means of a self-acting valve, regulated by means of a float in the sump.

Some time ago, either from the failure of the action of the float, or from some other cause, a pipe in connection with the pumps burst, and before the engine man was aware of it the water had risen up the pit above the pumps. He was thereby unable to get down to close the float valve. This happened in the middle of the night, and the engine man only discovered what was wrong by observing that the engine - the surface condenser of which is on the delivery from the pumps - was working non-condensing, the pumps having lost the water. The engine was not damaged in the slightest. Now, Sir, I will ask you to imagine what would have happened under similar circumstances to a Cornish engine. The load on the piston was 60 tons, and that was suddenly lost by the bursting of a pipe, and the engine man only know of it by finding that the engine was working non-condensing.

The conclusion jumped at was that the pumps had broken down, and a report to that effect got into the local papers. I could not contradict it, because the pit was full of water, and no one could see what had happened. If the commissioners had had the money and pluck to have put down reserve pumping power the accident would have been of no serious consequence, because the water might have been pumped out at once, and the broken pipe replaced. As it was they were compelled to have recourse to the old plant, quite inadequate for the purpose, and hence the delay and expense in getting the water out of the pit and starting the new engine again. The broken pipe in question was not supplied with the engine or pumps but whether it was broken because of defect or shock I am unable to say; the failure, however, did not arise from any defect in the engine or pumps.

Since I wrote the above, I have seen the report from which your correspondent has taken his information, and I see that he has been deceived by misleading statements.'

The Commission however, must have been happy with the Bradley engine as it worked virtually continuously until 1932, when it was replaced with electric centrifugal pumps. It was estimated that from 30th June 1890 to 30th June 1902 it had raised 11,686,000,000 gallons of water from a depth 384 feet, calculated to be equivalent to an average piston speed of 69 feet per minute!

The Moat Engine

The Moat engine (44"+76", stroke 10') was smaller than the Bradley. The engine together with associated pumping equipment was ordered on the 12th June 1883 (Orders 3798 to 3811). It was put into operation during 1885.

There were two lifts of pumps. The bottom lifts were 19½" bucket pumps that fed the upper lift, 19" plunger pumps. The system was capable of pumping 2.2 million gallons per day at seven strokes per minute.

In the 10 years up to 30th June 1902 it had raised an estimated 7,391,525,000 gallons of water from a depth of 620 feet, at a piston speed of 115 feet per minute.

The engine was finally stopped in October 1919 after a pump rod broke, and the pit became flooded.[12]

A contractor selected by the Commission maintained the Bradley, Moat and Stowheath engines. The Contactor, as the advert in Figure 73 indicates, was expected to keep the engines in thorough repair and working order, as well as being responsible for labour and stores. No doubt the supply of coal for the boilers was also included.

SOUTH STAFFORDSHIRE MINES DRAINAGE ACTS 1873 to 1882 - The Commissioners are prepared to Receive Separate TENDERS for the Working of the following PUMPING ENGINES, in the Tipton District:-

STOWHEATH PUMPING ENGINE, near Wolverhampton

MOAT PUMPING ENGINE, Summer Hill, Tipton

ROUGH HAY PUMPING ENGINE, near Darlaston

DEEPFIELDS PUMPING ENGINE, Coseley

BRADLEY PUMPING ENGINE, near Bilston

The Contractor will be required to find Labour and Stores, and to Be responsible for keeping the Engines in thorough repair and working order.

Payments monthly

The Commissioners do not bind themselves to accept the lowest or any Tender.

For further particulars and orders to inspect, apply to the General Manager, to whom the Tenders shall be sent.

EDMUND HOWL, General Manager
South Staffordshire Mines Drainage

25, Darlington Street, Wolverhampton   452

Figure 73. Advert (Copy): Invitation to tender for engine operation and maintenance
(Birmingham Daily Post January 25th 1889. p.1)

In a later paper[13], *'Some of the Considerations Affecting the Choice of Pumping Machinery,'* (1902), Messrs. Meysey-Thompson and Lupton were able to quote running costs for both the Bradley and Moat engines over a 13 and 10 years accounting periods, respectively. It is interesting to note that the smaller engine cost more to run.

	Bradley*	Moat**
Coal	£1,422	£1,410
Labour	£873	£918
Stores, including oil and packing	£99	£98
Repairs to engine and pumps	£110	£150
Repairs to boilers	£47	£23
Totals	£2,551	£2,599
Running cost per annum.	£196	£260

*   From accounting year ending 30th June 1890 to 30th June 1902 (13 years)

**   From accounting year ending 30th June 1893 to 30th June 1902 (10 years)

Unfortunately, after the engines went into service, there was still an outstanding amount of the cost that the Commissioners owed to Hathorn, Davey. By January 1886, this amount, together with interest, was in excess of £3,000. [14]

Inevitably, Hathorn, Davey wished to reclaim the outstanding amount from the Commissioners. A Statement of Fact dated 2nd March 1886 summarised the situation and showed that the original cost for supplying the engines was £11,095-9s-4d (apparently well below the original estimate) of which £8,046-3s-4d had been received, partly in cash, and partly in goods. That left an unpaid balance of £3,049-6s-0d, to which a further £293-11s-8d interest had been added.

Henry Davey had several meetings with the Commissioners and their Secretary all of whom admitted that the engines had done their work well and said they ought to have been paid for long since, and promised to pay the balance. The Commission intended to take out a loan, but there was some doubt about this happening under the terms of the 1873 Act. Consequently, in December 1885, Hathorn, Davey instructed their Solicitors to apply for the outstanding balance, warning the Commissioners that they would be prosecuted if the debt were not paid. No reply was received to this warning.

After further lengthy correspondence which examined the Commissioner's obligations under the terms of the Act, and whether the Commission could borrow money, or apply for a mortgage, to pay off a debt, or offer some other form of Security an Opinion was given by a legal official of the Inns of Court on the 10th March 1886. The Opinion indicated that the Commissions were empowered to offer security, in the form of a Bond, for example, to Hathorn, Davey for the outstanding amount. It would ultimately be a more rewarding solution than applying for the full amount through a Claims Court.

A security was formally agreed with Hathorn, Davey in the form of the rate raised by the Commissioners on the Tipton Drainage Area. In effect it was agreed to pay the outstanding amount, now £3,365-17s-7d, to Hathorn, Davey on 1st May 1886.

In fact the amount was paid on the 29th April 1886. The Law Clerk for the Commissioners asked for the Security Bond to be endorsed as a receipt.

However, 'once bitten twice shy' and so the next time the South Staffordshire Mines Drainage Commission owed money to Hathorn, Davey they were prosecuted.

That amount was small in comparison to the amount owed on the Bradley and Moat engines. It is unclear what the debt was for and seemed to be related to the installation of the engines. The amount owed was £597-2s-7d. The Defendants (SSMDC) admitted buying the goods, but simply said that they had no money at present to pay, and had no goods that could be seized. They admitted that they could levy a surface-drainage rate and a mines-drainage rate, but they said that the work for which these goods were ordered did not come for payment under either of these rates.

The goods were apparently purchased for the purpose of erecting a pumping engine for the underground drainage of mines in the Tipton district, and they said that the articles could only be paid for out of so much of the mines drainage rate as might be raised in the Tipton district. It is unclear why this amount couldn't be raised as in the previous manner, but certainly monies administered by the Commission at this time was under severe scrutiny, and ways to cut costs were being explored. Certainly the Court case ruled in favour of Hathorn, Davey.

It is inevitable that situations like this would occur because of the way the Commission was funded. These events don't appear to have soured the relationship that Hathorn, Davey had with the South Staffordshire Mines Drainage Commissioners. Hathorn, Davey continued to supply various parts for the pumps, as well as three inverted vertical compound engines in 1904. In fact it is probable that Henry Davey was on their list of Consultants.

Chapter 15 References

[1] South Staffordshire Mines Drainage. 1920. *Report of the Committee to Inquire into the Drainage of the Mines in the South Staffordshire Coalfield.* HMSO. p.13

[2] ibid. South Staffordshire Mines Drainage. 1920 pp. 13 to 14

[3] ibid. South Staffordshire Mines Drainage. 1920 p.15

[4] ibid. South Staffordshire Mines Drainage. 1920 pp.19 to 20

[5] Mines Drainage Commissioners. *Birmingham Daily Post*, 7th June 1883,

[6] ibid. South Staffordshire Mines Drainage. 1920 p.20.

[7] South Staffordshire Mines Drainage. *The Engineer.* 20th July 1883. p.56.

[8] Lones, T.E. 1930. The South Staffordshire and Worcestershire Mining District and Its Relics of Mining Appliances. *Transactions of the Newcomen Society.* VII. p.47.

[9] Compound Differential Pumping Engine, Bradley. *The Engineer*, 23rd January 1891. pp.74 and Supplement.

[10] Colliery Pumping Progress in Staffordshire. *The Engineer*, 5th October 1888. p.289.

[11] South Staffordshire Mines Drainage. *The Engineer*, 12th October 1888. p.303.

[12] Weaver, P. Some Interesting Data on the Birmingham Canal Navigation p278 to 303 (full reference not known)

[13] Meysey-Thompson, A. H. and Lupton, H. 1902. Some of the Considerations Affecting the Choice of Pumping Machinery. *Transactions of the Institution of Mining Engineers*, 24. pp.276 to292

[14] DLC Invoice - last date 31st January 1886

16. Henry Davey and the London Office

Davey's move to London in 1887 almost certainly gave him the freedom he probably desired. Away from the Sun Foundry, and the daily involvement in trivial decision-making, it allowed him time to think, and conceive even more novel ideas for engines, the majority of which he patented. By the end of the century he had developed both the vertical forms of the Triple Expansion and the Compound Engines, both of which were inverted, with the cylinders mounted above a crankshaft, or beam, respectively.

His proximity to the major engineering Institutions presented him with the opportunity to become involved with other projects. He was, a member of a Committee looking at the design and efficiency of steam-jackets for cylinders.

Equally important was the fact that it brought him closer to potential clients. London was certainly the first stop for a foreign company representative arriving in Britain seeking to purchase a steam engine!

The Hathorn, Davey and Company offices were located at 3, Princes Street, Westminster, London, S.W. just to the south of Oxford Circus, between Hanover Square and Regent Street. No 3 contained the offices, or 'rooms', for a variety of professions. One imagines the Company office to be a small affair with probably a desk for Davey, a table for meetings, and a drawing board. It is not recorded if Davey employed a secretary. Company accounts dated July 1888 show three items that relate to the London Office; Furniture valued at £40, and fees owed and paid from business transactions that were £371-10s and £102-15s-3d, respectively. No items refer to rental or to the purchase of the office, so possibly this was being paid directly from Davey's account. Davey was taking the biggest portion of the Company's profits.[1]

It would have been about this time that Henry Davey created an *'Album of Designs of Pumping Engines for Town Water Supply'* to promote the company.[2] The preamble explains that, *'The Davey Engines are well known, and the following illustrations are examples of different designs, showing some of the usual applications to Water Works purposes. Special designs are sometimes required to meet local conditions.'* The twelve pages in the album all bear Davey's signature and office address, and show a variety of Compound Engine designs. The first page compared different cylinder arrangements from Compound Engines as well as a balanced Cornish configuration. The following pages contained a series of plans and elevations of five Inverted Angle-Compound Engines with differing cylinder arrangements to operate force or well pumps. Figure 74 shows design 5 with high and low-pressure cylinders engines and a flywheel, which Davey described as a *'Davey Compound Beam Engine with Well and Force Pumps.'*

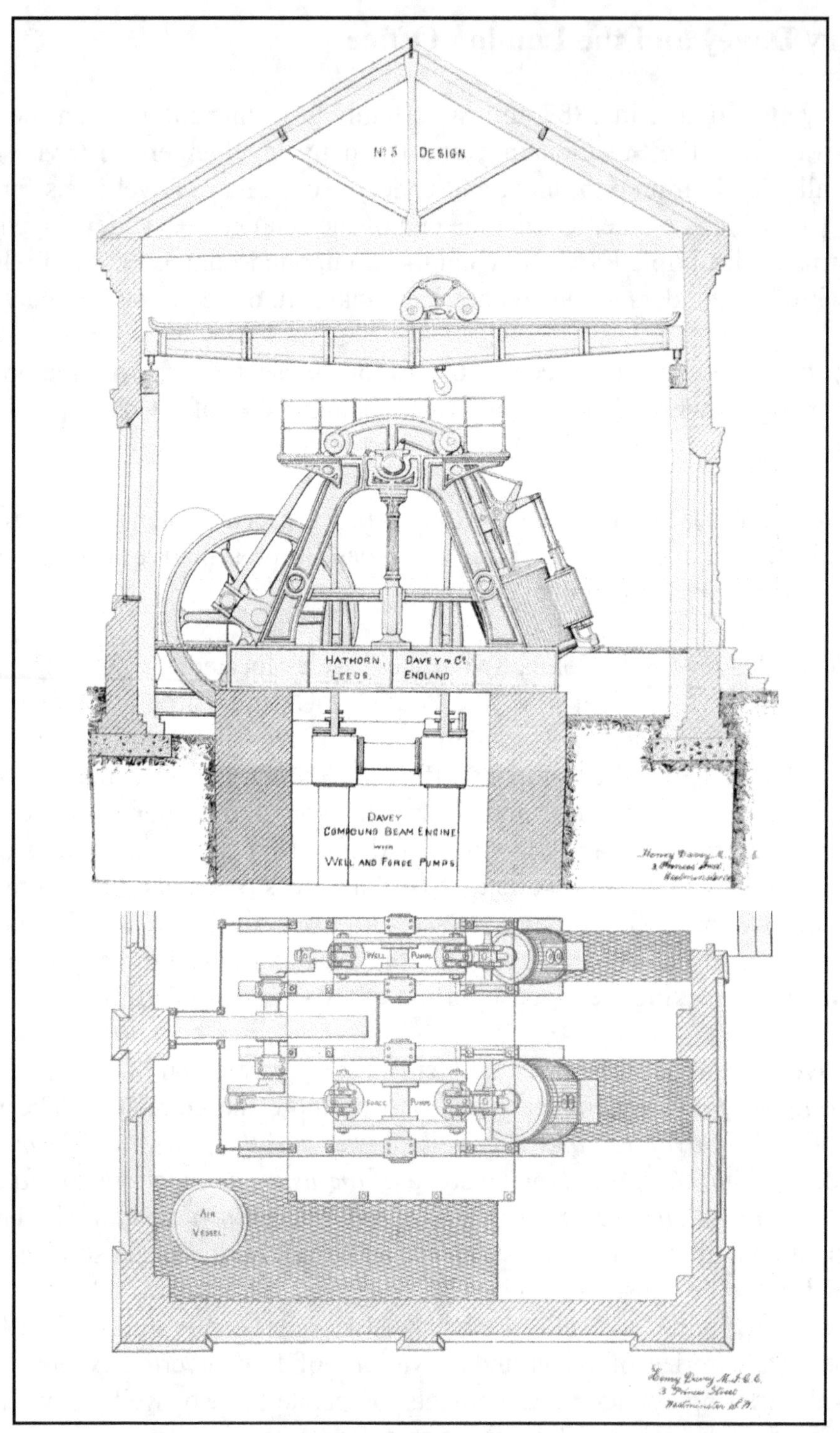

Figure 74. Design 5: Davey Compound Beam Engine with Well and Force Pumps
(Davey, Album of Designs of Pumping Engines for Town Water Supply)

The engines that Hathorn, Davey sold to the Weston Super Mare, and Wolverhampton Waterworks appear to have been manufactured to, or modified from, designs in the album.

i) Weston Super Mare Waterworks (Orders 4532, 4532 A to R, 1st November 1887)

The water for the town of Weston Super Mare, Somerset was originally supplied from a spring. In 1878, the Weston Super Mare Water Works Company that maintained the supply was put into liquidation and the responsibility of supplying water to the town was given to the Weston-Super-Mare Improvement Commissioners.[3] By 1886 the Commissioners had applied for a Parliamentary Act (1887), to enable them to purchase land to establish a new waterworks and a reservoir.[4]

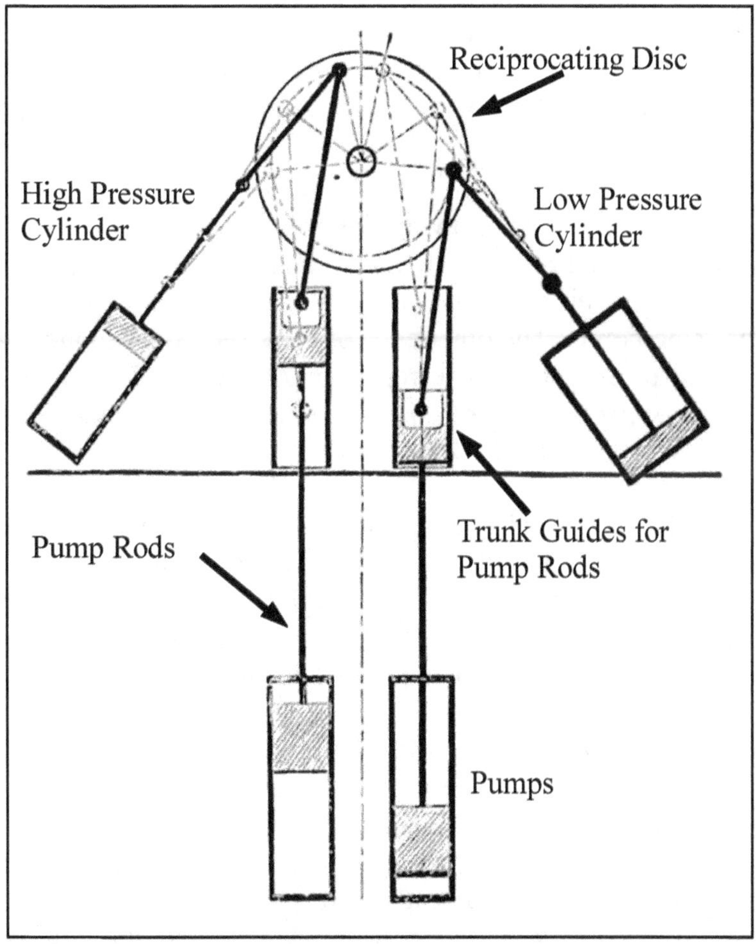

Figure 75. Diagram showing how the engine operated the two pumps.
(The Engineer. January 13th 1888, p.33 [GG])

To improve the water supply a 15 feet diameter well was sunk 27 feet into the Carboniferous Limestone at the eastern side of the town, and a heading was driven from it to connect to an existing well. At the time there was a drought, but nevertheless the large amount of water encountered caused some difficulties in driving the heading.

The new pumping plant consisted of two of Davey's patent Differential Compound Surface Condensing, Diagonal Pumping Engines, fitted with adjustable expansion valves. It was simply described as a Differential Compensating Pumping Engine.[5] For both the high and low-pressure cylinders, equilibrium-cushioning valves were fitted to the intermediate receiver, and an arrangement was provided by means of which both cylinders could be supplied with steam direct from the boilers when started. The high and low-pressure cylinders are 14" and 26" diameter, respectively, and both had a stroke of 3' 9".

The pumps were 15½" diameter with a 2' 6" stroke. There were two piston type pumps to each engine. Steam was supplied by two Galloway type steam boilers 28' long by 7' diameter.

Figure 75 is a simplified diagram showing the mode of operation. The high and low-pressure cylinders are inclined diagonally and opposite each other. The piston rods are attached to a reciprocating disc, and the sweep rods of the two pump rods are attached to the disc at the same points. The other ends of the sweep rods connect with the main pump rods seated in trunk guides. The report in *The Engineer* commented that, '*the engine under notice is well-designed, and appears to be a decided advance in the construction of direct-acting pumping engines.*'[6]

At a Weston Super Mare Council meeting in 1909 a query was raised regarding the pump capacity. It seems that the diameter of the pumps was half an inch more than specified, which had meant that for the previous 20 years, the amount of water pumped into the reservoir had been underestimated.[7] This discrepancy is not a surprise as others have also noted a difference in the quoted pump diameter![8]

ii) Wolverhampton Waterworks (Orders 4616, 4616 A to R, 4619, 13th October 1888)

The Wolverhampton engines are described as Double Cylinder Differential Pumping Engines. The first engine (which was referred to as No.4) was used as a lifting engine to pump water out of a well, and was capable of pumping 2,800gpm from a depth of 140 feet, into a straining tank. Exhaust steam was passed through the tubes of a condenser that was immersed in the cold water delivered by the pump. The pump was 27¼" diameter with a 4' 6" stroke.

The second engine (referred to as No. 5) was used as a forcing engine capable of pumping 1,400gpm into the water mains, against a pressure of 300 feet of water. The engines were so similar that their parts were interchangeable. The water from the suction pipe was used to condense the steam.

The cylinders for both engines were specified as 56" diameter with a 5' stroke. Both condensers would be constructed with gun-metal tubes with ample cooling surface to maintain, '*a vacuum within 1½ lbs of atmospheric pressure.*' The plunger pump was 20" diameter with a 4' stroke.

The engines, pumps, staging etc were for the Cosford Pumping Station and cost £7,500. The engines were started on the 4th November 1891 and were commemorated by a gathering of Council Officials, which Henry Davey attended. '*This makes the 5th engine at Cosford, two for raising from the well, and three for forcing the water to Tettenhall, where the Wolverhampton reservoir is situated. At the conclusion of the ceremony, the company sat down to an excellent dinner at the Jerningham Arms Hotel, Shifnal.*' [9]

The London office seems to have been very successful. The orders from mining companies were not as numerous as they had been in the previous two decades, but they were slowly being replaced by the pumping requirements of waterworks and Sanitation Boards of British Local Authorities, for example Barnstaple, Birmingham, Widnes. It was probably this trend that steered Davey to produce his Album of Designs, and led to him patent his version of the Triple Expansion Engine, a machine eventually favoured by a number waterworks companies.

Orders were also coming in from a variety of countries and foreign agencies. In the first half of the decade, engines had been supplied to Townsville, Queensland, Australia. The Crown Agencies of the British Empire also played their part by ordering engines for Malta, and no doubt encouraged the Indian State Railways to purchase Hathorn, Davey engines for their collieries.

John Taylor and Sons, who were London based, managed the mines on the Kolar Goldfield, India and ordered an engine for the *Mysore Gold Mining Company*. This was the first of a number of engines Taylors would order for their mining operations in India and elsewhere in the world.

Other Hathorn, Davey engines were exported to Brazil; Venezuela; Australia (Charters Towers, Queensland, that still survive *in situ*); Bridgtown Waterworks, Barbados, and closer to Britain, the Delft Waterworks in the Netherlands.

It is clear that Davey was enjoying his 'release' from Leeds, and the period from about 1887 to 1900 might represent the most creative phase of his life, but he did find time for other engineering distractions.

Figure 76. Henry Davey's design for the 'Great Tower of London.'
(The Engineer. May 23rd 1890 1888, p.427 [GG])

Possibly the first hint of a London Tower, similar to the Eiffel Tower, came in an interview that Sir Edward Watkin gave to a correspondent of the *Pall Mall Gazette*, that was published on the 22nd August 1889. Watkin was many things. He was a Baronet; and a Liberal Member of Parliament associated with railway projects and the Chairman of nine different railway companies. He was also regarded to be an ambitious visionary. In the interview, he proposed that a tower similar to the Eiffel Tower should be constructed from, 'British steel and British labour' close to the site of the Imperial Institute.[10] At that time, the Imperial Institute of London was being established at Exhibition Row, South Kensington, London. The building now forms part of Imperial College. So a tower at this location would have had some significance.

By the 26th October Watkin had formed, and become Chairman of the *London Tower Company, Ltd.* and had offered two prizes of 500 and 250 guineas for the best design of a 1,250 feet high tower. The panel of judges included eminent engineers, for example Baker (associated with the Forth Bridge) and Walker (noted for his work on the Manchester Ship Canal and Severn Tunnel).[11] There were a considerable number of entrants, and designs and drawings had to be submitted by the 1st February. It seemed only natural for Henry Davey to submit a design.

Sir Edward Watkin opened an exhibition of designs from some eighty entrants at the Hall of the Drapers' Company, London on April 29th April 1890. An article in the *London Evening News* (30th April 1890) described some of the best designs. Perhaps Davey felt he was 'home and dry' when in the final paragraph the report stated, '*Mr Henry Davey's design is one of the very finest and the most engineering-like in aspect. Twelve fine strong steel masts rise up, diminishing in size, and approaching each other as they reach a basket like gallery, above which rises a central column, surmounted by a lantern, crowned by a ball..*'

Figure 76 shows the design of Davey's tower. In plan it was hexagonal with the structure taking the form of a spire. He proposed to construct it from about 8,000 tons of steel. In the basement, there would be machinery for working the lifts and generating lighting. On the ground floor, there was a space of 70,000 square feet that would have housed a Winter Garden, with ornamental fountains.

Lifts would take visitors to the first stage 380 feet above the ground. The second stage, at a height of 980 feet, was reached by further lifts and consisted of four floors. From there, stairs would give access to the top sections. A report in the *Engineer* (May 23rd 1890) elaborated about the technical details that included lift capacity and speed.[12]

Unfortunately, Davey's design didn't win. First and second prizes were given to 1,200 feet and 1,300 feet, high towers designs, respectively. The Jurors summation concluded, *'the designs sent were of very varying merit and were disappointing as a whole, there being no single design which they could recommend as it stands for execution.'* [13]

Simply titled '*Game*', Davey's next patent (15, dated 1st January 1890) may be seen by some as a departure from character, and totally frivolous, but no doubt inspired by Norman, his two year old son. The patent for '*Game*' was also taken out in France, Germany and the USA. The drawings accompanying the patent are shown in Figure 77. The '*Game*' was similar to bagatelle, and was described by Davey as:

'My invention consists in certain new and useful improvements in a game comprising a table, slab, or board, over the surface of which balls are propelled by means of a spring or other suitable projector; and the invention also consists in a novel movable angular deflecting-cushion for use in connection with a game-table, all as will be hereinafter described and claimed.

In the accompanying drawings, Figure 1 is a perspective view of my improved game board or table with an angular deflecting-cushion, said view showing the table provided with two slots or tubes adapted for two projectors, and also with corner pockets, recesses, and also receptacles for toys, counters, balls, and angular movable cushion, toy figures, a ball, movable cushion, and projectors being represented.

Fig. 2 is a transverse section of a part of a table, showing more plainly the arrangement of the projector.'

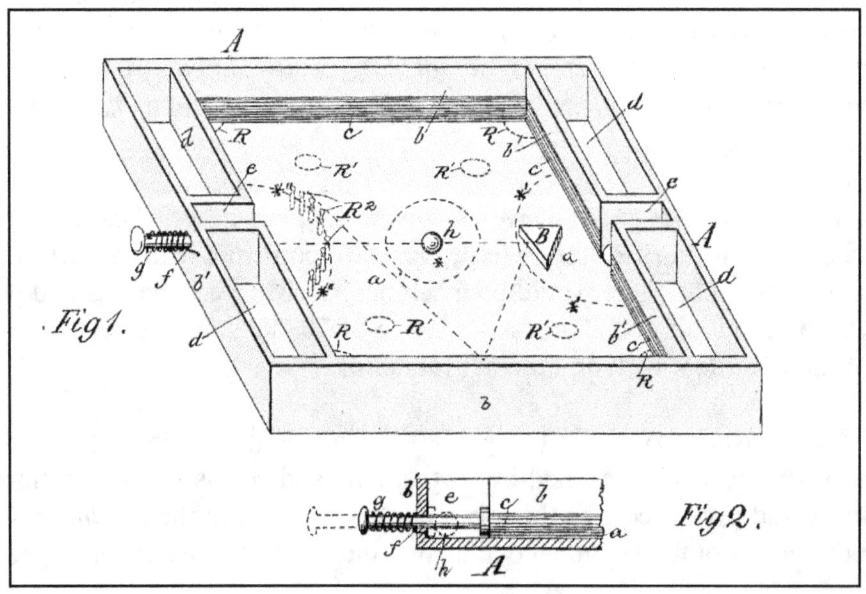

Figure 77. Henry Davey's patent for 'Game'
(Modified from: United States Patent No. 450,883 dated 21st April 1891)

Davey's interest also turned briefly to shaft sinking methods. In a paper that he gave to a Meeting of the British Association for the Advancement of Science held at Cardiff in 1891 he examined shaft-sinking techniques.[14]

The traditional method of shaft sinking was to drill a borehole to prove the ground where the shaft was to be sunk. A pump would then be placed in the borehole and as sinking progressed the water would be pumped from within the confines of the shaft. In Davey's method he proposed that pumps should be placed in boreholes around the area of the proposed shaft sinking, and that the ground was dewatered prior to sinking. For example, two engines working four pumps in strategically drilled boreholes could be used to dewater the ground.

Davey also took the idea further and commented about boreholes being used for town water supply, rather than the usual more costly well.[15]

'A simple borehole is made very cheaply and very expeditiously. Four 30-inch boreholes can be put down in a very small fraction of the time required to sink a 12-foot well. Instead of making a large well, the author [Davey] *put down four boreholes to accommodate the pumps for duplicate pumping engines - a pair of pumps to each engine. The boreholes being completed, the pumps are lowered into them, and coupled up to the permanent engines.'*

Davey then cited the Netherley Pumping Station (Widnes Water Board) where a Compound Engine (32"+60") (Order 4922, dated 17th June 1991) was pumping from two boreholes 20-feet apart. Davey was also able to devise a system using his parallel motion to pump from two different levels if the diameter of the borehole was varied.

For a brief period, Davey examined the automation of hydraulic sewage pumps using principles already devised for dewatering dip headings in mines. It used a system of floats with a vessel that could be made to fill or discharge. In other methods, the engine speed was varied depending on the volume of sewage. It also used a float system.[16]

Davey's Expansive Direct-Acting Accumulator Engine was also a novel idea that he was promoting at this time.[17] *'It is well known that a simple rotative engine employed to work pumps, and subject to variations of speed and to stoppages, is not capable of being worked with any considerable degree of expansion, because expansion depends in such engines on the momentum of the fly-wheel, and, therefore, when the engine stops, it requires full steam to start it again.*

The ordinary steam pump is an exceedingly convenient and effective machine for pumping water to work hydraulic machinery, especially when a pair of such pumps are arranged on the duplex plan; but the objection to the use of such pumps is the great waste of steam, no expansive working being possible in such engines. The Davey engine gets over this difficulty by a compensating mechanism of a very simple nature, consisting of a bell-crank lever, connecting together the pump plungers in the peculiar manner shown in the illustration.' [The bell-cranks are the triangular pivots located in the pits beneath the pumps.]

The two Expansive Direct-Acting Accumulator Engines shown in Figure 78 are Double-Acting, Non-Compound Engines, made by Hathorn, Davey for working hydraulic machinery.

Figure 78. Two Expansive Direct-Acting Accumulator Engines
(The Engineer, 12th April 1889. p.310 [GG])

'The effect of this arrangement is that the engines give a continuous flow in the pipes and are governed by the differential gear. Engines arranged on this plan, if large enough, need no accumulator, but may be employed to pump direct into the hydraulic main. Messrs. Hathorn, Davey, and Company are making a considerable number of

these engines for various pumping purposes. The engines illustrated have 40in. cylinders by 4ft. stroke, and together are capable of supplying 400-horse power to hydraulic machinery.'

There is no doubt that Davey spent much of this period designing various engines and their applications. Davey was constantly promoting the Company. He must have been well aware that the use of steam power was starting to diminish, being replaced with other sources of energy, like the internal combustion engine. However this did not deter him for pursuing his goal of producing efficient steam engines, which resulted in the Company developing the Inverted Triple Expansion Engine. It was an engine that gained favour with many waterworks companies.

Chapter 16 References

[1] SNA Balance Sheet dated 31st July 1888. GD455_77

[2] Photocopy of: Davey, H, c1888. Album of Designs of Pumping Engines for Town Water Supply, Courtesy of Mike Beevers.

[3] Notice. *Weston-super-Mare Gazette, and General Advertiser*. 1st December 1877. p..5

[4] Improvement Act. *Weston Mercury and Somerset Herald*. 20th November 1886. p.1

[5] Weston Super Mare Water Works Extension. *Weston Mercury and Somerset Herald*. April 20th 1889, p.5

[6] Differential Compensating Pumping Engine. *The Engineer*. 13th January 1888, pp. 27, 33 to 34 and Supplement

[7] Weston Water Works - Amazing Revelations. *The Weston Gazette*. 16th January 1909. p.5

[8] Chris Allen noted that a report in *The Engineer* stated that the steam cylinders were 14" & 26" x 3' stroke, and that the pumps were 16" x 2'

[9] Cosford - New Pumping Engines at the Waterworks. *The Wellington Journal and Shrewsbury News*. 7th November 1891. p.8

[10] An Eiffel Tower for, London. *Pall Mall Gazette*. 22nd August 1889. p.1

[11] The Proposed Great Tower in London. *Cheltenham Chronicle*. 26th October1889 (Supplement)

[12] The Great Tower for London - H. Davey. *The Engineer*. 23rd May 1890. p.427

[13] The Great Tower for London. *The Western Mail*. 26th June 1890. p.3

[14] Report - Sixty-First Meeting of the British Association for the Advancement of Science (Cardiff) August 1891. Office of the Association, Burlington House, Piccadilly, London. pp. 766 to 767.

[15] Davey, H. 1894-95. Bore hole Wells for Town Water Supply. *Cassier's Magazine*. 7. pp.244 to 248.

[16] Automatic Hydraulic Sewage Pump. *The Engineer*. 14th February 1890. pp.130 to 131.

[17] Expansive Direct-Acting Accumulator Engines. *The Engineer*. 12th April 1889. pp.310 and 316

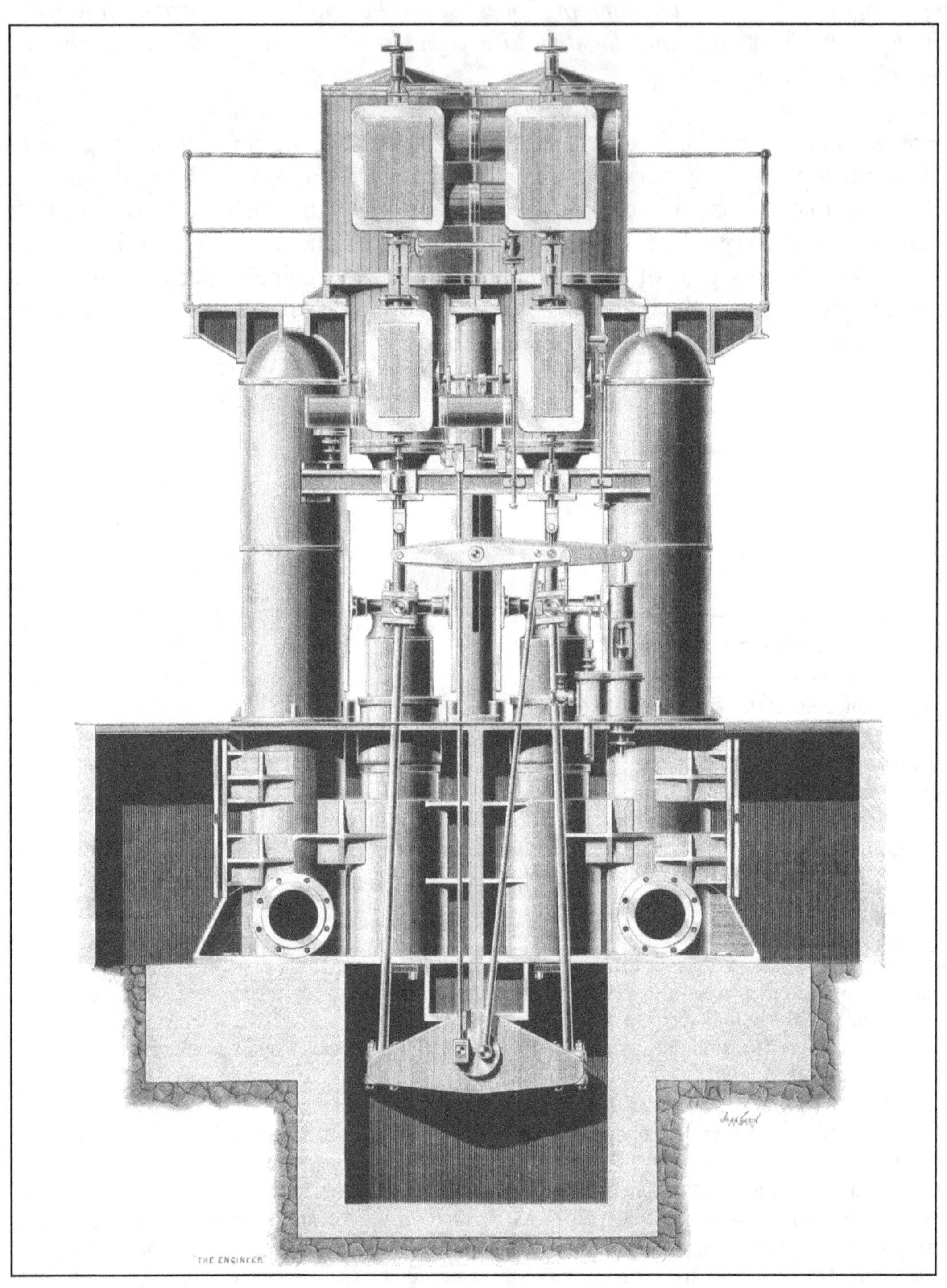

Figure 79. Triple Expansion Pumping Engine for Brazil
(The Engineer, January 13th 1893, p.31 [GG])

17. Triple Expansion Steam Engine

In the 1880s the compounding of steam entered a further phase of evolution, triple expansion. Firstly it was employed in marine engines, but then the principle was applied to stationary engines, with a number of companies promoting their own designs. Davey of course, designed several types of Triple Expansion Engines.

Hathorn, Davey and Company manufactured Triple Expansion Engines with a general configuration that included an intermediate cylinder between the high and low-pressure cylinders. The engines were commonly vertical and inverted. Power was usually applied directly to crossheads and a crankshaft running longitudinally beneath the engine. The shaft often had a flywheel, either located between the crossheads or at the end. The pumps, located in a pit beneath the engine, were directly coupled to the shaft. According to Hugh Lupton, this configuration gave a high degree of mechanical efficiency.[1] Ultimately these engines, which used superheated steam and worked on much higher steam pressures than the Compound Engine, would prove to be extremely reliable, efficient and economical. This efficiency was increased by the incorporation of an economiser. Other improvements included the use of steam jackets around the cylinders, and Corliss Valves to replace the slide valves.

For pumping from shafts, wells and boreholes, a cam arrangement at the end of the shaft generated a horizontal reciprocating motion. This arrangement worked horizontal con-rods and quadrants, and is referred to by some engineers as 'The Pitman.' However, Horizontal Compound Engines with two cylinders, more often than not, were still preferred for this type of operation. The Company also constructed Horizontal Triple Expansion Engines, with one exported to Japan to work an underground hydraulic pumping system.

Whilst high-intermediate-low pressure configuration of cylinders became the norm, there is one design of a triple expansion engine that remains a mystery. In the *Engineer*, 13th January 1893, there are illustrations and a description of a Triple Expansion Engine, where the steam is expanded three times in two cylinders.[2] The engine was designed by Henry Davey and manufactured by Hathorn, Davey for a waterworks in Brazil. (Figure 79) This engine is possibly listed as 4723, dated 2nd August 1889, for a Vertical Triple Expansion Waterworks Pumping Engine (two cylinder - 15"+30" later amended?) ordered by Walter Mansell who had previously bought engines from Hathorn, Davey for installation at Bahia, Brazil.

The detailed description stated that, *'the engine is of a new design, having several novel features. It is constructed under Mr. Davey's patented compensating device, by means of which high-pressure steam and a high degree of expansion is secured, without any rotative motion. It also has the new form of Davey's differential gear and cushioning valves. It will be seen on Figure 79, that there are two high and two low-pressure pistons connected together by trunks, and that there are two pump plungers directly connected to the piston rods; the one making the up-stroke whilst the other makes the down-stroke. The two plungers are connected by means of connecting rods taking hold of pins in a rocking-frame under the pumps, the rocking-frame constituting the compensating device.* [This is a similar bell-crank mechanism to that shown in Figure 78.] *This device has been largely applied to waterworks engines. It is a simple contrivance consisting of a bent lever for causing the pump resistance to decrease from the beginning to the end of the stroke, so that a high degree of expansion may be employed in the steam cylinder without the use of a flywheel and crank-shaft.'*

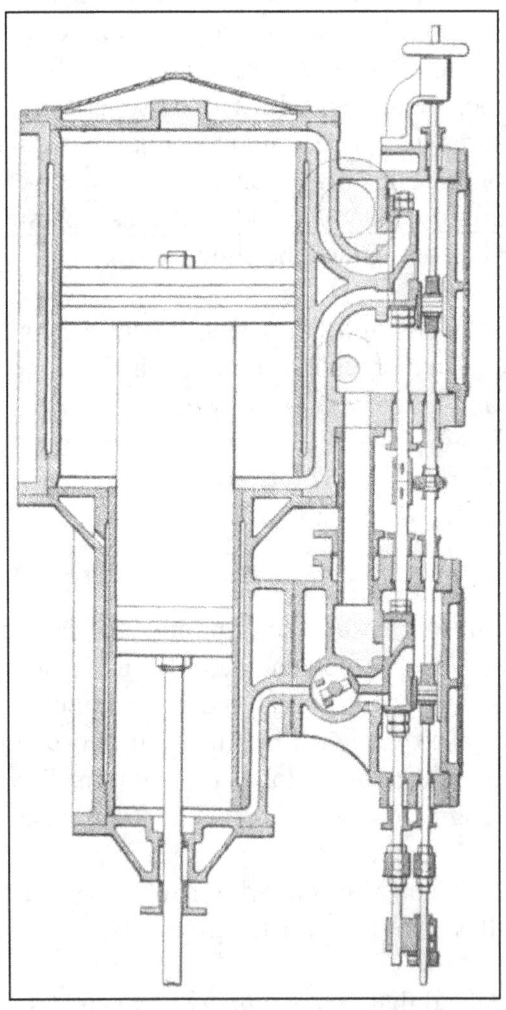

Figure 80. Triple Expansion Engine for Brazil: Details of steam cylinders and valves.
(The Engineer, January 13th 1893, p 26 [GG])

'The valve gear consists of main slide valves, moved by a subsidiary cylinder, controlled by a cataract. On the backs of the main slide valves are expansion valves, connected to the engine by levers, so as to receive the motion of the main piston on a reduced scale. The combination effects, in a more perfect way, the self-control of the engine, secured by Davey's original differential lever. When admitting steam to the engine, the main slide valve moves under the expansion valve in the same direction. Should the engine lose its load, the expansion valve closes immediately, and the main valve begins to admit steam to the cylinder.'

'The high-pressure cylinders are provided with cushioning valves, by means of which the length of stroke is determined and kept as nearly constant as possible. With a constant load there is practically no variation. The steam is first admitted to the engine under the small piston. It is then expanded into the annular space around the trunk, and from there it is finally expanded on the top of the large piston and then escapes to the condenser.' [Illustrated in Figure 80]

'The engine has 77½ in. and 30in. cylinders by 3ft. stroke, and 17in. plunger pumps. It will indicate 70-horsepower with 90 lb. steam pressure.'

Future Triple Expansion Engines would use Corliss Valves instead of slide valves and generally incorporate steam jackets.

Until the 1890s, the majority of valves used to control steam entering and exhausting cylinders were of the slide type, or its variants. The inlet and outlet valves were open and closed by a sliding block operated by connecting rods. This process was phased in with the engine's motion, and the valves operated at the appropriate part of the steam cycle.

In contrast, the Corliss Valve, originally designed in the USA, was semi-rotary. It could be opened and closed very quickly permitting steam to enter the cylinder without significant pressure drop, a process that would cool steam, with the consequential fall in efficiency, and thus loss of power. In addition, there was a separate valve for inlet and outlet, so the heat loss normally associated with the admission of steam, and the exhausting of cooler steam from the same slide valve was eliminated.

This was a major advance in steam power technology. Davey modified the Corliss valve system to his own design and placed the valves in the cylinder covers. They were operated by means of a counter-shaft that got it's motion by means of gearing from the engine shaft. There was a separate eccentric for each pair of steam and exhaust valves. The whole sequence of timing could also be adjusted by hand that meant that engines could be tweaked to produce even greater efficiency. [3]

The steam jacketing of a cylinder was not a new idea. Watt had pioneered the concept in the late 19th century, when he noticed that steam condensed without doing any actual work. Others had followed in his footsteps, but by the 1890s Watt's ideas hadn't really advanced significantly. The superheating of steam exasperated this problem significantly. Steam would readily condense on any 'cooler' surface. It was a problem that would have to be resolved if steam engines were to be made even more efficient.[4] The principal behind a steam jacket was to keep the cylinder warm enough so that steam did not condense. If steam condensed it would lose some of its expansive power.

It was a problem that occupied members of the Institution of Mechanical Engineers. Consequently, the Institution set up a Committee to examine the subject in 1886, and Henry Davey was appointed its Chairman. They published their, *'First Report of the Research Committee on the Value of the Steam Jacket'* in 1889.[5] The Committee compared the 'with and without' scenarios for a number of engines that included single cylinder and compound, from a variety of sources. Hathorn, Davey, manufactured none of the engines studied. In the first year the Committee just collected data.

The second report of the Committee was published in 1892.[6] In the intervening years since the first report, more data was collected, and Davey had conducted experiments with the Hathorn, Davey Triple Expansion Engine at Waltham Abbey (details below), and another engine at the Woolwich Arsenal. Comments from various observers on the subject were also published.

The third report was published two years later.[7] More data was tabulated, and it was noted that a final conclusion could not be reached without conducting some laboratory experiments for comparison. The reasoning behind this being that some element of consistency was required. Not all the tests were conducted over the same period of time, or similar conditions. Later tests were conducted on a steam locomotive.

An article appeared in *Nature* that summarised the findings of the Committee in 1892.[8] *'The general conclusion to be drawn from the experiments, as quoted, is that "the expenditure of a quantity of steam in an efficient jacket produces a saving of a greater quantity in the cylinder." It does not follow from this that the jacket is always desirable, as the saving may be so small as not to justify the additional complication and increased outlay at first cost. That, however, is a matter upon which steam users must themselves decide upon a commercial basis; and is, of course, outside the province of the committee, but what would have been valued would have been some critical remarks giving guidance as to what goes to constitute the "efficient jacket," what fresh engineering practice is opened up by the use of the efficient-jacket, and under what conditions it may be most effectually applied.'*

Basically, the findings of the Committee were inconclusive. One wonders how Davey would have felt about this. Had he felt that he had wasted his time? Certainly the author of the *Nature* article was not impressed.

After the Brazilian engine, no further orders were taken for Triple Expansion Engines until 1895. Between 1895 and the end of the century only five orders were placed for Triple Expansion Engines of which it is believed, only four were manufactured. They were:

i) Order 5268, dated 8th July 1895 for Liverpool Corporation. Priced at £4,015

The Triple Expansion Engine (15"+23"+38", stroke 3') was named the William Radcliffe Pumping Engine, and was sited at Aubrey Street, Everton, Liverpool. The engine was actually designed by James Alfred Towler, the works manager at Hathorn, Davey, and in 1897 he gave a paper about the engine to the British Association of Waterworks Engineers, and this was summarised in the *Engineer.* [9]

It was a relatively small engine compared to later models, and it was installed to pump water into an adjacent covered reservoir and overhead tank to supply the highest part of the City. It pumped a variable quantity of water at the rate of 600 to 5000gpm to a height of 102.39 feet.

The engine had a number of special features that included; a complete system of jacketed cylinders, covers (that also enclosed the Corliss Valves) and receivers, and a method for catching and re-evaporating moisture, and clothed and encased radiating surfaces. Clearances were small, but the steam valves were large. The engine was regarded as, '*quick in action and independent in adjustment, so as to effect a minimum loss of steam pressure and a maximum amount of work got out of a given volume pressure of steam.*'

Mr Parry, the engineer for the works, conducted a trial on the engine's performance and efficiency, which were compared with other makes of engine.

In Alfred Towler's paper the principle conclusion reached was that buyers of Triple Expansion Engines could confidently purchase one in Britain rather than from foreign competitors that were mainly American companies.

Figure 81. Leeds City Waterworks, Headingley: Vertical Triple Expansion Engine
(Hathorn, Davey & Co. Pumping Engines for Waterworks. 1899 Catalogue.)

ii) Order 5301, dated February 1896 for Leeds City Council - two engines

The two Triple Expansion Engines (15½"+25"+40", stroke 3') were again designed by Alfred Towler and installed at the Headingley pumping station, Leeds. *The practical result of the installation has been that, in spite of the growth in population, the coal consumption at the Corporation's Headingley pumping station has been reduced to about one-half of its former amount.'*

Referred to as Engines A and B, they were both thoroughly tested and several efficiency trials took place conducted by Professor William Cawthorne Unwin, an eminent engineer of the period, and are described in detail in *The Engineer*.[10] It was a general conclusion that the trials showed that the engines would have performed considerably better if the boilers had been more efficient.

There were two boilers of the Lancashire type, 28' in length by 7' 6" in diameter, each having two flues 3' in diameter, taper to 2' 6" at the back end, with four cross-tubes in the flues. Each boiler was fitted with Bennis' sprinkling stoker. The grate area was 36 square feet and the heating surface was 850 square feet. A Green's Economiser was arranged in connection with the boilers, consisting of 128 tubes that had a heating surface of 1,280 square feet.

Nevertheless, the engines (Figure 81) and the pumps (Figure 82) were illustrated for a number of years in the Hathorn Davey Sales Catalogues *'Pumping Engines for Waterworks'* and are listed as, *'No. 5. Leeds Type. Triple Expansion Vertical Rotative Pumping Engines.'* and are described in the catalogue as follows:

'Vertical, Rotative, Triple Expansion, Corliss, Pumping Engine, working single-acting ram pumps directly attached to the engine piston-rods.

These engines are intended for use in cases where the water to be pumped is within suction distance of the pump room floor, the work to be done considerable, and fuel economy a primary object.

The working parts are necessarily more numerous than is the case with the other classes of engine referred to in this catalogue, and the first cost therefore higher. On the other hand, the motion of the engine is very uniform, frictional losses and therefore wear are reduced to a minimum, and the steam consumption is less than in the case of any other pumping engine.

The main bed-plates are fixed at the engine-house floor level, and are supported on foundations which do not come above the basement. The crank shaft and fly-wheels are carried on these bed-plates, and they also support the cast iron uprights and steel pillars upon which the entablature and steam cylinders are placed. The engine-house walls are not, therefore, subjected to any strain from the working of the engine.

The cranks are placed at angles of 120 degrees, and the cylinders being double-acting, there are six points of maximum impulse in each revolution. The turning motion is therefore practically constant.

Figure 82. Leeds City Waterworks, Headingley: Three Throw Ram Pumps
(Hathorn, Davey & Co. Pumping Engines for Waterworks. 1899 Catalogue)

The engine is steam jacketed, both on the cylinders and covers. The former are protected by non-conducting material and planished sheet steel lagging, and the latter by polished shell covers or by steel lagging.

Steam jacketed receivers, covered with non-conducting material and lagged with sheet steel, are provided between the cylinders.

The piston-rod glands are fitted with metallic packing.

The condenser is of the surface type, fitted with solid-drawn brass tubes.

The steam and exhaust valves are of the Corliss type, and the steam valves of the high-pressure cylinder are connected to a sensitive governor, so that the point of cut-off is instantaneously and automatically varied to suit any alteration either in the steam pressure or head against the pump.

The reduction of clearance spaces has had special attention, and they do not in any case exceed 2 per cent. of the piston displacement.

The pumps, which consist of three vertical single-acting plungers, are placed upon the basement floor level, each immediately below one of the steam cylinders, and the plungers are connected directly to the engine piston-rods. The fly-wheels and connections are therefore merely required to average the work done by the engine throughout the stroke, and the strains upon the crank shaft and connecting rods are in consequence extremely small.

The dead weight of the plungers, pistons, and rods is adjusted so as to be approximately equal to one half the water load, so that, although the plungers are single-acting, the resistance is practically the same on both the up and down strokes.

Very ample valve area is provided, and the valve chambers are furnished with doors enabling the valves to be replaced without disturbing any portion of the pump work or pipes.'

Each valve box was fitted with an air vessel, and this, coupled with the steady delivery of the triple rams had a more constant displacement than any other type of pump. They produced a uniform flow, and reduced shock in the water mains to a minimum.

The engines operated three-throw single-acting ram pumps. The contract conditions with Hathorn, Davey specified that the engine under full load would pump $1^{1}/_{3}$ million gallons in 12 hours. As part of the trials, the engines pumped water through several of the water mains, although the engine had to be throttled back for safety reasons as it was presumed that higher pressures might have damaged the water mains.

The Leeds Waterworks engines were widely used to advertise the Triple Expansion Engine that was sold under license in Canada by Peacock Brothers, Montreal.

iii) Order 5384, dated 3rd December 1896 for East London Waterworks (Waltham Abbey). Priced at £5,473

Alfred Towler is also believed to have designed this Triple Expansion Engine.

The engine was also used in an experiment to access the effectiveness of the steam jackets with regard to increasing steam efficiency, and is described in detail.[11]

'The diameters of the cylinders are 18-inch, 30.5-inch, and 51-inch, respectively. The stroke of each of the three cylinders is 36-inch. Each piston cross-head is connected to its crank by a single connecting rod, and to a pump-plunger by a pair of pump-rods. There are thus three steam cylinders, three pump-plungers, three connecting-rods, and three pairs of pump-rods. Each of the three cylinders is provided with an ordinary slide-valve, actuated by a separate eccentric on the crank-shaft. The slide-valve on the high pressure cylinder is actuated direct by its eccentric, and is provided with a Meyer expansion valve, by means of which the speed of the engine was regulated during the

experiment. The slide-valves on the intermediate and low-pressure cylinders are actuated through variable expansion links, but the positions of these remained unchanged throughout the experiment.

The bodies and both ends of all three cylinders are steam-jacketed, the cylinders forming liners in the body jackets. Steam is supplied to these jackets' through a pipe connected to the main steam-pipe on the boiler side of the stop-valve. The steam to the jackets enters at the top on one side, and at the bottom on the opposite side; and, in ordinary working, the high-pressure jackets discharge directly into the boiler, while the intermediate and low-pressure jackets discharge through steam-traps into the hot-well. The jackets of the high-pressure cylinder receive steam at full boiler pressure; and, by means of reducing valves, the pressures in the jackets of the intermediate and low-pressure cylinders are maintained a little higher than the pressures in the irrespective steam-chests.

Each cylinder is therefore jacketed with steam a little above its own initial pressure.'

The results of the experiment were summarised as:

Triple Expansion Engine	Without steam in Jackets	With steam in Jackets
Duration of experiment in hours	8	8
Boiler Pressure psi above atmosphere	130	130
Number of Expansions	22	30
Indicated Horse-power	140	138
Feed water per indicated horse-power per hour, lbs	17.22	15.45
Feed water, % less with jackets	.....	10.3
Coal per indicated horse-power per hour in lbs.	2.09	1.79

It shows that there are reductions in consumables if the cylinders are jacketed.

iv) Order 5628, dated 16th December 1898 for the Mount Morgan Gold Mine, Queensland, Australia. Priced at £3,500. The order was cancelled.

v) Order 5687, dated 3rd August 1899 for the South Staffordshire Waterworks Company. Priced at £9,240

The Triple Expansion Engine purchased by the *South Staffordshire Waterworks Company* for their Trent Valley Pumping Station, Litchfield was different to those previously manufactured by Hathorn, Davey. It was a Horizontal Tandem Triple Expansion Engine, and was the first of this type the Company had designed. The engine (20"+30"+44", stroke 5') shown in Figure 83 actuated a pair of 15½" diameter pumps placed at a depth of 300 feet in two boreholes, and one 15¾ force pump. In 1901 when the engine was started, the water level was only 67 feet below the surface.

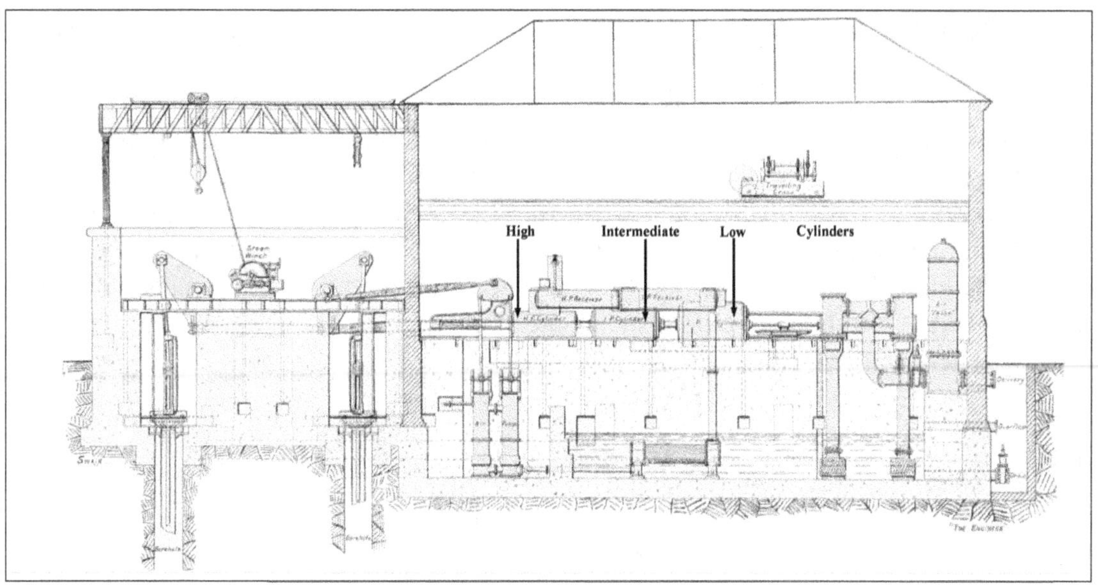

Figure 83. Trent Valley Pumping Station: Horizontal Triple Expansion Engine
(The Engineer, 26th September 1902, p.295 [GG])

The engine was fitted with Davey's Differential Gear that operated the steam valves of the engine, and it was driven by water taken from the rising-main, so that in case of a burst pipe, the gear is thrown out of action and the steam valves remain closed. As an extra precaution, a throttle valve was also placed on the steam pipe. This was usually fully open, but was arranged to close automatically, if the steam pressure varied. A pausing gear, allowed the speed to be regulated from two to fourteen strokes per minute to suit the water demand.

A few years later, the *South Staffordshire Waterworks Company* purchased a second Horizontal Triple Expansion Engine from Hathorn, Davey for the Trent Valley Pumping Station (Order 6058, dated 3rd December 1903).

Both engines were scrapped in 1959.

It seems to have been Davey's normal practice to file patents for the majority of innovations he designed. However, to date, no patent has yet been found for either his 1893 two-cylinder variation of the Triple Expansion Engine, or for a three-cylinder engine, if such patents existed. Certainly at this time, other firms were constructing similar engines, and would have filed their own patents, so perhaps all Davey could do would be to file minor variations to those designs, but sufficiently different enough to permit Hathorn, Davey to manufacture the engines without paying for a license or royalties.

Similarly, despite Alfred Towler designing Vertical Triple Expansion Engines, no patents have been found for them under his name, although he did later file patents for gearing.

Davey did however file one patent in 1893 for an *'Improved Triple Expansion Engine'* number 17,452, filed on the 16th September, and accepted on 21st October 1893.[12] The patent was for a method of coupling a Horizontal Triple Expansion Engine to a pump with a beam that would also actuate an air pump.

Chapter 17 References

[1] HL. 1940. *Hathorn, Davey and Company, Sun Foundry, Leeds.* p..5

[2] Davey's Triple Expansion Engine. *The Engineer.* 13th January 1893. p 26 and 31

[3] Davey, H. 1905. *The Principles, Construction, and Application of Pumping Machinery.* (2nd ed) p. 217

[4] Fletcher, W. 1895. *The Steam Jacket Practically Considered.* Whittaker and Co., London.

[5] Davey, H. 1889. First Report of the Research Committee on the Value of the Steam Jacket. *Proceedings of the Institution of Mechanical Engineers* (London). pp.703 to 745

[6] Davey, H. 1892. Second Report of the Research Committee on the Value of the Steam Jacket. *Proceedings of the Institution of Mechanical Engineers* (London). pp.418 to 513

[7] Davey, H. 1894. Third Report of the Research Committee on the Value of the Steam Jacket. *Proceedings of the Institution of Mechanical Engineers* (London). pp.535 to 597

[8] The Institution of Mechanical Engineers. *Nature.* 3rd November 1892. pp. 19 to 20.

[9] The William Radcliffe Pumping Engine, Aubrey Street, Liverpool. *The Engineer.* 15th October 1897. pp. 380-381.

[10] Triple Expansion Pumping Engines, Leeds. *The Engineer.* 9th February 1900. pp. 150 to 153.

[11] Fletcher, W. 1895. *The Steam Jacket Practically Considered.* Whittaker and Co. London. pp.184 to 185.

[12] Davey H., 1893, 'Improved Triple Expansion Steam Engine', Patent 17452.

18. Compound Engines in the 1890s and their remains

The 1890s was very much a decade of change for Hathorn, Davey. Several new designs for Compound Engines were introduced. Their international market expanded, and thoughts were turning once again towards becoming a Limited Company.

There were other concerns as well. There was, for example, considerably more competition in the market for the manufacture of heavy-duty steam engines, for example, Worthington-Simpson, Fraser and Chalmers, to mention two. In addition there was the gradually increasing threat from other power sources, particularly electricity, and later petrol and diesel engines. The age of steam was slowly, but stubbornly, drawing to an end. The decade also ended with a series of general engineering strikes that certainly sounded the death knell for some companies.

The Compound Engine was now in use worldwide for a variety of purposes. The horizontal tandem version was popular, perhaps because it may have been relatively easy to install. There are a few Hathorn, Davey Compound Engines and associated remains that have survived from this period.

Figure 84. Marine Colliery, Ebbw Vale, South Wales: Horizontal Compound Engine
(Courtesy of Alan Denney)

In Ebbw Vale on the South Wales Coalfield a Hathorn, Davey engine can be seen at the side of the main road (Figure 84). It is in a very poor state of preservation, and has been moved from its original location.

This example of a Horizontal Compound Differential Condensing Pumping Engine (36"+68") (Order 5102) came from the Marine Colliery, Cwm, Ebbw Vale. *The Ebbw Vale Steel, Iron and Coal Company Limited* placed the order at a cost £4,635 on the 9th September 1893. The engine was expected to pump 50,000 gallons of water per hour when working at seven strokes per minute.

When the Marine Colliery closed in 1989 and the site was being cleared, the engine was found in a covered sub-surface pit (undercroft) at the shaft top (Figure 85). A decision was made to rescue the engine and relocate it at the side of a new road junction. Although some of the engine parts are missing, it is possible to see the valve arrangements, as designed by Davey. Typical for the engines made at this time the order number is cast into the frame, a practice that presumably ensured parts did not get mixed up at the manufacturing and erection stage.

Figure 85. Marine Colliery: Hathorn, Davey Engine in downcast shaft undercroft
(National Archives, Kew. COAL 80/645)

Figure 86. Two engines had a combined water lifting potential of 1000hp
(The Engineer 21st December 1894, p.554 [GG])

A short illustrated article in *The Engineer* (1894) advertised that Hathorn, Davey had made three pairs of Compound Differential Pumping Engines with two of the engines having a combined power, equivalent to 1000 horsepower, in water lifted (See Figure 86).[1] The six engines (40"+76", stroke 10') were some of the largest yet built by Hathorn, Davey. According to the article, one pair of engines, complete with pumps, weighed 1000 tons. The pump rods were 20" square pitch pine. The six engines are listed as Orders 4942, 4994/4997 and 5363 for the following engines:

Orders 4994/4997, dated 1st June 1892 were for two Horizontal Condensing Differential Pumping Engines for the *Powell Duffryn Steam Coal Company*, South Wales. The orders also included the full pumping systems with connecting rods, quadrants, and pumps, etc.

The remaining two orders were for the Mitsui Company, Japan, for their collieries on the Miike Coalfield, Kyūshū. They had previously ordered an underground haulage engine (Order 4503, dated 31st December 1887).

Towards the end of the 19th century, the Japanese Government controlled mining operations on the Miike Coalfield, but in 1887 they sold their mining interests to the Mitsui Company. At one of the coalmines an influx of water had interrupted shaft-sinking operations, and so a young Japanese mining engineer, Takuma Dan, was sent to Europe and the USA to study pumping technology.[2] It is probable that during his visit to Europe he met, and befriended, Henry Davey. Eventually, Davey would act as a consultant for the Mitsui Company.

Orders 4942 and 4942A to T, dated 18th September 1891 were placed by the Mitsui Company. The cost for two Horizontal Compound Differential Condensing Pumping Engines (40"+76", stroke 10') including quadrants, spear rods, rising main, pumps and lowering apparatus, totalled £16,450. The two engines were to be installed at the Kachidachi (Katsudachi) Pit where an earthquake in 1889 had damaged the shaft lining and surrounding strata, and had resulted in the influx of water.[3]

A note at the end of the order stated, *'To be suitably packed for a long voyage.'* However, this was apparently insufficient, as Order 5011, dated 24th August 1892 was for, *'one Low Pressure Cylinder 76 inches diameter to replace one broken in discharging from ship.'*

Colyer (1892) describes the engines:[4] *'These are a pair of compound pumping engines made for a colliery in Japan, they were designed by Mr. Henry Davey, M.Inst.CE. The engines are made in the horizontal form and on the compound system, and are placed side by side. The high-pressure cylinders are 40 inches diameter, and the low-pressure cylinders 76 inches diameter, each having a stroke of 10 feet. Each of the engines is about 500 indicated horse-power, and they are together capable of pumping 360,000 gallons of water per hour to a height of 420 feet, at a speed of 7½ strokes per minute. Each cylinder is provided with four double-beat Cornish valves as shown, two for steam and two for exhaust. The steam valves of both cylinders also have a separate cut-off motion. The high- and low-pressure pistons are connected to the cross-head by means of two piston rods; the pistons being made very deep to provide a large area of supporting surface. The surface condensers are placed in a sump by the side of the engine house. The air pumps are worked off arms on the quadrant shaft.*
    There are two 24-inch plunger pumps and two 24½-inch bucket pumps to each engine, each having a stroke of 10 feet. The bucket pumps raise the water into a sump at a height of 120 feet from the bottom of the shaft, and in which the surface condensers are placed, and from this point the two plunger pumps force to a further height of 300 feet to the surface, through a wrought-iron rising main 26 inches diameter, the total lift therefore being 420 feet. The pump valves are of the double-beat type. Motion is given to the pumps by means of connecting rods and two powerful quadrants.'

Order 5363 and 5363A to D, dated 14th September 1896 was also for two Compound Surface Condensing Differential Pumping Engines, for the Miyanohara Pit. The cost was £6,090 (delivered to Hull) or £6,160 (delivered to London). The specifications for the two engines are detailed in Order Book 5:

Figure 87. Miyanohara Pit, Miike Coalfield, Japan. The surviving basement wall of the engine house constructed for the Hathorn Davey Horizontal Compound Engines.
(Author 2017)

'5363: Two Compound Surface Condensing Differential Pumping Engines, as per specification book (see copy specification book No. 18. Page 780). Each engine to be suitable for working a pair of 24" ram pumps, against a head of 500 feet with a steam pressure of 85 to 90 pounds. Each engine to have cylinders 40" and 76" diameter and 10' stroke of hard metal - flanges to be bracketed - valves of the Cornish type of gunmetal with phosphur-bronze spindles and nuts. Brass glands and neck rings. Cast iron guides carried upon mild steel pillars. Valves to be actuated in accordance with Davey's patents. Working parts of steel finished bright. Pausing gear and main pausing gear cataracts lined with brass and to have brass covered pistons, gunmetal adjusting plugs and hand wheels. Cylinder covers having stuffing boxes and glands bushed with gunmetal. Pistons to have Lancaster and Tonges patent metallic packing. ????? rings to be secured by bolts of best iron, having safety appliance to prevent them from working loose - piston rods of steel having deep nuts and copper pins. Crossheads of steel with slide blocks having large wearing surfaces. Bedplates of deep girder pattern

with slides cast on them to receive slide blocks - to have top slide bars complete in every respect including oil cups and holding down bolts.

5368A: Condenser of the surface type; vacuum gauge.

5368B: Air pump worked by an arm off the quadrant shaft and to have India rubber foot and delivery valves on removable grids.

5368C: Pipe connections between cylinders condenser and all necessary pipes within the limits of the stop valves of engine and the condenser, but no pipes beyond.

5368D: Spanners, counters.'

The Specification Book referred to has not survived. Later, specific items of machinery could be ordered directly by telegram, using various code words, from the catalogues.

The author visited the Miyanohara Pit in 2017. The head frame, and the house for the winding engine dominate the site. On the opposite side of the shaft to the winding engine (Figure 87) there is a tall brick wall adjacent to the shaft that formed the basement wall of the pumping engine house for the Hathorn Davey engines. The arched openings in the wall once provided access, as well as an opening for the con-rods that ran from the quadrants to actuate the air pumps.

The oldest surviving example of a Hathorn Davey engine in Australia can be found at Charters Towers, Queensland. Gold was discovered in the area in 1871, and the town grew quickly, and it became by far the largest gold mining town in the State. Water for mining purposes was taken from the local creeks, but the demand for a more reliable water supply increased with the expanding town. In the late 1880s, a scheme was proposed to pump water from the Burdekin River to a reservoir on Tower Hill, the highest point above the town. In 1888 the Burdekin Water Scheme Joint Board placed an order with Hathorn Davey.

Order 4552 (4th January 1888) was for one Compound Differential Condensing Pumping Engine (22"+38", stroke 5') to pump 20,000gph to a height of 570 feet through over 8 miles of 12" cast iron pipe. The engine was to look similar in appearance to another engine previously bought for Townsville (Order 4127) on the coast to the east, with eight ornate columns supporting the entablature.

The rest of the order (4552 A to Q) included, differential valve gear, surface condenser, air pumps and two plunger pumps (9½", stroke 5'), pump rods, two bucket pumps (10", stroke 5'), an air vessel with 12 times the capacity of one of the pumps, rising main and valves, filling tank, one Davey's Patent Engine Recorder, and a Richards Steam Engine Indicator. The order also included two Lancashire boilers (26' x 6'6") with two flues with mountings complete, and a donkey pump for filling the boilers.

Figure 88. Charters Towers: The two adjacent vertical compound engines sit on the entablature. The large beams that form part of Davey's parallel motion can be seen under the left side of the engines.
(Author 2014)

Pumping was a two-stage operation. The water was extracted from an inlet tunnel that had been driven 172' to the pump well from the Burdekin River, and in the second stage it was then pumped to the Tower Hill reservoir located above the town.

A major problem was discovered in April 1889 when it was noted that coal-burning Lancashire boilers had been delivered instead of the specified wood-burning boilers. Hathorn Davey agreed to pay for a suitable alternative. The engine was erected and commissioned in 1889-90 by an engineer employed by Hathorn Davey. The total cost was £3,460.

In the late 1890s, the goldfield was approaching its most productive phase with the consequence that in 1895-96, there was an even greater demand for water, and so a decision was made to purchase a second Hathorn Davey engine, again with a capacity to pump 20,000gph.

Order 5225 and 5225 A to N, dated 18th March 1895 was for one Compound Vertical Direct-Acting Differential Condensing Pumping Engine (22"+38", stroke 5') and pumps, etc, and had a similar specification as Order 4552, apart from the boilers. The order stipulated that, '*All parts when fitted must be distinctly marked, so as to facilitate the erection of the machinery*,' so it seems likely that perhaps the Water Board engineer felt competent enough to install the engine on this occasion.

The two engines worked until 1942 when they were replaced by electric pumps. Both engines still survive *in situ* (See Figure 88), and were added to the Queensland Heritage Register in 1995.[5] When the author visited the site in 2014, the engines were exposed to the elements, but seemed relatively complete. Davey's parallel motion could be seen beneath the engine, whilst both sets of Davey's Differential Gear were also obvious features (Figure 89). The potential danger of snakes in the pits under the engines did not permit closer inspection.

Figure 89. Charters Towers: One of the two sets of Davey's Differential Gears
(Author 2014)

Figure 90. Cheddars Lane: The oscillating disc that operated the pumps.
(Author 2019)

Two horizontal compound differential engines are still *in situ* at the Cambridge Museum of Technology, at Cheddars Lane, Cambridge (Figure 89). The Cheddars Lane Sewage Pumping Engines can still work under steam and are undoubtedly the best-preserved Hathorn Davey engines that have survived from this period.

The two Cheddars Lane engines were ordered (Orders 5148, 5148 A to M) during May 1894 and are described as Davey's Horizontal Compound Pumping Engines (22"+44", stroke 4'), and worked at a steam pressure of 80psi.

Each engine worked two sewage pumps with cylinders 34" diameter and 4' stroke, lifting some 250,000 gallons of sewage per hour 50 feet from a culvert to a delivery main, which carried the sewage a distance of 2¼ miles to a sewage farm.[6]

Figure 91 shows the pumping arrangements, together with an inset showing the system employed for connecting the pumps to the piston rod.[7] The pumps A and B are actuated by means of a rocking or oscillating disc (C). The oscillation is produced by a connection via a con rod and crosshead to the piston rod (D). The large oscillating disc is shown in Figure 90.

Davey designed a number of engines that pumped in this manner, and described the oscillating disc, as in effect a triangular lever, with the con-rods for the two pumps, to two corners. The con-rod from the piston crosshead connected to the third corner. (See Figure 90 insert) It was a method that reduced the pump's resistance as the stroke proceeded. Davey described this as a compensating principle, and was a method for turning a horizontal rocking motion to a vertical motion and was employed mainly for pumping from wells and boreholes.

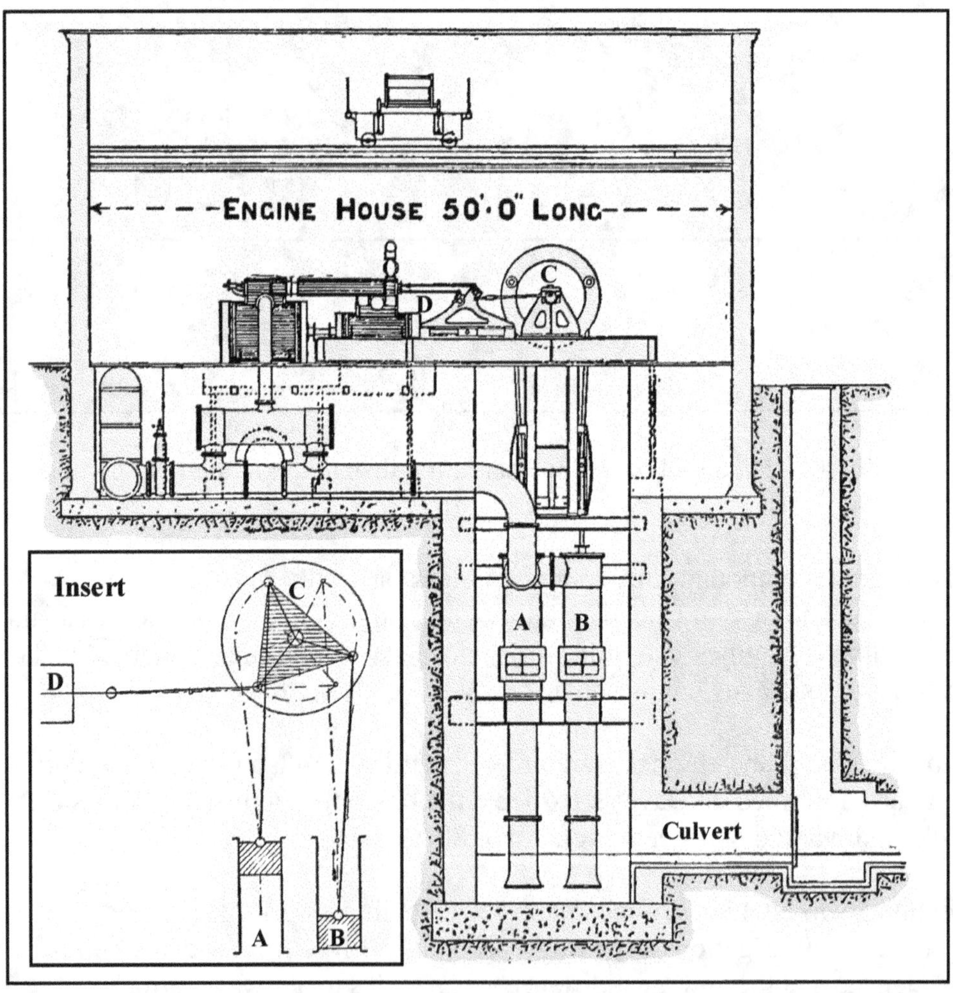

Figure 91. Cheddars Lane: Diagram showing engine and pump arrangements.
(Davey, 1905. p.269)

The cost of the two engines and associated equipment was £5,450. The two engines worked continuously until May 1966. The museum occupies a dominant site at the side of the River Cam, on the east side of Cambridge.

There is one more excellent example of a working Horizontal Compound Engine at the Mill Meece Pumping Station, near Eccleshall, Staffordshire. It was not built until the 1920s and will be featured in a later chapter.

In the 1890s Davey still flirted with other adaptations of the Compound Engine. His ideas for producing an Inverted Vertical Compound Engine, with the pumping beam located beneath the engine, resulted in some of the largest engines Hathorn, Davey ever constructed.

Chapter 18 References

[1] Davey Pumping Engine. *The Engineer.* 21st December 1894. pp.550 and 554

[2] Obituaries: Takuma Dan. *Transactions of the Institution of Mechanical Engineers.* 1932

[3] Morris, J. 1895. *Advance Japan: A nation thoroughly in earnest.* W.H. Allen and Company Ltd, London, p.314

[4] Colyer, F. 1892. *Pumps and Pumping Machinery* (Part 1) pp. 166 -167, drawing 42

[5] https://apps.des.qld.gov.au/heritage-register/detail/?id=601081

[6] Unwin, D. 2007. *The Tall Chimney.* Cambridge Museum of Technology

[7] Davey, H. 1905. *The Principles, Construction, and Application of Pumping Machinery.* (2nd ed) p.269

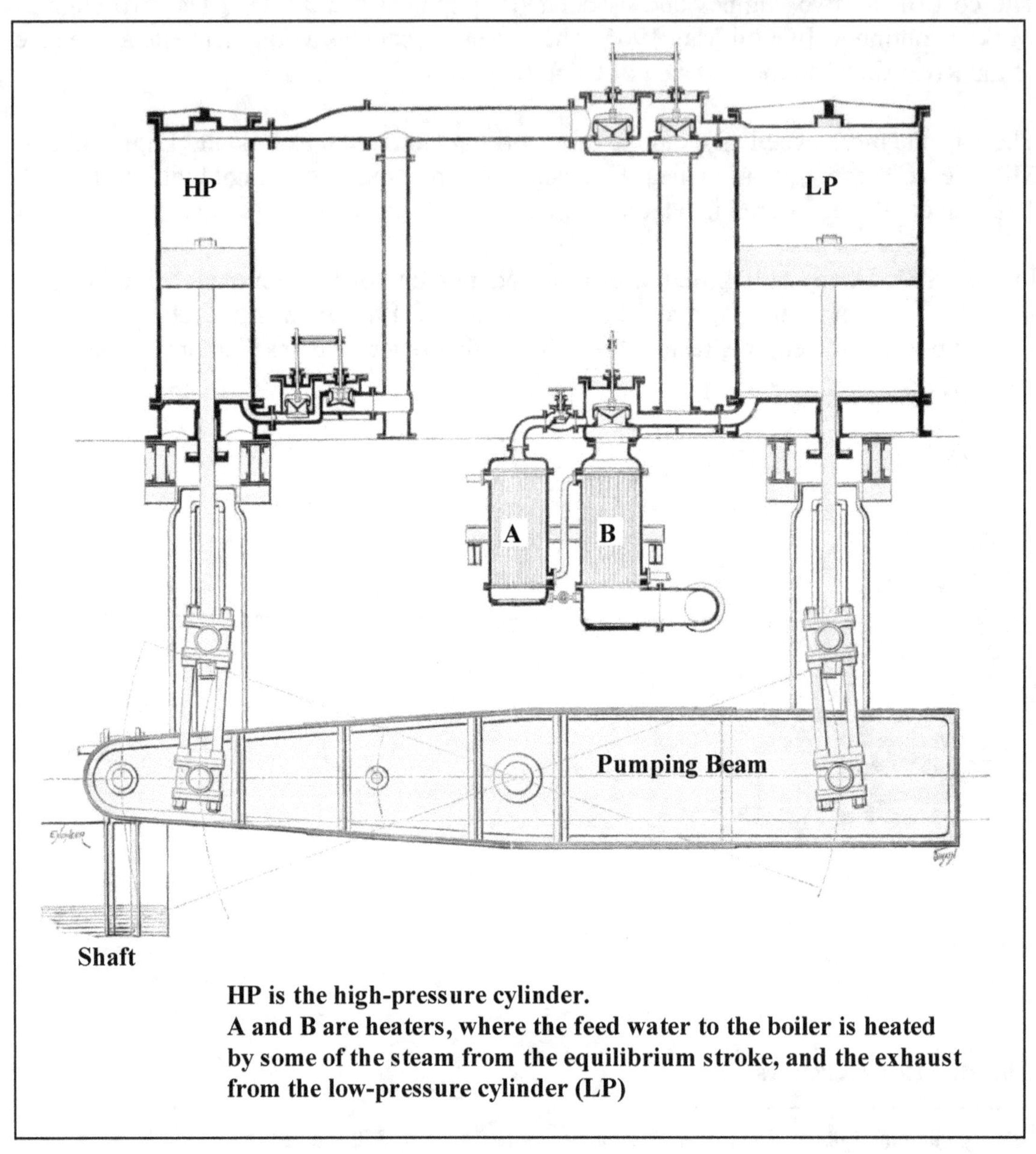

Figure 92. Davey's Vertical Inverted Compound Cornish Pumping Engine
(Modified from: The Engineer, 24th January 1896. p.85. [GG])

19. The Inverted Vertical Compound Engine

Davey's Inverted Vertical Compound Differential Pumping Engine was one of the largest engines manufactured by Hathorn Davey, and first made it's appearance in the 1890s.

On the 31st August 1895, Davey had applied for a patent for an *'Improved Steam Engine.'* His application was accepted on the 16th November 1895 as patent 16320, and was for an Inverted Compound Pumping Engine. Figure 92 is a simplified version of the drawing submitted with the patent application.

The vertical form of the compound engine consisted of inverted high and low-pressure cylinders with the pumping beam in a pit beneath the cylinders. The end of the beam protruded over the shaft and worked the pump rods. Unlike the horizontal form, it had to be placed on the immediate shaft top to function correctly.

Davey regarded it to be the only engine that Hathorn, Davey produced that worked on the true 'Cornish Cycle'. The cycle was more normally associated with the operation of a single cylinder Cornish Beam Engine. The cycle occurs in several stages where the piston is moved by both steam pressure and vacuum, but part of that action is produced by the weight of the pumping equipment in the shaft. The inverted vertical form of the compound engine was slower than the horizontal type, but was a very robust machine that could operate larger pumps, that could force water from greater depths.

An important new feature of the engine was the feed-water heater. Davey indicated that if exhaust steam had been used solely to raise the temperature of the feed, there would not have been much gain because the exhaust steam would only have had a temperature of 110°F or less. However, Davey designed the engine to also use steam acquired at the equilibrium phase of the Cornish cycle. This could raise the temperature to over 210°F or more.

In the patent (16320) Davey also illustrated how a Triple Inverted Compound Engine might work, although Hathorn, Davey never manufactured such an engine. A drawing shows the addition of an inverted intermediate cylinder, with the piston rod attached to a point on the beam between the high-pressure cylinder and the beam fulcrum.

The first engine to be built to this specification was ordered by the *Waihi Gold Mining Company, Ltd.*, New Zealand (Order 5284, 5284A to N) on the 23rd September 1895. Described as a Davey Patent Vertical Compound Differential Cornish Condensing Pumping Engine (45"+90", stroke 8') and cost, together with 19" plunger pump and sinking pump, etc, £6,100.

Sections and Plan of the Waihi engine, installed in the engine house, are shown in Figure 93. In addition to the feed water heater, located between the two cylinders on the front elevation, the side elevation and plan show that there was a significantly large cooling tank at the side of the engine house, to accommodate the air pump and the condenser. The up-down movement of a lever attached to the beam actuated the air pump.

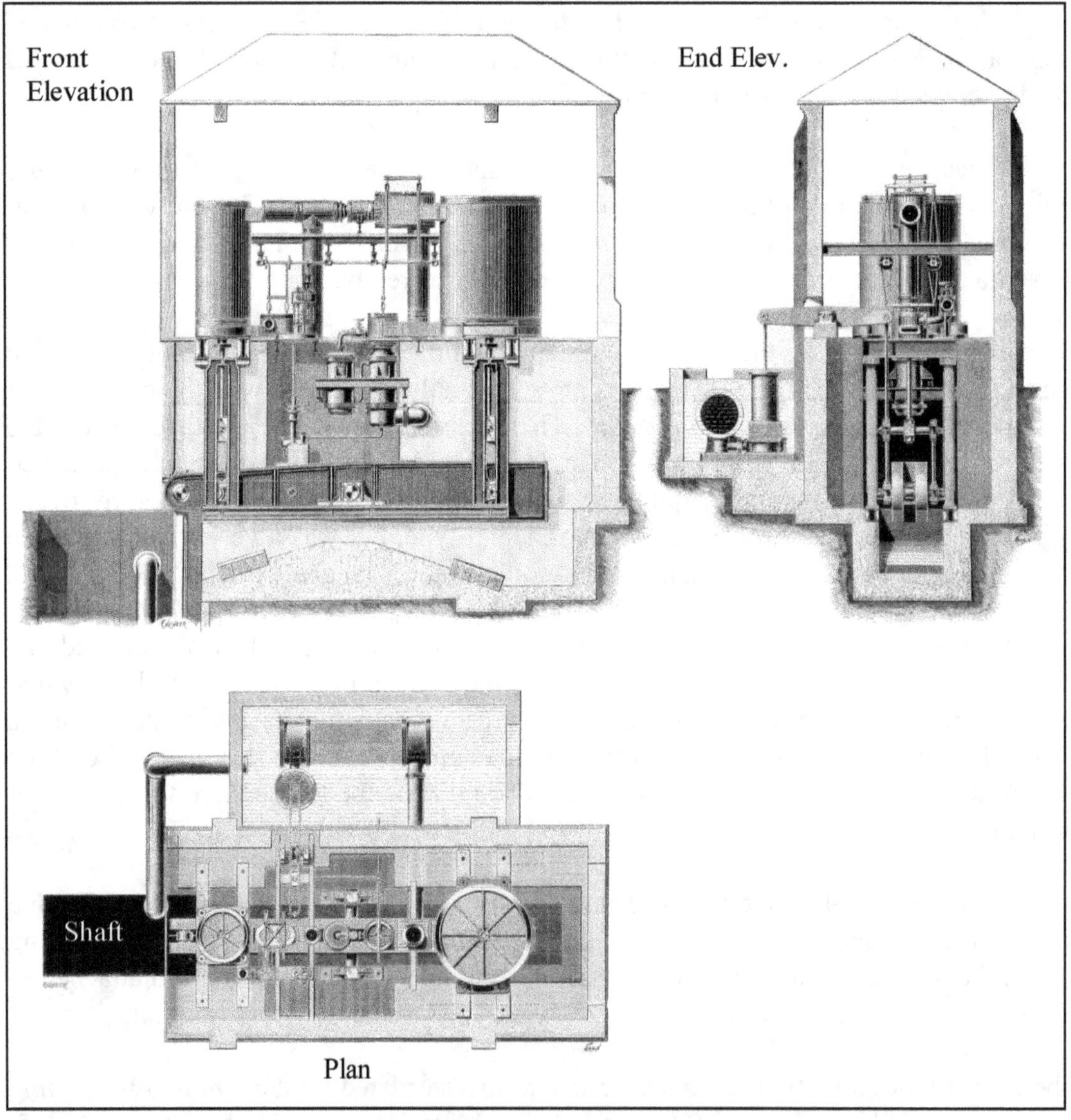

Figure 93. Waihi Gold Mine, New Zealand: Elevations and Plan of the Inverted Vertical Cornish Compound, referred to at the mine as 'A' Pump.
(The Engineer, 24th January 1896. p.84 and 88 [GG])

The engine is briefly described in *The Engineer* (1896).[1] *'The engine is for working the Cornish system of pitwork, viz., that of a single spear rod, and is particularly suitable for draining metalliferous mines, where the working of the mine is generally contemporaneous with the sinking of the shaft, and therefore the Cornish system or stage pumping is the most suitable. Sinking is performed with a bucket pump to a depth of 200 feet, when a plunger pump is put in, and the sinking is continued with the bucket pump.'* It was expected that the shaft would be sunk to a depth of 1000 feet.

On the 15th January 1898, Davey had another patent (29632) accepted for *'An Improvement in Compound Pumping Engine,'* for connecting both pistons to the same point on the pumping beam. This method was not used in any of his later designs.

The second Inverted Vertical Compound Cornish Pumping Engine can only be described as a hybrid machine. The engine is not listed in the Order Books except for the Differential Gear (Order 5372, dated 19th September 1896, costing £600), because parts for the engine had been salvaged from a single cylinder Cornish pumping engine.

On the 9th December 1895, a fire at Marriott's shaft, South Frances Mine, near Redruth, Cornwall had destroyed the engine house. However, the 80" cylinder of the Cornish Beam Engine survived. The following year, the mine was amalgamated with the adjacent Wheal Basset, and a limited liability company was formed: *Basset Mines, Limited.* The three existing Cornish pumping engines working on both mines were just about coping with water quantities entering the workings, so a solution to the problem was sought. Henry Davey was appointed Consultant Engineer to the new Company, and it was on his advice that an Inverted Vertical Compound Engine was installed on Marriott's Shaft. The engine was started on the 4th May 1899.

Completely surrounded by Cornish Beam Engines for many miles around, there is no doubt that the engine was a unique addition to the Cornish landscape. Barton (1969) described the engine and its history in some detail.[2] The engine utilised the 80" cylinder from the fire-damaged engine as the low-pressure cylinder. The high-pressure cylinder was 40" diameter. It is not known where this cylinder came from, as none are assigned to Basset mines in the Order Books, and more likely as not, it was acquired from a scrapped engine. The Cornish mining industry at the time was in a state of depression and Barton notes this point by remarking that the engine was christened with a glass of champagne rather than the customary bottle![3]

The high and low-pressure cylinders had 9' and 10' strokes, respectively, whilst the plunger pumps in the shaft had a 13' stroke. A report in *The Engineer* commented that, *'On the day of the opening it was an object of admiration to see the pump rods, which with their attachments weigh about 160 tons, swinging up and down with a 13' stroke, silently without jerk or vibration, throwing 1000 gallons of water per minute from a depth of 1000 feet.'* Certainly as Barton notes, with the high-pressure cylinder placed on the pump side of the beam's centre, this method of pumping puts less strain on both the beam and the centre point, the latter being a common point of beam breakage in Cornish engines. The engine's beam had been constructed from mild steel plates and angles, and boxed at the low-pressure cylinder end to form a balance box.

The engine cost £7,500, and worked 19" pumps. There were several other non-Cornish methods applied to the operation. The engine worked exceptionally long lifts of 70 fathoms (420 feet), and the pump rods were constructed of tubular iron with a cast iron base instead of timber. Four Lancashire boilers provided high-pressure steam. Davey apparently described the engine as *'the first true successful Cornish engine worked with high-pressure steam for a Cornish mine.'* A fact inscribed on the silver salver presented to Mrs Basset of Tehidy, Lady of the Manor, who had named the engine.

The beam was replaced in 1909. The new beam was composed of mild steel and manufactured by Findlay and Company, Motherwell, Scotland. It weighed 55 tons. The engine house, that still survives (Figure 94) was apparently built large enough to accommodate a second engine. The Basset mine ceased working in 1918 and the engine was scrapped after working for twenty years.

Figure 94. Marriott's Shaft engine house (left) & Cerro Muriano engine house (right).
(Author 1970s)

The third Inverted Vertical Compound Cornish Pumping Engine ordered from Hathorn, Davey was erected at the Cerro Muriano Copper Mine, Córdoba Province, Andalucía, Spain. The *Cordova Exploration Company* leased the mine in 1898, and ordered an Underground Horizontal Compound Differential Condensing Pumping Engine from Hathorn, Davey (Order 5823, dated 16th September 1901 and costing £724). The engine was capable of raising 400gpm from a depth of 600 feet. This was the depth that the Romans had reached nearly 2000 years earlier, which was a remarkable achievement, and is probably one of the deepest Roman mines in the world. However, even with the addition of a geared pumping engine with an 18" bucket pump, the water table could not be lowered below the depth of the Roman workings. Consequently, *John Taylor and Sons* were asked to examine the property, and a new company, *Cerro Muriano Mines, Limited*, was formed to work the mine, and a new larger pumping engine was ordered.[4]

The engine (54"+94", stroke 8'), one of the biggest yet built by Hathorn, Davey (Orders 6028 and 6028 A to V, dated 29th June 1903 at a cost of £5,525) was erected on the San Rafael Engine Shaft and was started during August 1904. It worked one set of 18" plunger pumps placed at a depth of 557 feet, with sinking bucket pumps fixed below. The shaft was then sunk to 695 feet, where good ore was encountered. The engine was capable of working the 18" plunger pumps down to a depth of 1,580 feet.

Other orders for the engine are also listed. Order 6032, dated 7th August 1903 for example, was for 550 feet of 18" Rising Main and 112 feet of 19½" steel pump trees (probably pump rods?).

In a report produced by the Córdoba mines inspectors dated 1st September 1904, it was remarked that two boilers provided steam (at 150psi). The engine was capable of pumping 6,000 cubic metres of water in a 24-hour period (1,319,815 gallons of water per day, or about 920 gallons per minute).[5]

Obviously, the Cerro Muriano engine was so unique it attracted much attention. Reginald Bonham-Carter, the manager of the Constancia Foundry, Linares visited the mine during December 1905.[6] He describes his visit in a letter to his mother. It is, *'a copper mine twelve miles north of Córdoba', that was, 'originally in the hands of Carrs of Córdoba, but now run by a group of well-known mining men and directed by the Taylor's firm.'*

'The mine is in its infancy but the Managers are hopeful of success, and are equipping the mine with first-rate English machinery. The country there is not yet spoilt and is covered in pine trees.'

'I only stayed there for twelve hours, arriving at four in the morning, and spending the first four hours in the pumping engine room studying an engine I had never seen before.'

The engine was scrapped sometime after 1919 when the mine was sold to a Spanish mining company. The engine house still survives (See Figure 94). The Author first visited the site in 1978, when it was not wise to linger, as it was adjacent to a military installation. The site was again visited in 2006. The engine house is now surrounded by a private garden!

Figure 95. Waihi: The 110 inch Low-pressure Cylinder, piston rings and crossheads.
(Mining Journal, Railway and Commercial Gazette. 4th June 1904. p.631)

In 1902 the *Waihi Gold Mining Company, Limited* of New Zealand ordered a second Inverted Vertical Compound Cornish Pumping Engine from Hathorn, Davey, and referred to it as 'C' Pump. It was larger that their first engine Pump 'A.'

'In order to facilitate delivery it was necessary that the whole plant should be constructed and got ready for shipment by the same steamer, and with this end in view, the work was divided among several firms.' The engine had 60" (6' stroke) and 110" (12' stroke) cylinders. The low-pressure cylinder weighed 21 tons. Figure 95 shows the low-pressure cylinder and piston ring. The first set of 23" pump plungers was placed at a depth of 700 feet from the surface. The engine was designed to raise 1,500gpm from a total depth of 1,550 feet. The work was divided among the following firms:

Messrs. Hathorn, Davey and Company was responsible for the pump plunger and plunger case. Messrs. Harvey and Company, of Hayle, supplied the high-pressure cylinder; the low-pressure cylinder was constructed by Messrs. McOnie, Harvey and Company. This firm also supplied the pipes. The large beam for the pumps, which weighed 50 tons, was made by Messrs. Findlay of Motherwell. The pistons were packed with the patent metallic packing made by Messrs. Lancaster and Tonge, of Manchester who also made the stuffing boxes for the piston rods of both cylinders, and also the piston rings. The whole of the work for installing the pumps was entrusted to Mr. John Henderson, the mine engineer of the Waihi Company. It is interesting to note that the photograph in Figure 95 is attributed to *Thames Engineering Limited*, who may well have been given the responsibility of collecting and checking the various components prior to shipping to New Zealand from the Port of London. [7]

Figure 96. Waihi; The Transportation of the Low Pressure Cylinder
(Courtesy of Peter Short, New Zealand)

An article in *Engineering* (18th March 1904) provides further information about the various components of the order.[8] The engine was to work at 7 strokes per minute. The beam, constructed from mild steel plate and angles, with massive cast iron centres for the shaft and pins, was 48 feet long overall, and 8 feet deep at the centre. The centre shaft was 22 inches diameter. The plummer blocks and sole plates that support the beam weighed 14 tons.

The condenser and air pump were placed in a sump at the side of the engine house, in a similar manner to that shown in Figure 93.

The low-pressure cylinder was 15' 4" long over the flanges and 2¼" thick. Together with the bedplates and cover it weighed 41 tons. Obviously a cylinder of that size and weight presented some transportation difficulties. Figure 96 shows how the low-pressure cylinder was transported to the mine site. It was presumably towed by a traction engine.

The *Mining Journal's* New Zealand correspondent described the starting of the engine on 28 April 1904.[9] *'The ponderous machinery, which is very much simplified compared with the smaller, more antiquated type (Presumably a reference to the first Hathorn Davey engine at Waihi), moved off into duty just as simply as the starting of an eight-day clock. Two or three strokes per minute of the large Findlay balance bob with the long plunge of the rods – viz., 700 ft., with a direct lift for the whole of the distance, very shortly found out by a few pump columns, which opened out and burst, necessitating a short stoppage to replace them.'*

    The probable cause of the defects may have been due to too rough handling in shipping and trans-shipping, and it is just as well to experience the temporary drawback at the first, as later on they may have proved inconvenient when the deepening of the shaft is resumed'.

There is very little detail in the Order Books for this second Waihi engine, with reference only being made to the ordering of the occasional parts. The engine was ordered at a time when Hathorn, Davey was still recovering from the effects of a major strike in the engineering sector that occurred at the end of the 19th century. In addition, the Company was in the process of becoming a private Limited Liability Company. Their Orders Books were also in a healthy state, so possibly to take on an order of this size at the Sun Foundry would have caused operational difficulties. Casting a low-pressure cylinder of that size may also have presented some problems. It would have been the largest that they would have ever made. But it seems reasonable to accept that the ability to fulfil the Waihi contract in the specified time was their major consideration in dividing the work up between so many companies.

In 1901, Henry Davey patented yet another variation to the inverted vertical Compound engine. Patent 23,905 *'Improved Pumping Engine'* was accepted on 16th January 1902. It related to an improved form of Compound Engine and method to actuate the steam valves. It was an Inverted Vertical Compound Engine but Davey had brought both cylinders closer together, and the piston rods were attached to the beam between the beam end and the fulcrum. Hathorn, Davey started to use this idea to illustrate their advertisements (Figure 97), and it was not too long before one organisation placed an order.[10]

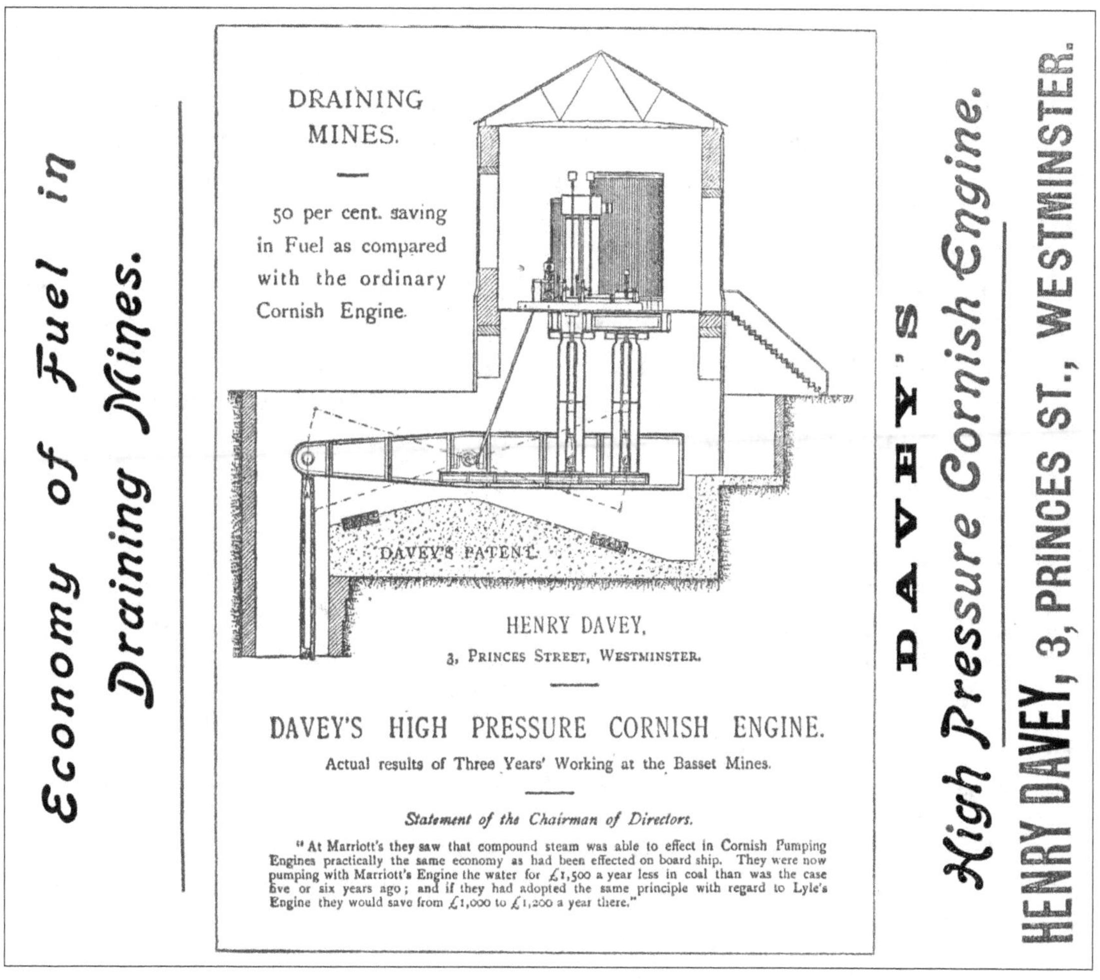

Figure 97. An Advertisement for Hathorn, Davey illustrated with
Davey's High Pressure Cornish Engine
(Mining Journal, Railway and Commercial Gazette. 25th October 1902. p.1437)

In October 1904, the *South Staffordshire Mines Drainage Commission* ordered three Inverted Vertical Compound Cornish Pumping Engines, built to patent 23905. They were the last Inverted Vertical Compound Engines to be manufactured by Hathorn, Davey. They were for:

Deepfield Pumping Shaft.   Order 6133 (25"+55")
Stowheath Pumping Station.   Order 6134 (25"+55")
Herberts Park Colliery. Darlaston.   Order 6135 (22"+47")

All three engines had ceased operation by 1948 and were scrapped.

Davey used the engine shown in Figure 96, for an illustration in the 1905 edition of his book, which is captioned, '*Compound Cornish Engine, South Staffordshire Mines Drainage.*' [11] One of the advantages of such a design was that the engine house was slightly offset from the shaft top. The pumping beam may even have been located in a short undercroft.

In 1902 Davey exaggerated his patent. It showed the two compound cylinders being replaced by two inverted vertical compound engines. Patent 21,708 was accepted on the 8th January 1903. Simply titled '*Pumping Engines,*' Davey states, '*My invention relates to an improved form of pumping engine of the quadruple expansion type...*' None, as far as the Author is aware, were ever constructed.

Chapter 19 References

[1] Compound Cornish Cycle Pumping Engine. *The Engineer.* 24th January 1896. pp. 84, 85 and 88

[2] Barton, D.B. 1969. *The Cornish Beam Engine.* D. Bradford Barton Limited. pp. 247 to 251

[3] Cornish Engine, Basset Mine. *The Engineer.* 12th May 1899. p.471

[4] Cerro Muriano Mines, Limited. *The Mining Journal and Railway Gazette.* 4th March 1905. pp.224 to 225.

[5] Cerro Muriano, *Libro de policía minera número 3.* Jefatura de minas de Córdoba. Córdoba Mining Police Archives, p.204.

[6] Vernon, R. and Vernon, M. 2016. *Don Regino.* Iberian Publications. p.214

[7] New Plant for the Waihi Gold Mining Company. *Mining Journal, Railway and Commercial Gazette.* 4th June 1904. p.631

[8] Waihi's massive pumping engine. *Engineering.* 18th March 1904.

[9] Mining in New Zealand: Waihi Gold Mining Company. *Mining Journal, Railway and Commercial Gazette.* 4th June 1904. p.631

[10] Advertisement for Hathorn, Davey. *Mining Journal, Railway and Commercial Gazette.* 25th October 1902. p.1437

[11] Davey, H. 1905. *The Principles, Construction, and Application of Pumping Machinery.* (2nd ed) p.140

20. The Sun Foundry: 1872 to 1900

One of the advantages of operating an engineering company was that you could always manufacture some of your own machine tools, and other heavy equipment. Hathorn, Davey was no exception. Whilst the size of the foundry didn't significantly expand in the first 25 years of the Hathorn, Davey Company, the range of equipment did. This was inevitable considering that their products were increasing both in range, and size.

It is not recorded how the business expanded to meet this demand, but it is possible to glean from the Order Books what machinery the foundry needed to conduct operations.

No items assigned to Hathorn, Davey appear in the first few Order Books. The Sun Foundry didn't appear to go through any significant changes, certainly up to Davey's departure from Leeds, and Hathorn's death. However, the appointment of Hugh Lupton as a partner may have been the beginning of significant change. Having started with the Company as an apprentice, Lupton probably knew the Foundry intimately. Change seems to have started slowly. Machinery was replaced, and some alterations were made to the shops. By 1900 additional shops had been built.

The first listing appears in 1880, and it is presumed that the Foundry was worked very much in the same fashion from when the Partnership took it over. Some of the first items listed under Hathorn, Davey were for engines. They were:

October 1880: One 16" cylinder (Order 3153); One Vertical Engine (3162); One Hydraulic Press (3306) / 1881: One Steam Simplex motor (3377) / May 1882: One Centrifugal Pump (3541); two Rotative Engines (3551); Clock counter and well-gauge frame (3554) / June 1883: one pair of Rotative Engines (3783) and in late 1883 a work table and apparatus (4033).

Lupton described the conditions in the Sun Foundry probably for the time since he had became an apprentice in the mid-1880s.[1] *'In those day there was, of course, no electricity available. The shops were driven by steam and lit with gas and all lifting was done by manual labour, with the exception of the Foundry, where there was a pair of steam-operated jib cranes. In the yard lifting was done with jacks and large castings were moved on rollers. In the shops there were hand-operated travelling cranes, which were known as "Goliaths" and which took about 15 minutes to travel from end to end of the erecting shop.'*

'Nor was there any system of shop warming, other than the heat from the engine exhaust, which was piped under the erecting shop for that purpose and in very cold weather open coke braziers were employed.'

'Coal was very cheap, the slack burnt under the shop boilers costing, if the writer's memory is correct, only 4/- [shillings] per ton delivered in the works yard. When in later years a heating system was installed the cost of the coal for this, but also including the comparatively small quantity used by the steam hammers, was very considerably more than the whole cost of driving the shop had formerly been.'

Between 1890 through to 1895, the majority of orders listed under Hathorn, Davey seems to be for alterations to existing equipment or occasional new items perhaps to replace those that had become worn, for example a boring bar, and included:

April 1890: alterations to Screwing Machine (4793) / October 1891: One 1½" Relief Valve and Pipe (4947) / July 1892: alterations to a Vertical Press (5004); one 10" Hydraulic Ram Press (5005); new Boring Bar (5047) / October 1894: one set of 20 ton Lifting Tackle (5170) / January 1895: Testing Pump for shop use (5208); one No.1 Motor (5215); new Crane (5216); Circular Motor for Slotting Machine (5226); Saddle and Rest (5230).

Figure 98. Sun Foundry: Probably the Interior of the Erecting Shop in 1891
(The Mining Journal. 10th January 1891. p.36)

The Sun Foundry was also a popular venue for various engineering groups to visit, and they would then publish occasional descriptions of the Sun Foundry. However, none of the reports actually described the Foundry, but would concentrate on listing some of the engines being manufactured.

Figure 98 is a lithograph showing what was probably the Engine Erecting Shop at the Sun Foundry in 1891. Certainly several Differential Compound Pumping Engines can be identified in a completed state, nearly ready for dismantling prior to packing and shipment. To the rear of the scene, several flues for forge hearths are set against the wall. There is also an obvious drive shaft and pulleys for working several machines.

One factor that may have limited any thoughts of modernising and extending the Foundry in the 1890s was the compulsory purchase of Sun Foundry land by Leeds Corporation. On the north and east sides of the Foundry some of the land was already owned by Leeds Corporation. An adjacent railway line was owned by the *Middleton Colliery Company*. Leeds Corporation wanted to build a new gasworks in this area. It would cover an area of over 24,000 square yards, so the Corporation compulsory purchased a portion of the land owned by the Colliery Company and Hathorn, Davey.[2] The purchase went ahead and the gasworks was eventually constructed. Plans of the Sun Foundry in the 20th Century show the foundry and the gasworks in very close proximity.

Orders listed under Hathorn, Davey from 1895 up to 1900 seem to suggest that several of the shops were being extended. Many of the orders in this period seem to be for heavy-duty lifting equipment and included:

November 1895: one 10 tons and one 20 tons Traveller (5288) / October 1896: one Steam Boiler and Connections (5374); Ironwork for extension of No.2 shop (5405); one Vertical Compound Engine, etc. (5412) / August 1897: Extensions to Iron Foundry (5444) and new Brass Foundry (5444A); Steel Girders and Cast Iron Columns (5509) / June 1898: two new Bogies (5586); Gear for applying power to a small Crane (5587); two Cast Iron Girders (5602) / October 1898: one Jib Crane (5621); one Crab for breaking metal in the top yard (5623) / January 1899: False Table (5633); one Cast Iron Marking Off Plate (5646); one 3 ton Crane (5649); Super-heater pipe connections (5652 and A); alterations (5666); one Tool Box (5671) / August 1899: one 2cwt Feeder (5685); Ironwork (5690) / November 1899: one 3 ton Jib Crane (5707); Shelving and necessary ironwork (5720); Material (5721) / May 1900: Jib Crane (5731); nine pairs of Steel Bushes (5732); one Base Plate and one Guide for Boring Bar, etc. (5760).

The above list indicates that in the mid-1890s the Sun Foundry was enlarged with the No.2 shop (probably for parts manufacture) being extended. There were also extensions to the iron foundry, and a new brass foundry was constructed. This was probably the result of an increase in the demand for heavy-duty pumping engines. In addition, the more complex Vertical Triple Expansion Engine was gradually replacing the Horizontal Compound. It had much larger castings for base structures and also had flywheels. Orders indicate that it was the preferred pumping engine for waterworks.

Until the 1890s, Hathorn, Davey produced mainly short information sheets about their products, or short catalogues about specific items, for example Davey's Domestic Motor. In 1894, the Company published what was probably their first major catalogue. It was titled, '*Practical Examples of Davey Pumping Engines for Town Water Supply.*' It was well illustrated with a small amount of description.

By 1899 this approach had changed completely, with very informative Catalogues produced for both waterworks (red cover) and mining (green cover). The 1899 Hathorn, Davey *Catalogue of Pumping Engines for Waterworks* also includes photographs of the Sun Foundry including one of the 'New Machine Shop' (Figure 99).

Figure 99. New Machine Shop
(Hathorn, Davey: Catalogue of Pumping Engines for Waterworks. 1899. Front)

A high-pressure cylinder of a Compound Engine occupies the foreground. A shaft for belting to drive other machinery including lathes runs along the length of the shop on the right-hand walling. The whole scene looks slightly disorganised. However, it was the practice to cast the order number into an order's main components, so that the various engine parts wouldn't get mixed. Other parts, that would include valve gear and operating rods for example, were often standard 'off the shelf' items. In Figure 99 there is an 'oscillating' disc propped up again a cylinder in the middle ground of the photograph, similar to that used with the Cheddars Lane engines.

Figure 100. The Erecting Shop
(Hathorn, Davey: Catalogue of Pumping Engines for Waterworks. 1899. p.2)

The final process would be to assemble the various orders in the Erecting Shop. Figure 100 shows the interior of the Erecting Shop. A Horizontal Triple Expansion Engine, probably destined for the South Staffordshire Waterworks, Trent Valley Pumping Station at Lichfield occupies the central area. An end plate for a cylinder stands vertically in the foreground (right).

After assembly and checking, an engine would have then been dismantled and crated ready for shipment. If for export, it would be taken by rail to Liverpool, Hull or London docks. Contract prices would often include transportation cost options for the different ports. Sometimes the orders would include a drawing of a shipping mark that had to be painted on the crates, to help in ensuring the complete order could be kept together during transportation and arrival at it's destination.

Another photograph in the Catalogue (included in the following Chapter) is of an interior scene in the iron foundry.

By the mid to late-1890s, probably as a result of the compulsory purchase of land for the gasworks, it is believed that the Sun Foundry had been extended to cover much of the east side of the Dewsbury Road, and certainly by the end of the decade the Company had demolished the housing along Higgins Street against Jack Lane.

The Company had a frontage along Jack Lane in 1900 and had constructed new offices. An Ordnance Survey map indicates that there was a travelling crane installed between the main foundry and the new entrance. The construction of the new offices must have been a relief to the non-manual workers at the Foundry, as space was probably at a premium.

When Hathorn, Davis and Campbell took over the Sun Foundry from Carrett and Marshall in 1872, it seems likely that most of the old employees transferred to the new business. Included was John Telford, one of the partners of Carrett and Marshall. Neither Hathorn nor Campbell had any engineering background, so the knowledge Telford possessed, combined with Davis's engineering background, were essential to keep the business operating in those early days.

In addition to the labouring workforce, clerks would be employed to look after paperwork and the day-to-day finances of the company. There was also a need for engineering draughtsmen who would transfer orders to paper, and produce the various drawings. The drawings were essential for the pattern and model makers who along with carpenters would prepare the wooden moulds for foundry castings, etc.

Everything, including the drawings, was linked to the order number. The Sketch Books contained a list of the drawings pertinent to each order, which will usually combine drawings specific to that order, as well as 'off the shelf' items. Some of the drawings exist at the Leeds Archives, albeit un-catalogued. These include detailed engineering drawings as well as instructive drawings showing the assembled machinery in its final setting. It can be assumed that the drawing office was probably the largest non-industrial department at the Sun Foundry.

Like most businesses at that time, a particular skill was handed-down from father to son. Quite often a whole families would work for one particular business. Consequently, only in exceptional circumstances would positions be advertised. Nevertheless, Hathorn, Davey did occasionally run advertisements in various newspapers in different parts of the country.

Advertisements for draughtsman regularly appeared in 1882, 1884, and 1891 for example. The 1880s were a time of increasing workload and no doubt other skills were also required. Perhaps the greatest loss to the Company was the death of John Telford on the 21st February 1881, although his son, John Schofield Telford, also worked for Hathorn, Davey in some managerial capacity. Alfred Davis had left some years previously, and Henry Davey was busy gaining customers, so in those early days of the business, there was probably a shortage of 'on the spot' engineers to manage the manufacturing process.

During January 1884, the Company advertised for both a General Foreman, and a Works Manager. It is not recorded who was fortunate enough to obtain either position, but there is circumstantial evidence in Company accounts to suggest that John Schofield Telford was promoted. Another possible candidate was James Alfred Towler. Certainly both names appear as subscribers to a proposal for the business to become a private Limited Liability Company in the 1890s.

Alfred Towler was born in 1859 at Eastby, near Skipton. It is speculation that he may have become an apprentice draughtsman within one of the many engineering companies in the Hunslet area of Leeds, in the 1870s. At his retirement from Hathorn, Davey when he held the position of Manager, in 1911, it was stated that this was after 26 years service.[3]

Towler was most certainly an engineer at heart, and promoted the Company very much at a local level. For a short period in the 1890s he was President of the Leeds Association of Engineers and gave talks to the Leeds University Engineering Society, and the Association of Water Engineers. He was manager of the Sun Foundry during what was probably the Foundry's darkest hour.

Work conditions at the Sun Foundry seem to have been acceptable. Whilst accidents undoubtedly occurred, very few were considered at the time to be major enough to be reported in the local Newspapers. In 1873, for example a 22-year-old employee was taken to the Leeds Royal Infirmary with severe lacerations to his arm that had been drawn into machinery.[4]

Like all the surrounding factories, Hathorn, Davey employees did make regular contributions to health and hospital funds that were duly acknowledged in the local newspapers. This action showed thanks and appreciation from the recipients, but also perhaps inspired competition between factories as to the generosity of the amounts donated.

In 1884 Hathorn, Davey employees donated £5-9s to the Public Dispensary, and £10-18s to the Leeds General Infirmary.[5] Thirteen years later, contributions were perhaps on a more formal footing, when it was reported that the employees of Hathorn, Davey had decided to double their weekly contribution to the Leeds Workpeople's Hospital Fund.[6]

In addition to the staff employed directly at the Sun Foundry, there were Erectors employed by the Company who travelled the world from contract to contract to supervise the installation of the engines. They would usually arrive on site at about the same time the engine was delivered. They would unpack and check the contents of the shipping crates and the condition of the parts, and possibly supervise the unloading and the transportation of the machinery to the site. Sometimes, parts would have been damaged in transit, and so this could cause lengthy delays for the Erector. Once the parts were assembled, months could pass before the engine was fully commissioned. Part of the Erectors job was to adjust the engine to ensure that it ran as efficiently as possible. In some instances this might cause the Erector to get replacement parts made locally.

Figure 101: Hathorn, Davey Advertisement for Non-Unionist Labour
(The Cornishman, 23rd September 1897. p.1)

Between July 1897 and January 1898, there was a general strike of engineers that was initiated by the *Amalgamated Society of Engineers, Machinists, Smiths, Millwrights and Patternmakers*. It was a Trades Union seeking to win an eight-hour working day for its members. The strike started in London and quickly spread throughout the country. By the 15th July it was reported that stoppages had occurred at Hathorn, Davey and Company. Two days later it was expected that 2,350 union engineers in Leeds would have ceased work.[7] An interviewed Union Official stated that the engineering dispute would probably result in a considerable proportion of them being thrown out of employment. That was exactly the reaction from some factory owners, although the response was mixed. Sometimes the labourer force was locked out of the factories. By the 19th July 1897, some 173 firms had given in to the striker's demands. In Leeds it was reported that an estimated 8,000 men were idle due to the strike.

It seems that Hathorn, Davey fared no better than any other company in the Hunslet area, and indications are that they dismissed a lot of the strikers. The consequence of this was that an advertisement (Figure 101) was placed in several newspapers stating that employment was being offered at the Sun Foundry to non-unionist Fitters, Turners, Machine-Hands, Drillers, Milling Machine-hands and General Smiths.[8] *'Comfortable quarters can be given on the premises out of reach of annoyance from strike-pickets.'*

It is assumed that the accommodation being offered was some of the old terraced houses along Higgins Street and the Dewsbury Road, which were due to be demolished, to make way for new offices (Figure 102)

Figure 102. The 'New' Hathorn, Davey Offices on Jack Lane in 1927
(Reproduced with the Permission of Leeds Library Service)

It was perhaps a contingency against the effects of another strike that provided the impetus for Meysey-Thompson and Hugh Lupton to purchase a piece of land in order to establish a foundry near Liege, Belgium in 1898.

Liege at that time was a centre of engineering excellence. The site of the proposed foundry was about 9 kilometres to the south west of Liege on the north bank of the River Meuse. It was located between the Quai du Halage and the Rue Spinette, near what is now Flemalle-haute Railway Station on the main Namur-Liege Railway Line.

The land was under several ownerships. The principal owner was the *Société Anonyme des Charbonnages des Kessales*, a coal mining company, who wanted to charge 45,000 Belgian Francs for the sale. The combined cost including other land was 54,000 Belgian Francs, which was calculated at the time to be £1,800.[9]

The sale must have proceeded, and the transaction was probably completed by April 1898 as the value of the land is incorporated into a later financial balance sheet. It is described as branch works at Liege, operated by the *Société Anonyme Hathorn, Davey Belgium*. Unfortunately, the Archives Department at Liege have no records that relate this company.

The engineer's strike did have a significant effect on the Company's profits. The result was a financial loss for the Company. It was undoubtedly this fact that probably encouraged the Partnership to convert the business into a Private Limited Liability Company to reduce further financial risk to the individual Partners.

Chapter 20 References

[1] HL. 1940. *Hathorn, Davey and Company, Sun Foundry, Leeds.* p.14
[2] Leeds Corporation. *The London Gazette.* 25th November 1890. p.6488.
[3] Presentation to a Leeds Manager. *Yorkshire Evening Post.* 18th December 1911. p.3
[4] Mill Accident at Dewsbury Road, Leeds. *Yorkshire Post and Leeds Intelligence*, 15th May 1873. p.2
[5] Local and Other News. *The Yorkshire Post*, 2nd January 1886, p.9
[6] Leeds and District. *The Leeds Times.* 25th June 1898. p.7
[7] The Engineers Dispute. *Huddersfield Daily Examiner*. 16th July 1897. p.4.
[8] Hathorn, Davey Advertisement for Non-Unionist Labour. *The Cornishman.* 23rd September 1897. p.1
[9] LMWS Box 21D2Bx1 There is a considerable amount of correspondence relating to the Belgian transaction.

21. Hathorn, Davey and Company, Limited and the Lupton Family

The idea of converting Hathorn, Davey and Company into a Limited Liability Company was first examined in 1885 when John Hathorn was trying to leave the Partnership. At the time it was questioned whether this route would lead to success. It wasn't until after the engineer's strike finished, that the idea was seriously considered again. Due to the strike, the Company had made a loss by the end of the Company's financial year (31st July 1898) of £1,290. (See Figure 103)

By the 24th March 1898 a proposal for forming a Limited Liability Company was being considered by Messrs. Lupton, Meysey-Thompson and Davey, and a draft proposal drawn up for consideration by all parties.[1]

'It is proposed to form as from _____ day of _______ 1898 a joint stock company limited by shares for the purpose of acquiring and continuing the business of mechanical engineering carried on by Messrs. Hathorn, Davey and Co. at the Sun Foundry, Leeds and carried on at branch works at Liege [Belgium].

1.   *Company name shall be Hathorn, Davey and Co. Ltd.*
2.   *Nominal capital £120,000 divided into 1000 shares of £100, and £20,000 of employee's shares*
   The first issue of £100 shall not exceed 750 of which £100 per share will be called, upon the formation of the company.
3.   *The Company shall purchase the whole real estate of the Sun Lane Foundry for £15,000*
4.   *Take over the firms liabilities valued at £3,900.*
5.   *Shares allocated in the first instance will be allotted as follows:-*

	No. of shares.	*Paid up Capital.*
Henry Davey	*240*	*£24,000*
Meysey-Thompson	*240*	*£24,000*
Hugh Lupton	*240*	*£24,000*
Towler	*10*	*£1,000*
Telford	*10*	*£1,000*
Un-named	*5*	*£500*
Un-named	*5*	*£500*
Totals	*750*	*£75,000*

6.   *Employees shares shall be £1 each and may be allotted by the Directors, as they feel fit, to employees.'*

The proposal was still short of two names to fulfil the legal obligation required under Company Law. It was a statutory requirement that to form a new Company there had to be seven subscribers, and this would be duly recorded and signed, together with shareholdings and profession signatories, in the proposed Company's Memorandum of Association.

	Profits after paying 5% interest on Capital but including Royalties and salaries to partners	Royalty and salary paid to Henry Davey	Salary paid to Meysey-Thompson	Salary paid to Hugh Lupton	Net profit after deducting salaries and royalties	Loss after 5% interest on capital	Nett loss after deducting from the apparent losses royalties and salaries to partners
1885	1758	1275	200	-	283	-	-
1886	-	776	200	-	-	2317	1341
1887	2558	984	200	-	1374	-	-
1888 a	2655	300	200	200	2055	-	-
1889	-	300	200	200	-	3268	2668
1890	-	300	200	200	-	1287	687
1891	3225	300	200	200	2625	-	-
1892	3631	300	200	200	3031	-	-
1893	2882	300	200	200	2282	-	-
1894	4394	300	200	200	3794	-	-
1895 b	4509 d	300	300	300	3609	-	-
1896	6323 e	300	300	300	5423	-	-
1897	4770	300	300	300	3870	-	-
1898 c	Strike	300	300	300	-	1290	390
1899	6909	Not known	Not known	Not known	Not known	Not known	Not known
1900	8880	Not known	Not known	Not known	Not known	Not known	Not known
1901	4153	Not known	Not known	Not known	Not known	Not known	Not known

a   Hugh Lupton became a Partner during the financial year 1887/88

b   Second partnership Agreement

c   Six month strike from July 1897 to February 1898.

d   Includes a percentage to Mr Towler from 31st July 1895

e   Includes a percentage to Mr Telford from 31st July 1896

Figure 103. Table showing Company finances from 1885 to 1901
(DLC Letter from Hugh Lupton 15th October 1900)

In a letter dated 15th October 1900, Hugh Lupton summarised payments made to the Partners. It also recorded the Partnership's profits and losses. The figures agreed with a later financial statement and showed that from 1885 to 1901 there were four years when the Company made a loss (Figure 103).

Although they were not Partners, the important role that both Alfred Towler and John Telford played in the running of the business was finally recognised in 1895 and 1896, respectively, when on top of their salaries, they were both awarded a percentage of the Company's profits.

In addition, Figure 103 shows the salaries paid to each Partner. Lupton was later to remark that when Henry Davey went to London his capital share was £1,914 and that he had a return from his patents as well.[2]

	Work in progress at end of year 31st July	Sales during year to 31st July	Purchases to stock for year to 31st July	Wages for year including salaries	Tonnage of castings from Foundry	Cost of model wood used to make patterns	Cost of patterns per ton of castings
1885			10215	14700	-		
1886			10015	14195	517	1180	£2-05s-07d
1887			13082	14823	653	1156	£1-15s-04
1888			13809	15363	730	1379	£1-17s-08d
1889			10120	15040	663	1290	£1-14s-0d
1890			13672	16996	662	1345	£2-06s-07d
1891			16101	16775	716	1220	£1-14s-0d
1892			22506	17149	786	1258	£1-12s-0d
1893			13812	15987	711	1227	£1-14s-06d
1894	11590	52064	24214	15581	723	1228	£1-14s-0d
1895	16581	37986	14034	15391	886	1350	£1-10s-06d
1896	13395	55282	18868	16871	779	1370	£1-15s-02d
1897	17300	48743	19544	18679	1076	1377	£1-09s-04d
1898	20885	31215	14705	11375	527	925	£1-15s-01d
1899	13253	67892	20409	18934	1006	1281	£1-05s-06d
1900	15827	65276	24158	19377	962	1435	£1-10s-0d approx*
1901	15586	67850	19907	19977	1018	1391	£1-08s-0d approx*

*Possibly too low

Figure 104. Table showing foundry costs between 1885 and 1901 together with 'work in progress' and sales between 1894 and 1901
(LMWS, 21D2BX2)

Figure 104 shows other aspects of the Partnership's activities from 1885 to 1901. The work in progress figures are only available for the period 1894 to 1901, but they do show significant increases for 1897 and 1898. This may be the result from a cumulative backlog of work caused by the engineer's strike when the work was disrupted. Sales figures show a different trend, as well as a healthy annual increase to over £65,000 for the three years to 1901.

The wages that also include salaries, rose about 25% over the period, and increased again following the engineer's strike.

Foundry costs were examined in detail and included annual income and outgoings. The purchase of consumables, for example brass, iron, copper, timber etc., rose steadily over the period.

The foundry may well have been one of the busiest shops at the Sun Foundry (See Figure 105). Pattern and model making were an important part of the casting process. The pattern makers were a specialist team. Any imperfection or error of their work would be reproduced in the final castings. While some parts could be cast using standard wooden patterns, many patterns had to be made from new. The cost of wood used to make the patterns is listed, with annual costs wavering between £1,100 and £1,400. In addition, the annual tonnages cast at the foundry are shown, as well as the calculated cost of a pattern per ton of casting, which was very consistent over the period.

Figure 105. The Foundry: The floor is composed entirely of casting sand
(Hathorn, Davey: Catalogue of Pumping Engines for Waterworks. 1899 p.4)

As Figure 105 shows, the interior of the foundry could be regarded as a 'health and safety' nightmare, with no one wearing safety equipment, just the ubiquitous cloth cap. The floor is composed of casting sand that may have been mined locally at either Garforth or Castleford, a few miles southeast of Leeds. There are two overhead gantries, one carrying a crucible, and to the rear, another hoisting a large receiver.

Nothing seems to have happened towards the aspiration to form a Limited Company until 1900, when a formal Memorandum of Agreement between Henry Davey, Meysey-Thompson and Hugh Lupton to form a Limited Company was produced.

By 1901, the intention to form a Limited Liability Company was well advanced and a draft Abstract of an Agreement had been drawn up between Henry Davey of the City of Westminster, engineer; Arthur Herbert Meysey-Thompson of the City of Leeds, engineer; and Hugh Lupton of the said City of Leeds, engineer (hereby called the vendors) on the one part and Hathorn, Davey and Company, Ltd (hereafter called the Company) on the other part. It was agreed that the Company would be formed under the Companies Acts 1862 – 1900 with a nominal capital of £120,000 divided into 12,000 shares of £10 with a view to acquire the business. The assets to be purchased by the new Company included:[3]

Cash represented by shares and debentures.	*£96,039-19s-9d*
Debts and liabilities.	*£13,668-9s-3d*
*Land and Buildings covered by stamp of conveyance**	*£18,159-6s-1d*
Fixed machinery covered by stamp of conveyance.	*£10,330-0s-0d*
Property not requiring stamp.	
*Shares in Belgium Company***	*£9,000-0s-0d*
Cash	*£2,781-0s-0d*
Stock	*£20,508-0s-3d*
Loose machinery	*£7,710-13s-4d*
Models and drawings	*£1,300-0s-0d*
Property for which agreement to be stamped.	
Book debts.	*£19,693-12s-8d*
Patents	*£2,900-0s-0d*
Debts owing to firm	*£13,668-9s-3d*
Goodwill	*£3,657-6s-4d*
Total	*£109,708-9s-0d*

*   Stamp, relates to the Stamp Act (1891) that imposed a tax on the transfer of assets and other property.
**   Shares in the Belgium Company refers to the 'Société Anonyme Hathorn, Davey Belgium.'

Figure 106. Hugh and Isabella (Ella) Lupton.
(Leeds Blue Book, 1927 pg 64-65)[4]

A draft Indenture soon followed. It confirmed the formation of Hathorn, Davey and Company, Limited, and was subscribed to by the following people:[5]

Henry Davey, 3 Princes Street, Westminster, London.	Civil Engineer.
Arthur Herbert Meysey-Thompson, Sun Foundry, Leeds.	Engineer
Hugh Lupton, Sun Foundry, Leeds.	Engineer
Ernest Claude Meysey-Thompson, Rokeby, Barnard Castle.	Esquire.
Arthur Greenhow Lupton, 11 Wellington Street, Leeds.	Aforesaid manufacturer.
Isabella Lupton, wife of the said Hugh Lupton.	
John Telford, Sun Foundry, Leeds.	Secretary
James Alfred Towler, Sun Foundry, Leeds.	Engineer

In addition to the five names on the 1898 proposal, Hugh Lupton's wife Isabella (Figure 106), Arthur Greenhow Lupton, Hugh's older brother, and Ernest Meysey-Thompson, Arthur Meysey-Thompson's older brother made up the full list of eight subscribers.

A letter dated 22nd August 1901, suggests that Ernest Meysey-Thompson was invited to be a subscriber so as to ensure that the family would be represented on the Board should Arthur Meysey-Thompson die. It is presumed that the same reason also applied to the Lupton family subscribers. Hugh married Isabella (née Simey) on 12th April 1888 shortly after joining Hathorn, Davey.[6]

On the 30th October 1901 the Private Company of Hathorn, Davey and Company, Limited was incorporated and became a registered company (Number 71739 on the Register). The Memorandum of Association states that the Company was formed, '*To acquire and take over as a going concern the business of Mechanical and Hydraulic Engineers and otherwise heretofore carried on at the Sun Foundry, in the City of Leeds, in the County of York.*' '*They would carry on the business of Mechanical, Hydraulic, Electrical, Water, Heat and Light Supply, and Sanitary and General Engineers in all their respective branches.*' Eventually all future advertising would denote that Hathorn, Davey was a Limited Liability Company.[7]

It is unclear how the Belgium Company was regarded but on the 3rd December 1901, an Indenture was drawn up to increase the share of the loan capital of the Belgium Company.[8]

On the 4th December 1901, Directors of the new Company had their first meeting at the Sun Foundry, Leeds the registered office of the company. The first General meeting was held on the 8th January 1902, and Henry Davey was the Chairman.[9]

The 20th century history of the Limited Liability Company was very much centred on members of the Lupton Family, as shown in Figure 107.[10]

Hugh Lupton's progress from apprentice to Partner was rapid. He was born on 11th May 1861 at Roundhay, Leeds.[11] After attending Rugby School, he read Modern History at University College, Oxford. His whole career was with Hathorn, Davey and Company, and ultimately he became the Managing Director. Along with other members of the family he was very much integrated in Leeds society and was involved with local politics, sitting firstly as a Rural District Councillor, and then on Leeds City Council. In 1926 he became Lord Mayor of Leeds. During November 1940 Hugh compiled a short fourteen-page history of the Company, which has been a valuable source of information.

In addition to Hugh Lupton, four other family members were involved with Hathorn, Davey and Company, Limited. Their names are shown in bold on Figure 107.

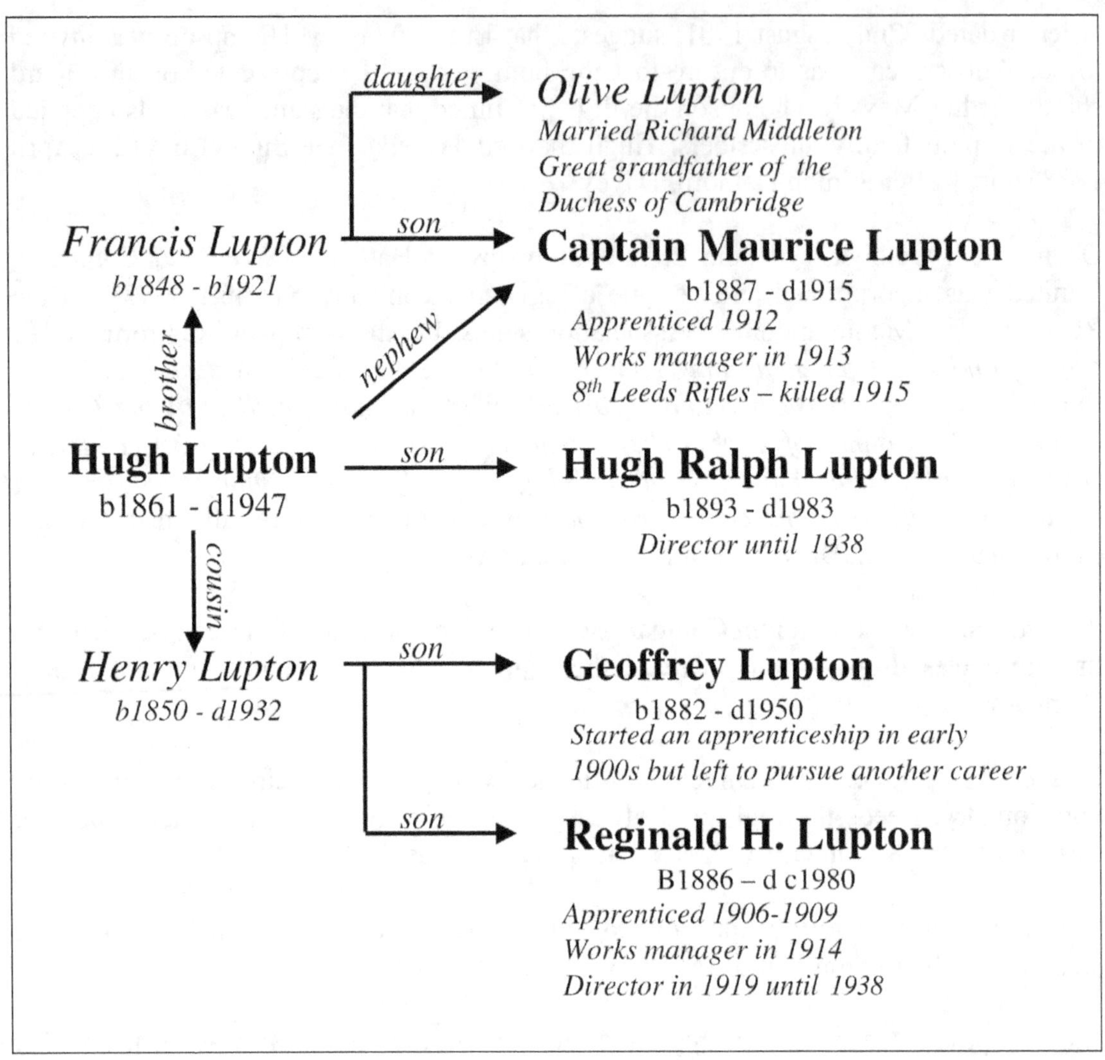

Figure 107. Chart showing the relationship of members of the Lupton family (shown in bold) who were associated with Hathorn, Davey.
(Author)

Geoffrey Henry Lupton was the son of Henry Lupton, Hugh's cousin. He was with the Company for a short while in the early 1900s. Geoffrey was born in West Yorkshire on the 2nd September 1882. It is presumed that he was in a regiment during World War I. Eventually he settled in Rhodesia, now Zimbabwe, where he bought a farm that he named Keld Head, probably after one of his favourite places in Yorkshire. He died from injuries received when a bull gored him on the 30th December 1950, at Marondera, Zimbabwe. [12]

Maurice Lupton, the son of Hugh's brother Francis. Maurice eventually replaced Alfred Towler as Works Manager in 1913.

Maurice was born at Potternewton, Leeds on the 4th June 1887 and educated at Rugby School from 1902 to 1906, and gained a BA degree from Cambridge in 1909.

While Maurice was completing his degree, his father wrote to Hugh Lupton on the 13th June 1909 with apparently a follow up to a suggestion from Hugh.[13] *'What I understand you to ask is whether Maurice is willing to finish a term with you as pupil, with the view, should you and he both wish for such an arrangement when his time of learning is over, of his remaining with you, and eventually becoming one of the partners. Maurice is quite willing to do this, and hopes to satisfy you sufficiently to be kept by you.*

It seems to me unwise to bind either side more than this; unless he makes himself useful enough, there is no use in your having him.'

On the 5th August 1911, Francis again wrote to Hugh concerning Maurice's position in the Company.[14] Hugh had been considering transferring over some of his shares in the Company to Maurice. In addition he had taken Maurice on as an apprenticeship waving the normal fee. *'You have given Maurice, without any fee, his apprenticeship and great opportunities, and he has now the chance of becoming, if he proves himself useful, one of the managers, eventually, of your concern; and it is only reasonable that if he takes up shares with you, he should do so at par.'* Some of the quota of shares were coming from Alfred Towler's allocation, probably because he left the Company in 1911. *'I hope therefore that you will agree altogether your very kind and liberal proposals and let him carry out the purchase from Towler, as soon as he has paid you for your £100 of shares.'* It was clear from the remainder of the letter that Francis was extremely grateful for Hugh's help in giving Maurice a career.

According to Hugh Lupton's notes about the history of the Company, Maurice was apprenticed to Hathorn, Davey and Company, Limited in 1912. Together with his cousin Hugh Ralph Lupton, he spent three months at the Messrs. Brück Kretschels works at Osnabrück, Germany, manufacturer of turbines.[15]

On the outbreak of World War I Maurice enlisted in the 8th Leeds Rifles, and eventually became Captain in the 7th Battalion West Yorkshire Regiment (Prince of Wales's Own). He was killed in action by a sniper near Lille, France on the 19th June 1915. [16,17]

Reginald Hamilton Lupton (Figure 108), also the son of Henry Lupton, was born in Leeds on the 31st July 1885. He was apprenticed to the Company between 1906 and 1909. Between 1909 and 1914 he was the Company's representative in India. In 1914 he replaced Maurice as Works Manager. He became a Director in 1919.

Reginald made five commercial voyages to India and Australia on behalf of the Company between 1919 and 1930. He retained an executive position when the Company was taken over by Sulzer Brothers in 1936. He resigned in 1938 and spent his retirement at Budleigh Salterton, Devon where he died on the 27th April 1975.[18]

Figure 108. Reginald Hamilton Lupton at Brisbane (Mnt. Crosby) Waterworks in 1927
(Courtesy of the Lupton Family)

Hugh Ralph Lupton, Hugh's son, was born 14th April 1893 at Leeds. He was educated at Wellington College and Trinity College, Cambridge. During World War I he initially served with the Leeds Rifles and was awarded the Military Cross. He finished his military service as a Captain in the Northumberland Fusiliers and had become an Instructor in Mathematics at the First Army Infantry School, British Expeditionary Force.[19]

He joined Hathorn, Davey in 1919. He also became a member of the Institution of Civil Engineers the same year (Proposed by Meysey-Thompson) and was awarded the Institutions Watt Gold Medal in 1926. He became Technical Director, and stayed with Hathorn, Davey until 1938. He was appointed Mechanical Engineer to the Metropolitan Water Board, London in 1939. He died on the 8th March 1983.[20]

As an aside, Olive Lupton the daughter of Francis married Richard Middleton, the Great Grandfather of the Duchess of Cambridge, thereby establishing a Royal connection with the Company!

At the time of becoming a Limited Liability Company, the order books were looking healthy, and some large and valuable orders were shortly to be placed.

Chapter 21 References

[1] DLC Proposal for forming a Company dated 24th March 1898

[2] DLC Letter from Hugh Lupton 15th October 1900

[3] DLC Assets purchased by the new Company 1901

[4] I would like to acknowledge the help provided by Josh Flint, Assistant Librarian Manager (Local & Family History/ Leodis), Leeds Libraries in locating the photographs of Hugh and Isabella Lupton

[5] DLC Draft Indenture regarding the formation of the Company 1901

[6] Marriage of Miss Simey. *Sunderland Daily Echo and Shipping Gazette*. 13th April 1888. p.3

[7] DLC Memorandum and Articles of Association. 30th October 1901

[8] DLC Indenture to increase the share of loan capital of the Belgium Company. 3rd December 1901

[9] DLC First Meeting of Directors and First General Meeting 3rd and 4th December 1901, respectively

[10] The Author is grateful to members of the Lupton family for providing this information.

[11] Hugh Lupton, England, *Oxford Men and Their Colleges*, 1880-1892

[12] https://www.findagrave.com Find a Grave Memorial - Geoffrey Henry Lupton (1882 -1950)

[13] LMWS - Box 21D2BX2 Item 19931.3C Letter regarding Maurice 13th June 1909

[14] LMWS - Box 21D2BX2 Item 19931.4C Letter regarding Maurice 5th August 1911

[15] HL. 1940. *Hathorn, Davey and Company, Sun Foundry, Leeds*. p.6

[16] https://www.geni.com/people/Capt-Maurice-Lupton/6000000009407117238

[17] Leeds Rifles Officer. *Leeds Mercury*. 23rd June 1915. p.2

[18] Probate Records

[19] Institution of Civil Engineers - Proposal for membership - H.R. Lupton - received 7th March 1919.

[20] Hugh Ralph Lupton. 1983. Obituary Notices, *Transactions of the Newcomen Society*, 55:1, p.249

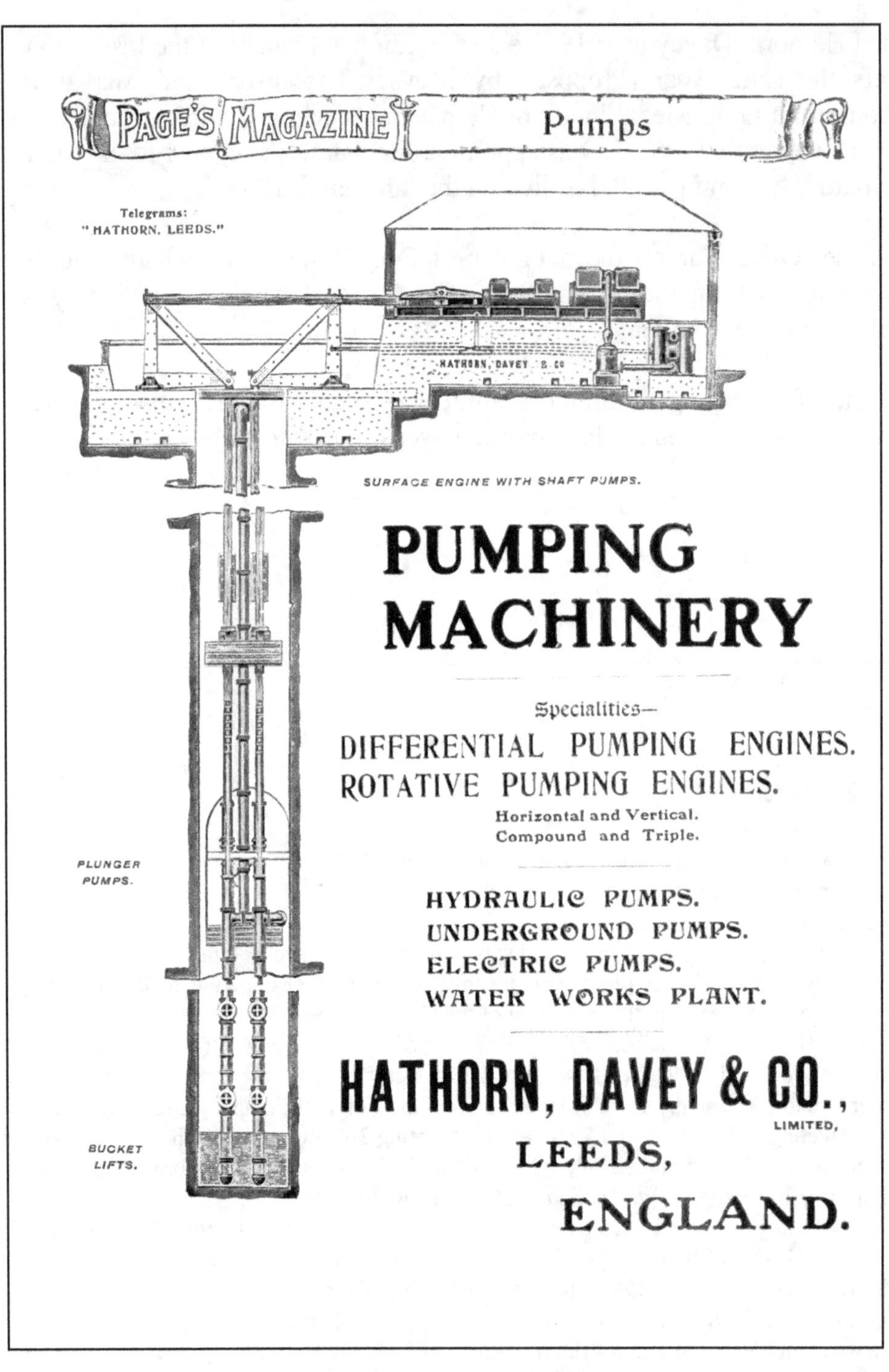

Figure 109. A 1904 Advertisement for Hathorn, Davey still featured the Horizontal Compound Engine
(Page's Weekly, 28th October 1904. p.9)

22. The Limited Company: 1900 to 1914

The period leading up to World War I may well have been the most profitable for the Company. This was a period when steam power was being replaced by both the oil engine and the electric motor, and so there was a certain degree of adaptability by the Company to the relatively new technologies. According to Hugh Lupton, the Davey Domestic Motor was superseded by the oil engine, which in turn was almost certainly replaced by the electric motor that would have offered a convenient, low maintenance, and certainly cleaner means of driving equipment.

Urban expansion at both home and abroad had placed greater pressure on the need to supply fresh water to ever expanding towns and cities, so new waterworks were being built, or extended, to increase pumping capacity. The Vertical Rotative Triple Expansion Engine fulfilled those needs, and by the turn of the century they had almost replaced the once popular Compound Engine at waterworks, and even on some mine sites. Nevertheless, the Compound Engine was not being phased out altogether. It was still being advertised (Figure 109) and sold as one-off items to smaller mining companies. Some of the largest orders for Compound Engines were placed at this time for exportation to Australian and Japanese mine sites.

Lupton briefly refers to the Rotative Triple Expansion Engine as using much higher steam pressures and superheated steam.[1] In this type of engine, '*the pump plungers were directly coupled to the crossheads and placed immediately below the cylinders, an arrangement that gave a high degree of mechanical efficiency. The gearing used for the control of these engines was of the Craig Trip Gear type, introduced from the Cotton Mills of Lancashire by the then Works Manager, Mr. Alfred Towler.*'

The patent Lupton referred to, was applied for by Towler, and accepted on the 1st May 1897. Patent 1775 was titled '*Improvements in Valve Gear for Pumping Engines Driven by Steam or Compressed Air*'. Towler described the patent as:

'*My invention relates to valve gear for pumping engines driven by steam or compressed air having the valves of the main cylinder actuated from a subsidiary cylinder which has its distribution valve or valves operated from the piston of the main cylinder.*
The chief object of my invention is the application of gear hereinafter described to the actuation of valves for the main cylinder of the Corliss kind with provision of trip gear for rapid closing of the supply valves and an arrangement of the exhaust valves so as to provide adequate cushion on the main cylinder.'

Towler's gearing for use on the Triple Expansion Engines was essentially similar in application to Davey's Differential Gear for Compound Engines.

Higher steam pressures and the use of superheated steam were complimented by changes in boiler construction that occurred about this time. Improvements to steam generation technology included; i) The tube-boilers manufactured by Babcock and Wilcox, that were frequently used to provide steam for Hathorn, Davey engines, and ii) Economisers, a method where flue gases were used to heat the water prior to it feeding into the boiler. The latter was used more frequently with Triple Expansion Engines. The Green's Economiser[2] (Figure 110), manufactured at Wakefield, West Yorkshire was often the preferred model. A Green's Economiser for example, was installed with the two Hathorn, Davey Triple Expansion engines (Order 6314, 10th May 1907) at the Onga River Pumping Plant, Nakama City, Kyūshū, Japan.

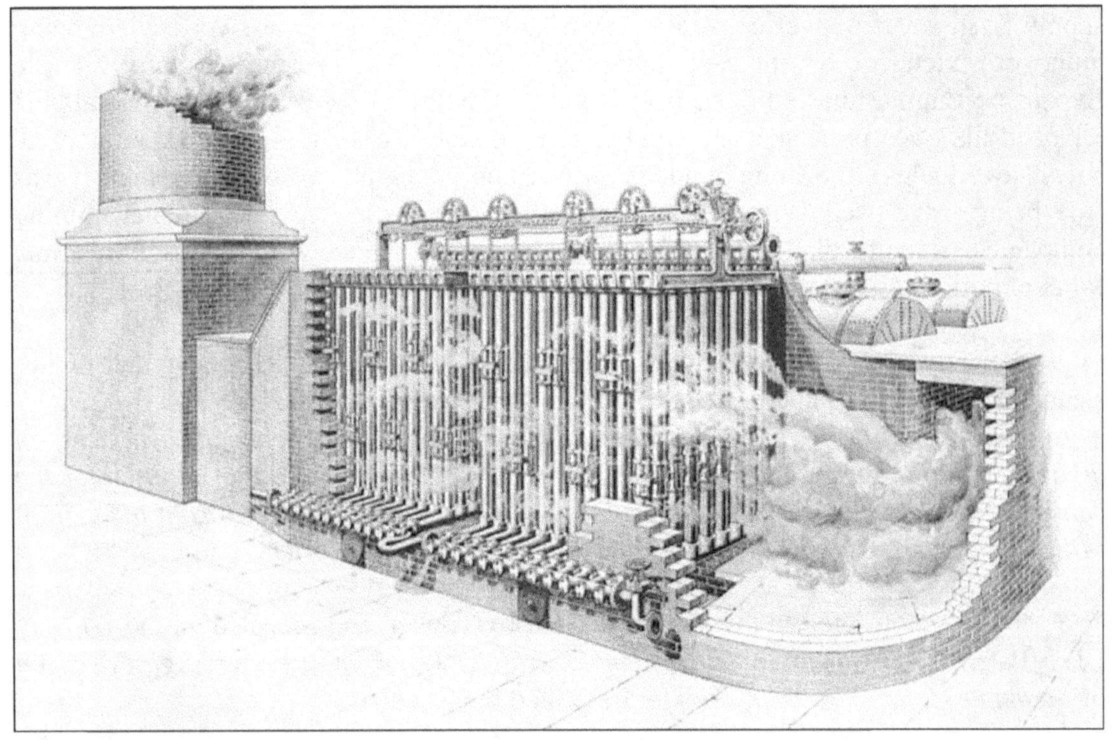

Figure 110. A Green's Economiser located between the boilers and the chimney
(Fifty years' history of the development of Green's Economiser. Frontpiece)

Although Henry Davey was still involved with the new Company, he remained London based, working from the Princes Street office. He was Chairman of the Board of Directors, and remained so for a number of years. Meysey-Thompson and Hugh Lupton managed the business and Davey was very much left to his own devises.

During 1900 Davey published, '*The Principles, Construction, and Application of Pumping Machinery (Steam and Water Pressure)*' widely regarded as a classic work, in

which he used for his examples, many of the engines manufactured by Hathorn, Davey.
Davey published a second edition of the book in 1905.

Davey also started to examine the history of steam technology in detail and published
several articles and papers about the atmospheric engine invented by Thomas
Newcomen.

Figure 111. Henry Davey. 1904
(Page's Weekly, 4th November 1904. p.541)

This may have been one of Davey's most prolific periods for registering patents. Most were related to improvements in steam technology, but Davey's son Norman was also showing interest in engineering, almost certainly due to Henry's encouragement. Consequently several patents were eventually registered under joint names. Both Henry and Norman worked on aspects of weir design and tidal energy, as well as gas and water turbines, and Norman ultimately published books on the two subjects.

Henry Davey at this time was also enjoying a certain degree of popularity for his many achievements and was featured in a brief article in *Page's Weekly* in 1904.[3] The article included his photograph (Figure 111). It mentioned his improvements to the Compound Engine, and provided examples of the system's capabilities. It also referred to his awards that included the Watts and Telford Medal presented to him by the Institution and Civil Engineers. It was also mentioned that Davey was a British Association general committee member, and a member of Council for the Institution of Mechanical Engineers.

A few years after the Partnership was converted into a Private Limited Company, the old four-storey building facing the Dewsbury Road, which had been the original works for Carrett and Marshall, was pulled down. The erecting shop that ran along the east side of the Dewsbury Road, was extended. They also bought some more land on the east side of the Sun Foundry adjacent to Leeds City Gasworks.

According to Hugh Lupton's notes, *'later a further considerable extension of the works was undertaken. Land which had formed a part of the Leeds Pottery, was bought from the Middleton Colliery Company and a new foundry, with a connection to the colliery siding was built.'* This must have been in the area immediately south of the Gasometer, towards Jack Lane.[4]

Davey seems to have been involved with the designs of a new machine and erecting shop. A plan prepared by him shows that the building would cover an area of 100 feet by 100 feet.[5] It would consist of two central shops, 25 feet wide, with about 25 feet at one end entirely for large machines, whilst the rest of the space would be devoted to erecting. Rails for wagons would run down the centre of each building. Overhead each building would be furnished with a travelling crane. Running the whole length at the side of the structures was angled girder work that would act as buttresses for stability. Small machines, such as lathes, would occupy the two spaces formed beneath the buttress girders. The building would be covered with wood or galvanised iron and light would enter through skylights in the roof.

Much of this work was put out to tender and an advertisement was placed in *The Leeds and Yorkshire Mercury* in 1904 by Stephen Ernest Smith, Architect, who invited tenders for the erection of new offices, model shop etc.[6]

During this period of reconstruction, in what was probably a major upheaval for the Company, it is known that some orders for engines are not listed in the Order Books. Reference has already been made to the fact that the components for the large Vertical Compound Engine for the Waihi Gold Mine, New Zealand were made by a number of other companies on behalf of Hathorn, Davey. Similarly four Compound Engines made for the Mitsui Company for the sinking of the Manda Pit, Miike Coalfield, Japan are also not registered. Despite an extensive search of the Order Books no listing has been found for those engines. It can be assumed that similar scenarios may exist for other orders of this period and that only orders put through the Sun Foundry were entered into the Order Books.

Securing a contract is important for any business, and whilst no one in Hathorn, Davey was appointed specifically with this task in mind, it seems that Meysey-Thompson was the person who generally carried out this function.

In early 1900 Meysey-Thompson made a visit to Odessa, in Russia. Previously the Partnership had sold a number of engines to British Companies trading within Russia. One of the main entry ports into the country at this time was Odessa on the Black Sea, and whilst some of the past orders may have been for mining concerns, more recent orders were connected with supplying water to the city. Four pumping engines were ordered on 11th March 1898 (Orders 5541 to 5568) and two more on 2nd August 1899 (Order 5684). Meysey-Thompson's visit would secure a further order for two more engines for the Chumka Pumping Station costing £10,350 (Order 5806, 25th June 1901).

Meysey-Thompson may have also visited South Africa, and perhaps secured an order for four engines from the Rand Water Board; and to India, although Lupton states that later, between 1909 and 1914, the Company was represented in India by Reginald Lupton.

There were many other large orders placed with the Company in the years leading up to World War 1 and some of those will be detailed in the next chapter.

During 1902, Meysey-Thompson and Lupton gave a paper, *'Some of the Considerations Affecting the Choice of Pumping Machinery'* to the North of England Institute of Mining and Mechanical Engineers.[7]

The paper summarised the advances in steam technology concentrating on engine efficiency, particularly those manufactured by Hathorn, Davey. It also made the point that an equally efficient steam generating plant was just as important as the engine. Another observation was that, '*At many collieries, the weight of coal raised is far less than that of water pumped, and as pits become deeper the horsepower necessary to raise the water to the surface must proportionately increase.*'

The fact that the Compound Pumping Engine could be sited away from a cluttered shaft top was also reiterated. The Bradley and Moat engines operated by the *South Staffordshire Mines Drainage Commission* were used as examples to demonstrate the engine's efficiency. After some seventeen years of use, it was stated that all parts of both engines were in excellent order, '*no important parts of either engines or pumps have at any time been replaced, with the exception of the pump clack-boxes - seven of which have been renewed.*'

Some emphasis was put on the problems associated with the manufacture of pump valves. The castings for large diameter pumps forcing against a heavy pressure had to be thicker. '*The flanges of the valve-box must be made enormously thick, even more so than the rest of the casting. Consequently, unequal strains are set up in cooling after casting; and the box is weak.*' '*Cavities are apt to form under the skin in casting, and are exceedingly hard to discover.*' Due to the many problems experienced, Hathorn, Davey had a standard pattern of valve-box, 6 inches in diameter, the number varying depending on the size of the pump. They emphasised that such valve boxes could be sited in a shaft recess, where a workman could work safely and normal shaft operation would not be interrupted.

They concluded that, '*In the writers' experience, where circumstances allow of it, the method of pumping by rods actuated by a Compound or Triple Expansion Engine on the surface, and when required to do so, also driving hydraulic pumps for draining dip-workings, is the most economical for heavy pumping, both as regards consumption of fuel and of repairs.*'

Up to this point, the paper didn't reveal any new information, so perhaps the real reason for presenting it is exposed in the final pages. '*Electricity, as a means of transmitting power for pumping purposes, has come into use in later years, and seems destined to play a larger part in future. The lightness and portability of the cable as well as the small loss of power in conveying the current to a distance, are important points in its favour.*' Whether Meysey-Thompson and Lupton were seeing effects on sales figures due to electrical innovations, or whether they could see the future implications of the relatively new power source is not known. Nevertheless electric power may have been perceived as a threat to their business.

Problems with electric power were then cited in the paper including the risk of an explosion due to sparking from dynamos and the unreliability of electricity due to breakdown of supply; its low efficiency. The experience of an Insurance broker was quoted, *'One out of every nine [electrical] machines had broken down in the past year... if electrical-driven machines above ground give so much trouble, it is hardly to be expected that the attention and cleanliness required........ would be forthcoming to a greater degree at the bottom of a pit.'*

Needless to say, the paper encouraged much discussion. Professor Henry Louis (Durham College of Science) didn't think that the authors had treated the application of electric power to pumping machinery fairly. Another person in the audience questioned the examples they had used to berate the use electric power for underground purposes. The paper did achieve it's objective and perhaps gave them the publicity they desired, and the discussion was later continued in the next two volumes of the *Transactions*.[8,9]

Although recognised as a source of power, Hathorn, Davey only seemed to apply electricity to work pumps after about 1910. Whether this was due to the reluctance of Henry Davey, is not recorded. Certainly Davey never applied for a patent involving electricity!

However, by 1914, Hathorn, Davey was reported in the *Daily Mail* as having provided an electrically driven pumping system for the New Joint Dock at Hull. [10]

'The firm of Hathorn, Davey, and Co., Ltd., Leeds, have a reputation for electrically driven hydraulic pumps. They are the contractors for this equipment at the Joint Dock. The plant consists of three sets of horizontal, electrically driven, three-throw pumps, each of which is capable of forcing 250 gallons per minute against an accumulator pressure of 850lbs. per square inch, and will supply power water for operating the dock gates, bridges, and other hydraulically driven plant at the dock. The power water, after being used for this purpose, is automatically returned to a tank above the pumps and used over again. The accumulators automatically start and stop the pumps when at the bottom and top of their stroke respectively. In order to relieve the shock at starting, Messrs Hathorn Davey, and Company have patented an arrangement by which the electric motor is first set in motion by the accumulator, and the pumps are started against no load. When the electric motor has attained its speed, the pumps gradually begin to pump and when the accumulator begins to stop the pumps, the quantity of water delivered by them is gradually reduced to zero, and the motors are then stopped. This method is found not only greatly to prolong the life of the hydraulic plant, but also obviates any danger of excessive flow current in the motors at starting. Each of the pump rams is of hard manganese bronze, 5½in. in diameter, and 18in. stroke, and runs at 55 revolutions per minute. The two accumulators the pump-house are 16in. in diameter, with a clear stroke of 18ft. A third accumulator, which is

placed on the pipe system, near the dock entrance, serves to regulate the flow of water at the end of the dock furthest from the pumps. This accumulator has a ram 18in. in diameter and stroke of 23 feet. Each the three return water tanks above the pumps has a capacity 5,200 gallons. In the event of this supply being interrupted through rupture of main or other cause, an alternative supply is provided by means of electrically driven centrifugal pumps, which draw water from the dock. Each of these three pumps is capable of raising 300 gallons a minute. The electricity for driving the pumps is obtained from Joint Dock's own generating station, the current being continuous at a tension of 440 volts.'

The 'reputation' Hathorn, Davey had gained for electrically driven pumps must have been recently acquired. Six years previously a pumping system at Plymouth for a Graving Dock, for the Great Western Railway, was steam powered.[11]

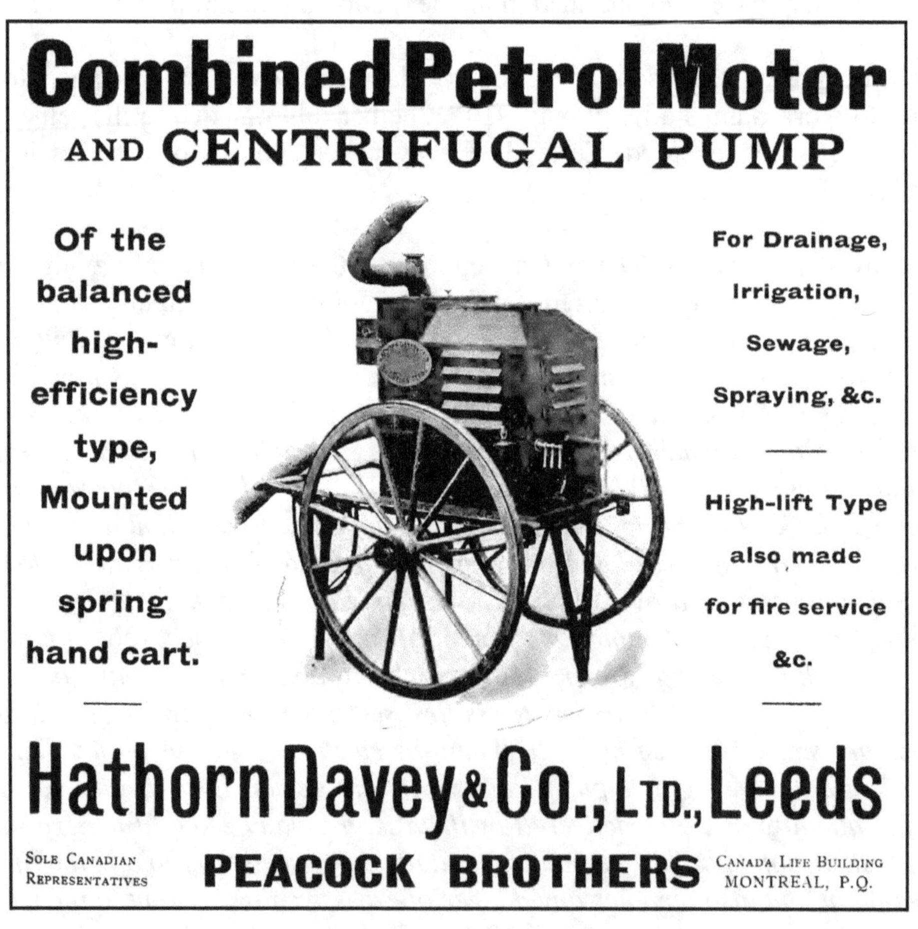

Figure 112: Advert for the Portable Combined Petrol Motor and Centrifugal Pump
(Canadian Mining Journal 1907 p.xxiii)

In 1907, Hathorn, Davey produced a portable petrol driven pump that was described as a '*Combined Petrol Motor and Centrifugal Pump in Portable Form, with Cover'*. The Petrol Motor driven Centrifugal Pump could be bought as an open unit or covered, high or low lift system, and mounted on a two or four wheel handcart depending on the size. Peacock Brothers of Montreal, Canada also manufactured it under licence (Figure 112).

It was produced for both low and high lift pumping, for different applications.

The low lift units came in nine sizes suitable for drainage and irrigation and they were advertised as being capable of dealing with water or sewage. Types AI to AIII (100 to 300gpm); BI to BIII (140 to 400gpm), and CI to CIII (180 to 500gpm) and were capable of lifting water from 10 to 30 feet. CIII with the greatest capacity was sold for £130. It was unprotected and not mounted on a handcart.

The price included: '*Magneto ignition, petrol tank, lubricator, petrol, oil, circulating water pipe connections, spanners and a screwdriver.*' An extra £3 was charged to fit shafts to the cart for it to be pulled by a pony.

The high lift unit came in ten sizes and was suitable for fire-fighting and other purposes. The delivery varied from 50 to 750gpm to a maximum height of 120 feet. The prices varied depending on delivery and type of mounting (two or four wheeled). The different sizes are detailed below together with the price.

F.I	50gpm delivery	£134 (open)	£152 (closed 2 wheel)	
F.II	75	£195	£215	
F.III	100	£245	£270	
F.IV	150	£290	£350	£400 (4 wheel)
F.V	200	£365		£470
F.VI	250	£450		£550
F.VII	300	£520		£630
F.VIII	400	£590		£750
F.IX	500	£695		£850
F.X	750	£860		£1020

Between 1902 and 1913, the Company made a profit. The largest amount was £13,488-19s-4d for the financial year ending 31st July 1904. This was about the time that the Company had manufactured some of their most valuable orders. Up to 1905, the dividend distribution to the shareholders was 6%, but in 1906 it fell to 4%. No dividends were paid at all in 1907 and 1908, reflecting a small profit of £962-15s -8d, an amount lower than that usually carried forward from the preceding years.[12]

From 1910 onwards, although profits continued to decline, smaller dividends were paid, for example, up to 5% in 1911.

There was a need to raise capital with the sale of other assets, and in 1913, the Belgian property was sold.[13] Apparently there was no major development carried out on the Belgian site, although the Company does appear to have collected a rental from some of the property that occupied it. With capital brought forward from the previous year, and the Belgian sale, the Company was able to declare a profit of £7,345-6s-6d, although no dividends were declared.

The year 1914 has gone down in history as the start of World War I, but for the Directors of Hathorn Davey it was a year of change and may be remembered for other reasons.

Chapter 22 References

[1] HL. 1940. *Hathorn, Davey and Company, Sun Foundry, Leeds.* p.5

2 *Fifty years' history of the development of Green's Economiser, with notes on other economiser inventions and early tubular boilers* E. Green and Sons, Wakefield. Edited by Fowler W.H. 1897

[3] Mr. Henry Davey, M.Inst.C.E., FGS. 1904. *Page's Weekly.* 4th November. Pp.540-541

[4] HL. 1940. *Hathorn, Davey and Company, Sun Foundry, Leeds.* p. pp. 6 and 7

[5] LMWS HD 19933 Unsorted Letters

[6] Persons desirous of tendering. *The Leeds and Yorkshire Mercury*, 19th March 1904. p.3

[7] Meysey-Thompson, A.H. and Lupton, H. 1902-03. Some of the Considerations Affecting the Choice of Pumping Machinery. *Transactions of North of England Institute of Mining and Mechanical Engineers.*v.24. pp. 276 to 292

[8] Discussion on Messrs. Meysey-Thompson, A.H. and Lupton, H. 1902-03. Some of the Considerations Affecting the Choice of Pumping Machinery. *Transactions of North of England Institute of Mining and Mechanical Engineers.* v.25. pp.175 to 189.

[9] Discussion on Messrs. Meysey-Thompson, A.H. and Lupton, H. 1902-04. Some of the Considerations Affecting the Choice of Pumping Machinery. *Transactions of North of England Institute of Mining and Mechanical Engineers.* v.26. pp.87-90

[10] Engineers and Contractors for New Joint Dock. Hathorn, Davey and Co., Ltd. *The Daily Mail.* 26th June, 1914. p.6

[11] White, A.J.L. n.d. Concerning the New Plymouth Graving Dock. A reprint from the *Great Western Railway Magazine*

[12] LMWS Box 21D2 Bx1 19865. Balance sheets 1902 to 1934

[13] LMWS Box 21D2 Bx1 19683.10 to 19863.9 Belgium land sale. 5th February 1898

23. Sales of the Compound Engine: 1900 to 1914

There were two major achievements for Hathorn, Davey in the period leading up to the Great War. Firstly, the Company constructed their largest engine, the vertical compound engine bought by the *Waihi Gold Mining Company, New Zealand,* previously mentioned in Chapter 19. Secondly, Hathorn, Davey was awarded one of their largest contracts by the *Tasmanian Gold Mining Company, Limited.*

The Tasmania Gold Mining and Quartz Crushing Company, Limited had been working the Beaconsfield gold mine, Tasmania profitably since 1888, and had distributed some £769,000 in dividends until the early 1900s. However, as the mines were deepened, there was a great inflow of water into the lower levels that rendered it necessary to provide further expensive pumping machinery.

A new company, *The Tasmania Gold Mining Company, Limited* was registered in October 1903 to purchase the property and the assets of the previous company. The capital of £500,000 raised from the flotation of the new company would finance the purchase of the pumping equipment. When *John Taylor and Sons* were appointed both managers and mining engineers for the new company, they immediately drew on previous experience and purchased the necessary equipment from Hathorn, Davey.[1]

At a general meeting of the *Tasmania Gold Mining Company* in January 1904, a shareholder asked the Chairman who was supplying the pumping equipment. He replied, '*Hathorn, Davey and Co. of Leeds; and I believe they stand in the front rank as manufacturers*'.[2]

There seems to be no doubt that *John Taylor and Sons* had considerable experience with Hathorn, Davey's Compound Pumping Engines, and probably regarded the engines to be reliable and efficient. Taylors had previously ordered a number of Hathorn, Davey engines for the their mines on the Kolar Goldfield, India. These included the *Oriental Gold Mining Company of India* (Orders 5307 and 5358), *Champion Reef Gold Mining Company* (5348), *Balaghat Mysore Mines* (5394), *Mysore West Gold Company* (5596) and the *Ooregum Gold Mining Company of India* (5663).

Orders 6059, 6060 and 6061 for the *Tasmania Gold Mining Company*, placed on the 16th December 1903, were for three similar Horizontal Compound Differential Condensing Surface Pumping Engines (50"+108", stroke 10'), one for Hart's Shaft, and two for Grubb's Shaft. Orders were also placed for surface quadrants (a total of six), pumps (eight pairs of 20" plunger pumps with a 10' stroke), rising main (960 feet of 16" steel rising main) and pump rods (5,640 feet of spear rods). Together with the engines, the total cost was £48,784. The detailed specification in the order book gives some idea of the expected water problems. The three engines working together would be capable of pumping 8 million gallons a day from a depth of 2,000 feet.[3]

Payment was made in five instalments of 20% representing different phases of manufacture. However, penalties could also be incurred for late delivery. The first engine was to be ready for delivery (completed at Leeds) five months from the order being placed, otherwise a penalty of £5 to £20 per a day would be incurred. There were no penalty clauses attached to the second or third engine. '*No penalty would be forced for delay resulting from the "Act of God", such as strikes or the lock out of men or the failure of an important casting.*'

The engines and equipment were shipped to the small landing jetty at Beauty Point, on the west bank of the River Tamar estuary, Tasmania, and were transported to the mine site on robust logging wagons, towed by traction engines (Figure 113).

Figure 113. Beaconsfield Gold Mine 1904: Transporting the low-pressure cylinder and its end plate from Beauty Point to the mine.
(Courtesy of the Beaconsfield Mine and Heritage Centre, Tasmania)

There is an excellent photographic record of the mine being constructed, lodged in both the Tasmanian Archives and the Beaconsfield Gold Mining Museum, some of which show the engines being installed. Figure 114 shows the setting up of one of the low-pressure cylinders with end plate. Figures 115 and 116 shows the completed engine installation on Grubbs Shaft. The engines were started on the 4th February 1905.

Figure 114. Beaconsfield Gold Mine 1904: Fitting a low-pressure cylinder end plate.
(Courtesy of the Beaconsfield Mine and Heritage Centre, Tasmania)

Each engine was fitted with a separate surface condenser, having brass tubes and brass tube plates and independent air pumps. The pump work proper was of an exceedingly massive and interesting character. The quadrants, for instance, had arms each 15 feet centre to centre, and were built up of steel plates blocked with pitch-pine, the main quadrant bearings was 18" diameter and 24" long. The spear rods themselves were 22" square, and in lengths of about 47 feet, the joints being made with steel spear plates and bolts in the 'ordinary manner'. The pumps were fitted with outside packed ram plungers, each 20" diameter by 10' stroke. The valves were of the double-beat type, but arranged in multiples in separate boxes. There were eight suction and eight delivery valves to each plunger. (See Figure 117)

Figure 115. Beaconsfield Gold Mine 1905: Inside the Grubbs Shaft Engine House
(Courtesy of Tasmanian Archives: LPIC147/1/79)

Figure 116. Beaconsfield Gold Mine 1905: Compound Engine, Grubbs Shaft
(Courtesy of the Beaconsfield Mine and Heritage Centre, Tasmania)

Figure 117. Beaconsfield Gold Mine 1905: An underground pump lodge
(Courtesy of the Beaconsfield Mine and Heritage Centre, Tasmania)

The rising main, which is of steel, is 16 inches internal diameter, and has welded flange joints. As the weight of the spear rods and plungers is much in excess of the water resistance, balancing beams are arranged on each level, and relieve the quadrants of unnecessary strain; the working efficiency of these engines is 83 per cent, in relation of indicated horse-power to the actual or water horse-power.

With the main shafts being deepened, the *Tasmania Gold Mining Company* placed further orders, for pump rods (1,000 to 1,500 feet) and four complete sets of 40 tons balance beams at a cost of £2,565 (Orders 6229 and 6229A).

During more recent mining operations parts of the Hathorn, Davey pumps were discovered and are now on display including valve chests, clack valve chambers, and sections of the pump rods (spears) and quadrants (See Figure 118). The top of the massive pump rods can still be seen protruding from Grubbs Shaft. The Grubbs Shaft engine house and the adjoined boiler house now form the mining museum.

Figure 118. Beaconsfield Gold Mining Museum. The photograph shows a section of the 22" pump rods (spears) and part of an underground balance quadrant (right). *(Author 2011)*

In the early 1900s the pumping-system designed by Henry Davey for sinking the Manda Shaft, Miike Coalfield, Kyūshū, Japan was probably the largest Hathorn, Davey pumping-plant ever erected on one shaft.

The four engines are described in a number of publications, and are illustrated by Henry Davey in both editions of his book.[4,5] However, none of the engines appear to have an order number in the Hathorn, Davey Order Books. They would have been ordered at a time when the Sun Foundry was undergoing changes. Consequently, another engineering company appears to have manufactured the engines on behalf of Hathorn, Davey, but to Henry Davey's specifications.

The complete pumping-plant was capable of raising 9,000gpm from a depth of 900 feet when the engines were running at six strokes per minute. This equated to 58,000 tons of water in twenty-four hours, or 13 million gallons. The total horsepower in water lifted was 2,500.

Figure 119. Miike, Japan: The interior of the Manda Pumping Engine House.
The engine house was probably larger than that shown in Figure 115.
(Source not known: Adaby is a corruption of 'A Davey.')

The Manda shaft was rectangular 41 feet long by 12 feet wide, and was made to accommodate four sets of twin pit-work and two pairs of winding cages.

The complete system consisted of four pumping-engines that were arranged side by side in one large house. (See Figure 119) The surface condensers for the engines were placed in tanks outside the engine house. The cylinders were 45" and 90" in diameter and had a 12' stroke. Each engine actuated two sets of pump rods by means of a pair of quadrants.

The pithead frame was built of steel and had a total height of 101 ft. Over the winding pulleys were placed girders for carrying the capstan pulleys. The winding pulleys were 17 feet and the capstan pulleys 5 feet in diameter. Each capstan pulley could carry 30 tons.

A 30-ton steam capstan was placed at each end of the winding engine house for the purpose of lifting and lowering the pump work. Guide pulleys were fixed on each side of the pithead frame for leading the capstan rope to the overhead pulleys.

Each engine was provided with a pair of sinking pumps (23" diameter with a 12' stroke and suspended on steel rods), for the purpose of sinking in stages of 300 feet. On the completion of the first stage, plunger pumps were put in to take the place of the sinking pumps, and then the sinking pumps were used for another stage of 300 feet, when the operation of putting in plunger lifts was repeated.

Figure 120. Llanover Colliery, South Wales: The late George Watkins
views the 33"+66" cylinders Compound Engine.
(National Archives, Kew. COAL 80/613 - Photograph by the late John Cornwell)

The remainder of the period up to World War I saw a slow decrease in the sale of Horizontal Compound Engines. The large numbers sold in a single order, as per those to Tasmania and Japan above, were never to be repeated.

There were steady sales of individual engines to colliery companies. On the 9th October 1913 (Order 6676), the Bargoed Coal Company ordered a Horizontal Compound Tandem Differential Jet-Condensing Pumping Engine (33"+66") for the Llanover Colliery, South Wales. It cost £6,000. It had a similar situation to the Marine Colliery engine at Ebbw Vale in that it was sited in a undercroft at the side of the shaft, as shown in Figure 120.

The iron ore mining industry also ordered a number of Compound Engines. For example, the Lumpsey iron mine, near Middlesborough (Order 6320 dated 19th July 1907) bought one engine (20"+44") for £1,965.

Probably the last Compound engine to be bought for an iron mine was for the Llanharry mine, South Wales operated by the *Cardiff Haematite Iron Ore Company Ltd*. Order 6764, dated 22nd July 1915 (38"+72") cost £6,863, but it didn't actually start working until after the War on 11th January 1920.

Other Compound Engines were still being sold to water companies despite the more efficient Triple Expansion engine being very much in vogue. Established customers of Hathorn, Davey like the *South Staffordshire Waterworks Company* bought many engines including one for the Brindley Bank pumping station near Rugeley (Order 6118, dated 24th August 1904). This order also included two Lancashire boilers.

Compound engines were also ordered before the outbreak of World War I by the *Cambridge Water Company* for the Fleam Dyke pumping station. Order 6662, dated July 1913 was for a pair of Steam Compound Horizontal Rotative Pumping Engines (23"+51"). Each engine drove one horizontal force pump directly, and two bucket-lift pumps via quadrants. They pumped from a 162 feet deep well that had been sunk into the Chalk Marl horizon. Due to the War, the station didn't actually open until 1921. As R.C. Hodrien stated in 1976, *'The engines at Fleam Dyke represent virtually the ultimate development of their type, but even before they started pumping they were being overtaken by later developments - firstly by steam engines working at higher pressures and speeds, then by diesel and electric pumps.'*[6]

The last Horizontal Compound engine constructed by Hathorn, Davey was for the Mill Meece pumping station, near Eccleshall, Staffordshire, and it is included in a later Chapter.

Chapter 23 References

[1] Tasmania Gold Mining Company Ltd. New Issue and Prospectus. *Mining Journal*, 17th October 1903 (ii). pp. 433 and 435.

[2] Tasmania Gold Mining Company Ltd. *Mining Journal*. 2nd January 1904 (i). pp.10 and 11.

[3] Tasmania Gold Mine, Beaconsfield near Launceston, Tasmania. *Mining Journal* (Supplement), 1st July 1905 (ii). pp.i to iv

[4] Davey, H. 1900. *The Principles, Construction, and Application of Pumping Machinery*. Charles Griffin and Company Limited, London, pp. 110-117.

[5] Davey, H. 1905. *The Principles, Construction, and Application of Pumping Machinery*. (2nd Ed). Charles Griffin and Company Limited, London. pp. 118 to 125

[6] Hodrien, R.C. 1973. Fleam Dyke Pumping Station. *Cambridge University Engineering Society Journal*. v.43. (Reprinted by Cambridge Society for Industrial Archaeology 1976)

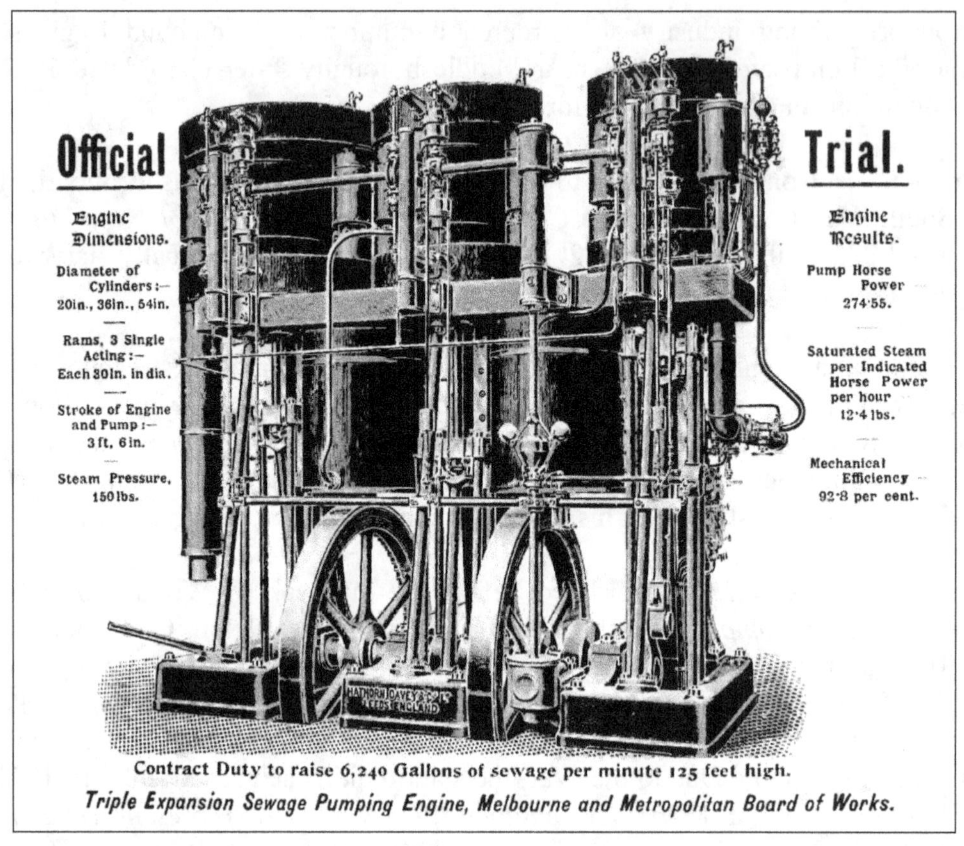

Figure 121: Hathorn, Davey advertisement featuring the Spotswood Pumping Engine
(Page's Weekly 17th March 1905 p.11)

Figure 122 A: The Spotswood Engine 2011   B: Counter (top) and Gauges
(Author 2011)

24. Triple Expansion Engines: 1900 to 1914

There is no doubt that by 1900 Hathorn, Davey's Vertical Triple Expansion Engine was held in high regard, both for it's design and efficiency. It was widely advertised and was in the *Catalogue for Water Works* as the 'Leeds Type' that was based on the original installation at Headingley, Leeds. The advantage of the engine was that pumps, sited in pits beneath the engine, could be driven directly from the engine's crossheads.

The first Triple Expansion Engine to be ordered in the 20th Century was for the Spotswood Sewage Works, Melbourne, Australia, and was widely used in Hathorn Davey advertising (Figure 121). Listed as 5729 in Order Book 8, dated 3rd July 1900. The engine is described as, *'One Pumping Engine including all fittings and accessories, with steam and delivery pipes...(20"+36"+54", stroke 3' 6"). Engine of the Inverted, Vertical, Triple Expansion, Corliss, surface condenser, rotative three-crank type - steam pressure 150 psi. Pumps placed vertically under cylinders and worked from engine crossheads by side rods and columns.' 'To raise from the suction wells through the delivery mains to the head of the outfall sewer, not less than 1000 cubic feet (6240 gallons) of sewage per minute, under an equivalent head of 125 feet from the invert of the suction well to the invert of the outfall sewer.'*

Details were also provided of the expected efficiency of the engine and pumps, with a penalty if it did not fulfil those requirements. *'We guarantee when the engine is working at full load a duty of 120 million pounds of sewage (calculated at the rate of 65 pounds per cubic foot) raised 1 foot high for each 1120 pounds of dry saturated steam supplied to the engine at a boiler pressure of 150 psi. Should the engine fail to give this duty, £150 to be deducted for each million pounds the duty falls below the rate specified. Previous to the engine commencing the duty trials, the engine shall be employed in pumping and for such time as may be necessary for thoroughly testing it. The test to be made by, and at the expense of, the purchasers or their representatives. We however* [Hathorn, Davey] *reserve the right to be present or represented at the official tests to check over the various readings or measurements recorded during the trials.'* Figure 122 B shows the array of gauges that would have been observed in the trials.

The specification for the contract is very detailed and covers twelve pages of the Order Book. The two flywheels for example, were 10' 6" diameter, each to weigh about 6 tons, cast in two halves with a polished outer rim. One flywheel was to have a barring rack on the underside of the rim.

The engine operated, *'Three Main Pumps of the vertical single acting ram type. Rams of cast iron 30" diameter and a 3' 6" stroke.'*

The cost of the order totalled £11,708-10s-0d, that included delivery and erection at Spotswood. Mr Nicholson who was an Erector, presumed to be employed by Hathorn, Davey installed the Engine. He was assisted by one fitter and three or four labourers. A detailed diary of their work exists at Spotswood, and is summarised as:[1]

28th Oct. 1901	Work started fixing penstock to the pump well.
9th to 22nd Nov. 1901	Hathorn, Davey parts being delivered on site.
16th to 31st Dec. 1901	Bolt holes for pumps bored by Calyx drill.
2nd Jan to 19th Feb. 1902	Pump chambers being fitted out - girder work & pumps
20th Feb to 19th Mar. 1902	Lower section of engine installed - engine bed plates / plungers / crank shafts / flywheels / entablature columns
20th March 1902	Placing entablature on columns.
22nd to 27th Mar. 1902	Placing high-pressure cylinder & base on entablature etc
1st Apr. 1902	Placing intermediate-cylinder and con rods.
2nd to 10th Apr. 1902	Completing intermediate-cylinder and placing low-pressure cylinder etc.
11th Apr. to 29th May 1902	Installing all pipework and Corliss valves etc.
30th May to 23rd July 1902	Floor plates / lagging / steam piping / cylinder covers etc
24th July to 5th Aug. 1902	Expansion joints fitted / general lubrication / adjustments
6th Aug. 1902	Steam turned on by mechanical Engineer at 11:50am. Engine started at once and easily worked with by-pass opened till 12:10. Engine stopped to adjust Governor gear. Started again at 3:30pm and ran till 4:30 pm.
15th Oct. 1902	Engine trial.

The engine must have passed the trial (referred to in Figure 121). Its success encouraged four more Triple Expansion engines to be ordered for the works from the *Austral-Otis Engineering Company*, Melbourne. The design for the four engines was based on Hathorn, Davey's engineering drawings.

Extensive duty trials were conducted for comparison between the Hathorn, Davey engine, one of the Austral-Otis engines, and an earlier sewage pumping engine manufactured by Thompson's Foundry, Castlemaine, Victoria. The results of the trials, *'proved that the Hathorn, Davey was capable of performing at a duty of 163 million foot-pounds of sewage lifted for each 1,120 pounds of steam consumed.'* This exceeded the guaranteed rate of 120 million foot-pounds by 36%, and was 28% better than the Austral-Otis engine.[2]

It has been calculated that the Hathorn, Davey engine pumped 91,850 million litres (20,204 million gallons) of sewage until it stopped working in about 1925. The Spotswood engine still survives *in situ* (Figure 122A). When visited by the Author in 2011 the intention was to get the engine working again.

In addition to Australia, Triple Expansion engines were exported to South America, South Africa, and India. In most cases they were used to supply water.

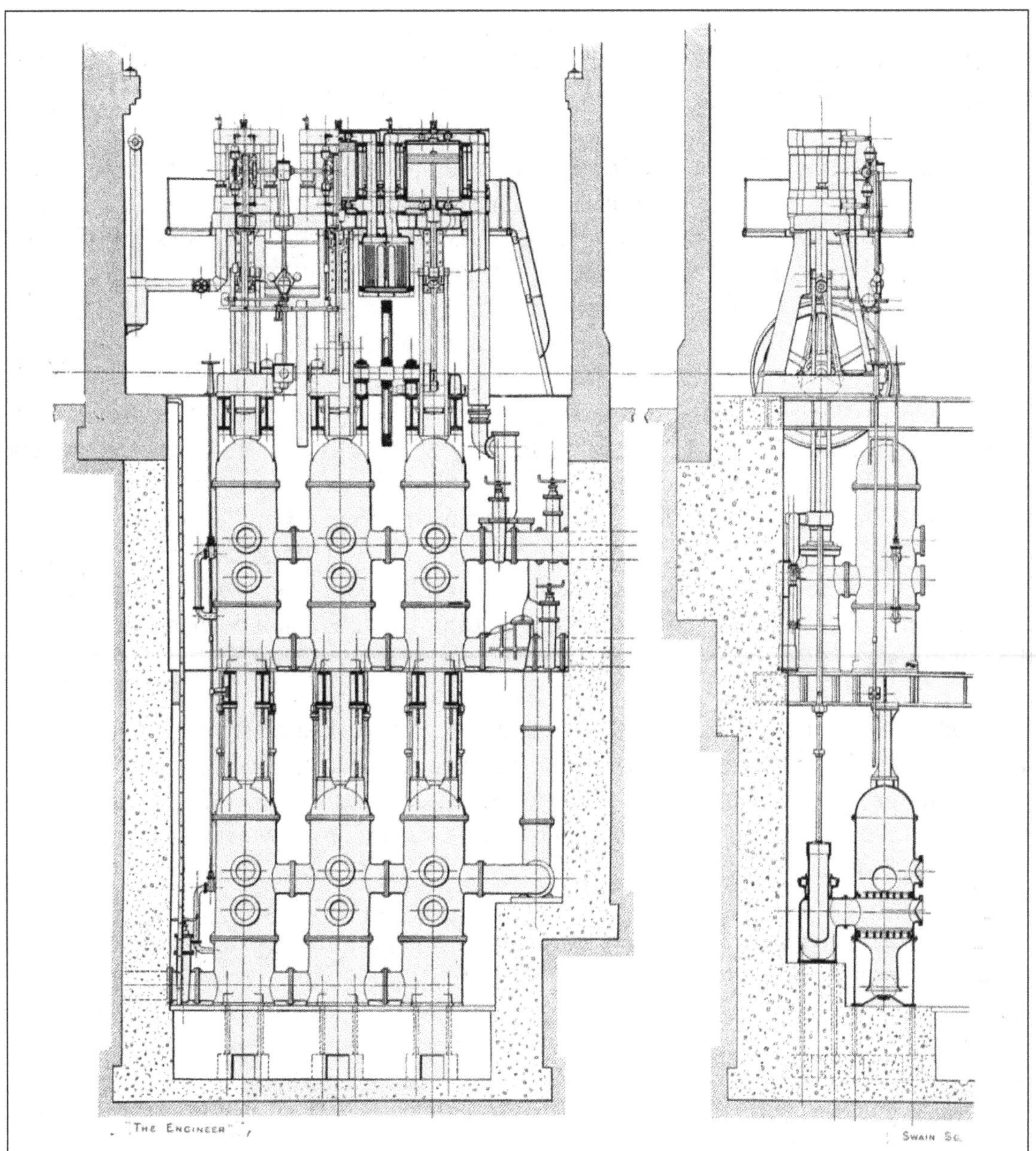

Figure 123. Rosario, (Order 5988): Elevation showing the
Triple Expansion Engine (15"+25"+40") and the two lifts of pumps.
(The Engineer. August 18th 1905. p.156 [GG])

231

Several water companies were formed in London to supply water to Argentinian and Uruguayan cities and bought from Hathorn, Davey Triple Expansion Engines for their waterworks. They were:

Argentina - Buenas Aires: San Fernando Waterworks.
6389   - 14th October 1908   - Buenos Aires, Argentina (12"+20"+31")

Argentina - Rosario: Consolidated Waterworks Company of Rosario de Santa Fe
5988   - 18th February 1903 - Rosario, Argentina (15"+25"+40")
6252   - 7th June 1906   - Rosario, Argentina (15"+25"+40") - Cost £4730
6322   - 25th July 1907   - Rosario, Argentina (15"+25"+40") - Cost £6566
6409   - 11th March 1909   - Rosario, Argentina
6474   - 14th July 1910   - Rosario, Argentina (20"+36"+54") - £10,146

The engine (5988) supplied water to Rosario from the River Parana.[3] It drove three single-acting ram plungers 16" diameter (high-lift), and a further 16" diameter pumps for the low-lift (all having a stroke of 3') at a speed of 36 revolutions per minute, which corresponded to a plunger speed of 216 feet per minute. (See Figure 123)

The pumps were designed to pump at the rate of 4,000,000 gallons per 24 hours against a head on the high-lift pumps of 170 feet, including friction, and a head of 55 feet with the low-lift set of pumps.

The boilers were supplied by Babcocks and produced superheated steam at 200psi. The high and intermediate pressure cylinders and their respective receivers are jacketed, with boiler pressure steam, while the low pressure cylinder is jacketed with steam at a reduced pressure. The position of the receivers is such that all water drains into them and is re-evaporated.

Butler (1922) provided a brief description of the engine's mode of operation.[4] *'The valves are gun-metal, faced with rubber, and are of the multiple type for both sets of pumps, and of a waterway capacity exceeding the area of the plungers. The three cranks are set at 120°, the two outer cranks being overhung; and on the shaft are keyed two built-up flywheels, one situated between each pair of bearings and on each side of the middle crank. The distributing valves in each steam cylinder are of the Corliss type, operated by a Craig trip motion, and are opened positively by hardened steel dies fixed on small eccentric clips that engage with other corresponding dies attached to levers on the valve spindles; which eccentric clips receive their motion from the valve shaft, and, as seen, are placed in front of the cylinders, where they are driven by cut gear wheels from the crank shaft. By means of this improved form of a trip gear a very rapid steam opening is obtained, and the dies keep in good condition for years without renewing, owing to the sliding motion of one die upon another tending to keep the cut-off edges sharp. The exhaust valves are all operated by cams, and each steam valve is separately adjustable for lap, lead, cut-off, cushion, and release while running. An automatic safety controlling device operated by an hydraulic governor is fitted to cut off the steam supply, and destroy the vacuum in case of loss of load, resulting for instance from a fractured pipe.'*

Uruguay - Montevideo: Montevideo Waterworks Company, Limited.
6258   - 6th July 1906   - Montevideo, Uruguay (23"+43"+64") - Cost £10,597
6453   - 26th February 1910 - Montevideo, Uruguay (23"+43"+64")

Figure 124. Hathorn, Davey engines, Montevideo Waterworks, Uruguay
(https://www.youtube.com/watch?v=HmJUmNl1A1I)

The *Montevideo Waterworks Company, Limited* was formed in 1888 to supply water from the River Santa Lucia, at a point about 33 miles northwest of Montevideo. New waterworks were ultimately constructed at Aguas Corrientes, and a pumping station built to accommodate the engines. Water was pumped to a reservoir constructed at Las Piedras, Montevideo. The engines still survive, and some are believed to be still in working order. (See Figure 124)

The Crown Agency for the Colonies in the British Empire was distinct from the British Civil Service, but carried out similar functions. However, they do not appear to have to placed orders for India perhaps because it had a different status within the Empire. The Agency occasionally placed orders for Triple Expansion engines that included:

Figure 125. The Zwaartkopjes Pumps: Single-acting ram, with 12" plungers, stoke 36"
(The Engineer. 22nd November 1907. p.520 [GG])

A number of engines were ordered for the Rand Water Board, South Africa. Order 6226 dated January 1906 and 6245, 11th May 1906 were for four Triple Expansion engines. The latter order cost £22,782 and included boilers etc. It was order 6321, dated 24th July 1907 for Zwaartkopjes, (23"+43"+ 63") (Cost £8,327), that seemed to receive the most attention. This engine underwent extensive trials and it was mentioned in various articles. [5] Figure 125 shows the set of pumps supplied by Hathorn, Davey.

Other Triple Expansion engines (12"+23"+34") ordered by the Crown Agency were Orders 6604 and 6604A dated 10th July 1912 for Lagos Water Supply. Two were built to the same specification, one for high lift, and the other for drawing water from a well.

The Japanese coal industry was still showing an interest in Hathorn, Davey engines. One engine in particular should be mentioned.

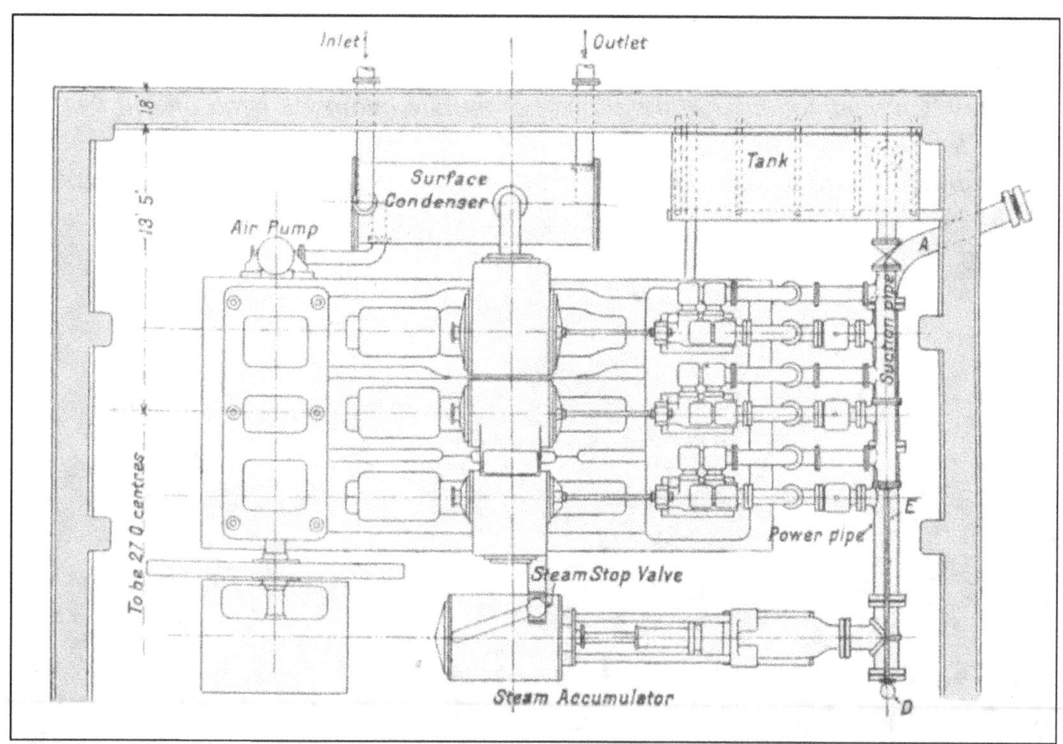

Figure 126. Miike: Plan showing the surface layout of the engine and accumulator
(The Engineer. 15th June 1900. p.614 [GG])

Hathorn, Davey order number 5678, dated 10th July 1899 was for a Horizontal Three Cylinder Compound Condensing Rotative Pumping Engine and pumps that cost £6,835, for Miike. In a report dated 1909, '*Mining in Japan, past and present,*' there is confirmation that this engine was installed at the Miyanoura Pit, owned by the *Mitsui Company*, and managed by Takuma Dan.[6] There is a description of the engine in *The Engineer* (15th June 1900) that includes a plan of the surface installation (Figure 126) and photographs of the Triple Expansion Engine (Figure 127) and the pumps (Figure 128).

The description of the equipment published in *The Engineer* is as follows.[7]

'*Hydraulic Plant. The plant is for the purpose of draining the dip or inclined workings of the mine.*'
> '*The coal is worked from the bottom of the shaft to the dip, and the water follows the working of the coal. It therefore becomes necessary to have the pumping*

plant to raise the water from the coal workings to the sump at the bottom of the shaft, where there are situated plunger pumps worked by a surface engine, and employed in pumping all the water of the mine to the surface.'

'The coal dips at about 1 in 10 and the workings at present extend to about 3500 ft from the shaft bottom. At this point, the two hydraulic pumps will be fixed, where they can pump about 1,400 gallons per minute against a head of 350 feet Pilot pumps will be employed to follow the coal workings for the purpose of lifting the water to the main hydraulic pumps. The consumption of steam is, however, very great, and there are obvious practical objections to taking steam into the workings so far from the pit bottom. '

Figure 127. Miike: The Triple Expansion Rotative Power Engine (20"+29"+41").
(The Engineer. 15th June 1900. p.613 [GG])

'The plant consists of a Triple Expansion Rotative Power Engine and steam accumulator placed on the surface of the mine, as well as the two duplex pumps placed underground 3,500 feet from the pit bottom, together with the pilot pumps, and necessary connecting piping. The rams of the duplex hydraulic pumps are 14inches diameter, and capable of running at fourteen double strokes per minute.'

'The system of working can be described. The power water is taken from an overhead tank in the engine room, and forced by means of the power engines into 7 inch pipes leading down the shaft to the hydraulic pumps, and is returned by the hydraulic pumps to the tank, so that the power water is used over and over again.'

Figure 128. Miike. The underground duplex hydraulic pumps with 14" rams.
(The Engineer. 15th June 1900. p.613 [GG])

'There are two sets of power water pipes, one for supply and the other for return. The power engine pumps the water under a pressure of 1000 lbs per square inch., and that pressure is kept constant by means of the steam accumulator, or regulator.'

'The accumulator consists of a steam cylinder 40 inches diameter, and a hydraulic ram 12½ inches diameter. The hydraulic cylinder of the accumulator is in free communication with the power pipe, whilst the steam piston is acted on by steam at a constant pressure, this pressure being adjusted by means of a reducing valve so as to maintain a constant pressure of 1000 pounds per square inch on the hydraulic ram. The steam accumulator forms a governor to the power engine. If the power engine runs too fast, then the ram is pushed out, and made to actuate the throttle valve.'

'The power engine has cylinders 20", 29" and 41" diameter, with a common stroke of 2 feet 4 inches. The pumps are of the ram type, worked directly from the pistons. Each ram is 5¼in. diameter by 2ft. 4in. stroke. There are two suction and two delivery valves for each pump.'

'The power water will be used over and over again, and will have a little oil added to it for the purpose of keeping the working parts in good order.'

Henry Davey appears to have been a Consultant for the *Mitsui Company*, and although Davey doesn't appear to have visited Japan, he clearly had met and befriended Takuma Dan on one of Dan's five visits to Europe. After the installation of the Miyanoura Pit pumping system Dan wrote to Davey on the 18th June 1902.[8] *'I am pleased to say that the pump fully satisfies our expectation, and I can confidently predict a great future for it, at least for shaft work, if not for all services. It works smoothly without any shocks, as I first feared, and comparative little care and experience required in running it with absolute safety and comfort of the attendants. In regard to economy, the advantage is also very conspicuous. It requires about 11 pounds of bad coal per delivery horse power per hour compared with over 30 pounds of coal for Duplex steam pump at the same distance from the boiler. In fact, it compares very favourably with your differential pumps working in the shafts.'*

Dan however was not too impressed with the pipe work to the pumps. *'The great defect is in the cast iron pipes which altogether too unwieldy for dip work. The position of dip pumps must naturally, be occasionally shifted and never be so permanent as at shaft bottom, so that the pipes must be of such weight and size as to be easily handled and fixed without extensive preparation in the pipe way etc; therefore unless lighter material be used this system, even if it is otherwise perfect will not be widely introduced for dip working.'*

Dan concluded that electric pumps would supersede this system, but perhaps to soften the impact of his comments, *'However, for heavy shaft work (not for sinking) the hydraulic possesses all the advantages of your differential engine without its inconvenience or danger, and no doubt be the pump of the future.'*

He finished the letter, *' I send enclosed a picture of my little girls as promised to Mrs Davey.'* suggesting that the Davey's and Dan must have met socially on at least one occasion.

Dan may have also influenced the purchase of two engines for the Onga River pumping station located near Nakama City, Kyūshū, Japan. The pumping station supplied water by an 11.4 Km pipeline to the Imperial Yawata Steel Works located further to the east. In 1905 Dan was appointed Director of the Japan Steel Works, and as a member of the Government Commission, did much towards the shaping of Japanese national policy with regard to the iron and steel industry.

Figure 129 Onga River Engine House (Left) and Boiler House, Nakama City, Kyūshū, Japan. The arc cut into the brickwork accommodated the roof covering for the Green's Economiser. It was once sited in the area to the left of the people. The water pipes and the flue gases passed through the two walled up arches.
(Author 2017)

Order 6314 dated 10th May 1907 was based on a quotation given on October 1906, for two sets of pumping engines. The original quote must have been for 15"+25"+40" cylinders, but this was increased to 16"+28"+44". Each engine would be capable of pumping at the rate of 2500gpm, and the engines would operate on a steam pressure of 140psi. The two engines were very much the 'Leeds' type conforming very much to that illustrated in the 1899 Hathorn, Davey catalogue.

The main pumps were vertical single acting ram types 14½" diameter with a 3' stroke to match the engine.

A Green's Economiser was also purchased and installed in an open building between the boiler house and the chimney.

The two engines were scrapped sometime in the 1950s when they were replaced by electric pumps.

The engine and adjacent boiler houses still survive (See Figure 129), and have now been included in the Japanese list of UNESCO World Heritage Sites devoted to monuments that represent the industrialisation of Japan during the Meiji Restoration (1868 to 1912). [9]

In addition to the orders listed above, other foreign orders included:-
Port of Spain, Waterworks, Trinidad (5798)
Odessa, Russia (5806)
Bloemfontein, South Africa (6027)
Badami Bagh, Bombay, India (6469)
Ahmedabad, India (6493)
Newcastle and Hunter Rivers water supply, Sydney, Australia (6549)

The period 1900 to World War 1 was a time when Triple Expansion Engines were very much in vogue, particularly for waterworks, and Hathorn, Davey had many competitors in this field. It is pleasing to note that such Hathorn, Davey engines still survive in Australia and Uruguay, to compliment those that still work in England, at the London Museum of Water and Steam, Kew Bridge, and at Twyford, Hampshire, for example.

Chapter 24 References

[1] The Author would like to express his thanks to Matthew Churchward for this information.

[2] Facts and Figures on No.5 Pumping Engine, Spotswood. Museum Victoria information sheet. Scienceworks Museum, Spotswood, Melbourne, Australia.

[3] Pumping Engine - Rosario Waterworks. *The Engineer*. 18th August 1905. pp. 156 and 157

[4] Butler, E. 1922. *Modern Pumping and Hydraulic Machinery*. 2nd Edition. Charles Griffin and Company, London. pp.27 and 28

[5] The Rand Water Board Plant and Works. *The Engineer*. 22nd November 1907. pp. 516 to 518, 520

[6] Anon. 1909. *Mining in Japan, past and present*. Report by Japanese Bureau of Mines, Department of Agriculture and Commerce of Japan. pp.291 to292.

[7] Hydraulic Power Plant. *The Engineer*. 15th June 1900. pp. 613 and 614

[8] LMWS. Box. 21/D2 BX3 Letter from Dan to Davey. 18th June 1902.

[9] Kato, K. n.d. Site of Japan's Meiji Industrial Revolution. Department of Industrial Heritage, Cabinet Secretariat, Government of Japan.p.45

25. Engines and Pumps sold in Britain: 1900 to 1914

To compliment the worldwide sales of the Triple Expansion engine, there was also an increase of orders in Britain. By far the largest numbers of sales were for water supply companies. The engine didn't prove popular with British colliery companies perhaps because of its size and space required to accommodate it at a shaft top. The vertical form was certainly too big for installation in an undercroft, and perhaps the horizontal version was too long. However, those potential problems did not deter the iron mining industry of northwest England from purchasing several.

Figure 130. Triple Expansion Engine: The Newmarket engine reinstalled at the London Museum of Water and Steam, Kew Bridge.
(Photograph by Gordon Ridgewell)

The Hathorn, Davey engine (Order 6421, dated 28th June 1909) is described as an Inverted Corliss Vertical Triple Expansion Surface Condensing Pumping Engine (12"+20"31", stroke 30"), and was installed at the South Field Pumping Station, Newmarket. It is one of several Hathorn, Davey engines that can still be seen working in England, albeit not in its original location (Figure 130).

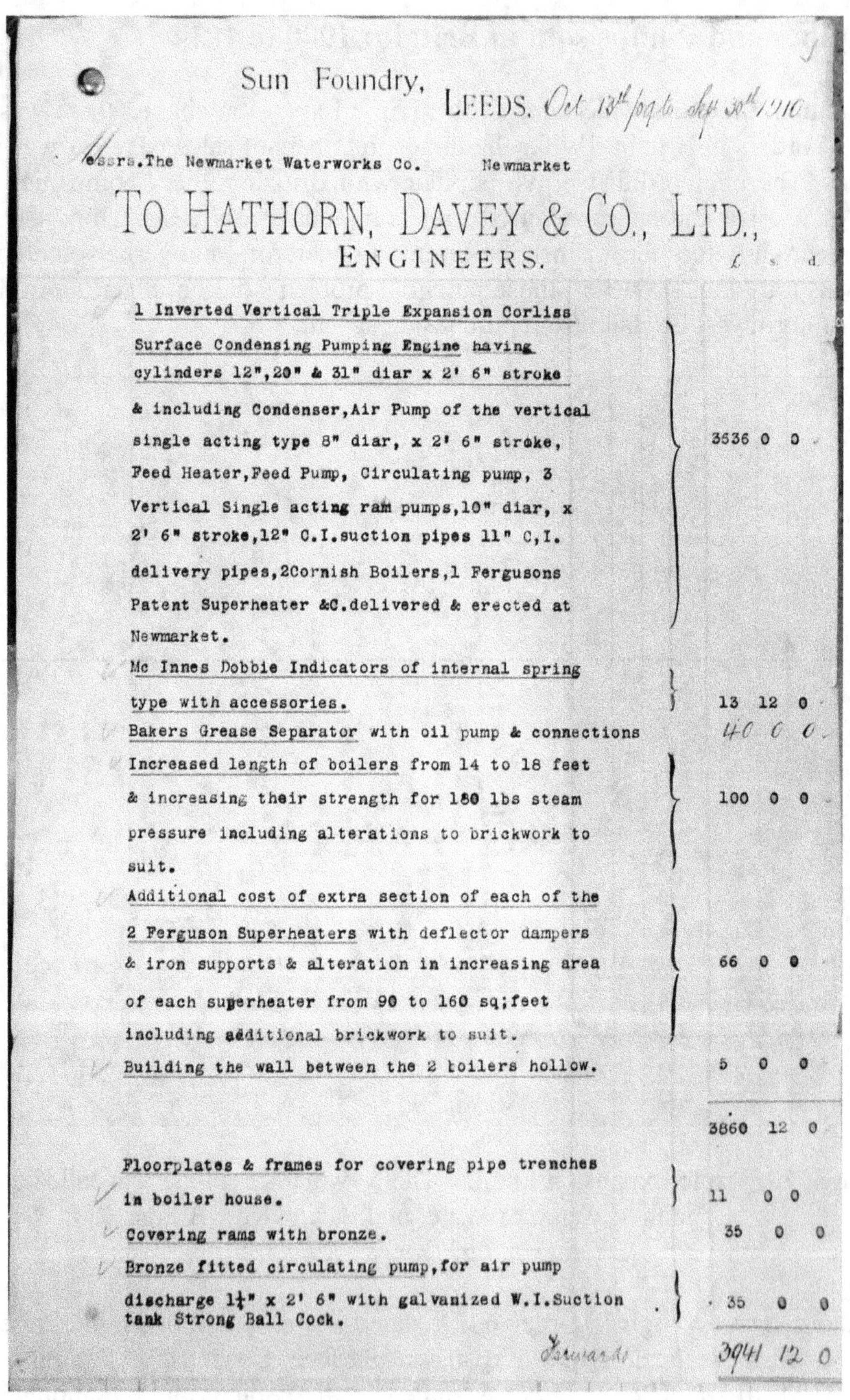

Sun Foundry,

LEEDS. *Oct 13th /ogt6 dep 30th 1910*

Messrs.The Newmarket Waterworks Co.   Newmarket

To Hathorn, Davey & Co., Ltd.,
Engineers.

	£	s	d
1 Inverted Vertical Triple Expansion Corliss Surface Condensing Pumping Engine having cylinders 12",20" & 31" diar x 2' 6" stroke & including Condenser,Air Pump of the vertical single acting type 8" diar, x 2' 6" stroke, Feed Heater,Feed Pump, Circulating pump, 3 Vertical Single acting ram pumps,10" diar, x 2' 6" stroke,12" C.I.suction pipes 11" C,I. delivery pipes,2Cornish Boilers,1 Fergusons Patent Superheater &C.delivered & erected at Newmarket.	3536	0	0
Mc Innes Dobbie Indicators of internal spring type with accessories.	13	12	0
Bakers Grease Separator with oil pump & connections	*40*	*0*	*0*
Increased length of boilers from 14 to 18 feet & increasing their strength for 150 lbs steam pressure including alterations to brickwork to suit.	100	0	0
Additional cost of extra section of each of the 2 Ferguson Superheaters with deflector dampers & iron supports & alteration in increasing area of each superheater from 90 to 160 sq;feet including additional brickwork to suit.	66	0	0
Building the wall between the 2 boilers hollow.	5	0	0
	3860	12	0
Floorplates & frames for covering pipe trenches in boiler house.	11	0	0
Covering rams with bronze.	35	0	0
Bronze fitted circulating pump,for air pump discharge 1¼" x 2' 6" with galvanized W.I.Suction tank Strong Ball Cock.	35	0	0
Forwards	*3941*	*12*	*0*

Figure 131 Hathorn, Davey: Quotation for the Newmarket engine.
(Courtesy of LMWS Item 21-D2-BX3)

Notes kept by the engineer who inspected the engine in 1936 still survive, and as would be expected, it contains information on valve settings that are described, and illustrated, in detail.[1]

The total cost of the order, as per Hathorn, Davey's quotation (Figure 131) was £3,941-12s with additional charges for the pumps (three Ram Pumps 10" diameter) and boilers (two Cornish Boilers).[2]

The engine was officially started during March 1911 and could lift 55,000gph, but was expected to work at about 50% of this amount, as the engineer stated, *'Can raise 720,000 gallons/day easily.'* The engineer's notes also record that the pumps had a mechanical efficiency of 92%. Various parts were occasionally replaced, particularly valves.

The Newmarket engine finally stopped pumping in 1962. In the late 1970s, the owners of the engine, the *Anglian Water Authority* donated it to the *London Museum of Water and Steam,* Kew Bridge, London where it was re-installed and returned to steam in 1980. The well-maintained engine, now 110 years old, is in remarkable condition and still works.

Figure 132: Twyford Pumping Station, Hampshire. The remains of the 'pitman.'
(Author 2020)

There is also a working Hathorn, Davey Triple Expansion Engine at the Twyford Pumping Station, Hampshire (Order 6579 dated 2nd February 1912). The engine (16"+28"+46", stroke 36") worked two pumping systems. Firstly, pumps located beneath the engine forced the water via a pipeline to a header reservoir. Secondly, a crank at the end of the engine's main crankshaft drove a short line of reciprocating con rods that connected to a pair of pumping quadrants (Figure 132) located at the top of the water supply well.[3] The latter arrangement, known as the 'pitman', is a term commonly used for a coal miner. The term may be derived from the fact that this pumping arrangement was more commonly associated with Horizontal Compound Engines frequently used for pumping coalmines.

Figure 133. Ilford Waterworks, Essex. Spur and pinion gearing to work the 'pitman'
(Courtesy of Chris Allen)

One of the best examples of the 'pitman' arrangement worked by a Triple Expansion Engine (15"+25"+40", stroke 36") was at the Ilford Waterworks, Essex (Order 6177). Alongside the flywheel, the large spur wheel / pinion system converted rotary to reciprocating motion via a crank, that worked two pumping quadrants (See Figure 133).

Four Triple Expansion engines were purchased by mining companies working the north-west Lancashire / Cumberland iron ore deposits. They were the *Bigrigg Mining Company* (6566, 22nd November 1911), *Ullcoats Mining Company* (6646, 9th May 1913), *Beckermet Mining Company* (6703, 16th July 1914), and the *Millom and Askam Haematite Iron Company* (6709, 18th August 1914). The pumping system employed was probably similar to that used at Twyford.

Figure 134. The Kesteven Asylum Engine
(Courtesy of LMWS Item 21-D2-BX3 - Pamphlet)

In addition to the major water suppliers that utilised the large Triple Expansion engines, smaller stationary engines were often purchased for independent water supplies at factories, and for example, mental health asylums.

The Kesteven Asylum, Lincolnshire (order number not known) bought a relatively simple engine that was described in an article in *Engineering*.[4] It is shown in Figure

134. Most asylum water supplies were usually extracted from boreholes, which were cheap to construct, compared to a well. At Kesteven, '*Owing to the length of rods in the borehole, quick running and short stroke is inadmissible. To get a fairly high pump speed, it is necessary to adopt a long stroke, which also involves less wear and tear and consequent breakdowns in the hands of comparatively unskilled men employed about small installations of this kind; whose attention, moreover, can often not be exclusively devoted to the pumping machinery.*' Basically, this was a low maintenance engine a few steps up from Davey's Domestic Motor, but working on a steam pressure of 80psi.

The engine was in essence a self-contained 'beam' engine (10" with a 24" stroke) that worked both a borehole pump designed to raise 5000gph, to a tank 20 feet above the engine house floor, and a force pump that raised the water to a reservoir 80 feet above ground level. '*The delivery from the force pump passes through the dome topped-casting which supports the beam. The upper part of this casting serves as an air vessel, while the surface condenser is placed in the lower half. The valves are worked by Davey's Differential Gear. The pumps deliver on both the up-and-down stroke, and the ram at the top being half the area of the bucket. It will be noted that the engine is self-contained, on a cast-iron bedplate which rests on a concrete bed; the cost of the installation is reduced to a minimum.*'

Figure 135. West Sussex Asylum Waterworks Engine
(Hathorn, Davey Catalogue for Waterworks 1899 pp.24 to 27)

An illustration in the Hathorn, Davey Catalogue for Waterworks 1899 [5] shows another example of an asylum installation that was probably specially manufactured for the West Sussex Asylum Waterworks. (See Figure 135)

Described as the 'West Sussex Type', a Rotative Geared Engine for well and borehole pumps, it consisted of a vertical single cylinder from which motion is transmitted by means of a spur gear or belting to the bucket pump crank at the top of the well or borehole. A single pump was used, and a counterweight was required to ensure a constant discharge. As in the Kesteven example, it was advised to deliver the water from the source into a tank and then re-pump by another means to the required height.

The Catalogue also illustrated a variety of heavy-duty pumps that included direct acting differential steam pumps (Kettering type), and three-throw single-acting ram pumps (Leeds type) sited directly beneath a Triple Expansion Engines. A small rotative steam pump (Lord Durham Type) was a double-acting ram pump attached directly to the engine piston rod. This was more suitable to be used as a boiler feed pump and could deliver heads of water up to 400 feet.

Figure 136. Types of two and three throw pumps
(Hathorn, Davey Catalogue for Waterworks 1899 pp.66 and 67)

Other types of boiler feed pumps were duplex, described as horizontal, direct-acting duplex, with two cylinders placed side-by-side and working two double-acting plungers directly attached to the engine piston-rods.

One section of the Catalogue is devoted to pumping machinery driven by belting, ropes or electric motors. They could raise small quantities of water to a moderate height, and were compact inexpensive pumps with few working parts, but were strong and moving parts were easily accessible. Figure 136 shows two examples.

However, various changes in pumping technology occurred in the early 20th century. There was a move away from steam power to petrol, diesel, and electricity driven installations. The change did not happen overnight. It was gradual and the transition period probably lasted until the 1950s, interrupted by the two World Wars.

In the first decade of the 20th century, hydraulic pumping still occupied a niche in the market and Hathorn, Davey sold several systems in this period.

The Company did try to embrace the new technologies as their portable petrol motor driven centrifugal pump demonstrates (Chapter 22). It is also apparent from the Order Books that centrifugal pumps (usually driven by an electric motor) were starting to gain a foothold in the market.

Figure 137.Plymouth: Centrifugal pumps driven by steam cross-compound engines
(Plymouth Graving Dock. A reprint from the Great Western Railway Magazine)

A centrifugal pump is designed to move liquids using rotational energy by means of rapidly rotating rotors, or impellers. Liquid enters the rotating impeller along its drive axis and is thrown out by centrifugal force. This increases the liquid's velocity and pressure and directs it towards the pump's outlet.

Hathorn, Davey evidently conducted various experiments with centrifugal pumps. Order 6274, dated 25th October 1906 for Hathorn, Davey was for four 5" centrifugal pumps, and a bedplate on which to fix them, and was stated to be for experimental purposes.

Other orders for centrifugal pumps are also listed. Order 6283, dated 7th December 1906 was for two steam-driven centrifugal pumps with impellers 30" and 24" diameter. It was placed by Leeds Corporation for the Knostrop Sewage Works. Shortly after an order (6293, dated 11th February 1907) was placed by the *Great Western Railways* for an '*Electrically Driven High Efficiency Centrifugal Pump*,' capable of pumping 150 to 250gpm a height of 14 feet through a 6" rising main. The pump would be driven by a three-phase electric motor manufactured by Ernest Scott and Mountain. It was intended to install the pump in the City Subway at Paddington Station.

Sales of the centrifugal pump increased for a variety of uses. Professor Thompson, Leeds University Mining Department, placed an order (6357, 24th February 1908). The *Great Western Railway* bought several more centrifugal pumps from Hathorn, Davey for various applications. They installed two pumps (Figure 137) that were driven by Cross-Compound Steam Engines at their Plymouth Graving Dock. The system was designed to lift 16,000gpm of seawater a total height of 32 feet. [6]

Two years later the railway Company purchased another for their Power Station at Fishguard.[7] '*A single-stage centrifugal pump has recently been constructed by Hathorn, Davey and Co., Leeds, to work under unusual conditions. It has been built to supply condensing water for the Fishguard Power Station of the Great Western Railway Company, and has to be capable of delivering varying quantities of seawater against heads which are liable to alteration with the tides. In order to comply with these conditions, the impeller has been designed to give a slow velocity to the water which passes through it, and the vortex chamber has no guide vanes or other obstructions. The chamber is built up of renewable manganese bronze rings. The impeller is balanced hydrostatically for end thrust by means of a by-pass arrangement, and the bearings are of ample length, well lubricated and automatically cooled by a constant stream of water. The stuffing-box is water-sealed. The pump is driven by a direct-current motor, to which it is coupled by a flexible coupling. It has a 7" inlet and 6" delivery branches. The following are particulars of tests carried out with the pump at the makers' works:-*

Static head (feet)	Gallons per minute	Revolutions per minute
25	408	553
21	750	540
16	930	550
26	1350	790
10	1620	790
56	300	790

These heads were obtained by adjusting a valve on the delivery branch, on which was fitted a pressure gauge, and the quantity of water passed was carefully measured.'

It is evident that Hathorn, Davey was trying to increase their range of pumps. Order 6358 (21st February 1908) for Dowlais, South Wales, was for a *'Patent Three-Stage Electrically Driven High Lift Balanced Turbine Pump'* capable of raising 334gpm at a speed of about 950 revolutions per minute.

Whilst the majority of the centrifugal pumps were used for raising water, others were being developed for sewage treatment. In 1913 and later, Hathorn, Davey produced the Stereophagus pump. Lupton described it as, *'an eater or rather chewer of solids. It is a centrifugal pump, but fitted with a cutting blade, so that it not only pumps the sewage, but also divides it into small pieces not liable to clog the delivery mains.'* [8]

Figure 138. The Stereophagus Pump
(*Engineering. 5th April 1912. p.444*)

R.C. Parsons first designed a special form of the centrifugal pump, driven by electricity. The Stereophagus pump was designed to meet the demand for a cheaply operated installation capable of dealing with unscreened sewage at small sub-stations.

Parsons filed his patent [9] in 1910 and set up the *Stereophagus Pump and Engineering Company, Limited* in December of the following year to acquire the right for manufacture and trade. In 1913, Hathorn, Davey manufactured a number of the Stereophagus Pump under license.

Figure 138 shows the components of the pump.[10] The conical impeller is provided with a number of spiral blades running out of the point. A fixed plate (which Parsons in his patent calls a shearing plate) nearly rubs up against the rotating impeller blades and cuts up solid matter. The impeller is carried on a horizontal shaft with a long bearing to give support, and reduce wear. The system is such that the blades can be adjusted, and are in effect self-sharpening.

There is no doubt that towards the commencement of World War I the change to new forms of motive powers were affecting the sale of Hathorn, Davey's steam engines. Only a few were produced in 1914, and the Company's business seemed to be gradually changing to pump manufacture.

The financial loss of £3,826 incurred by the Company in July 1914, may well have been repeated the following year, but World War I started, and the Company's business was turned to supporting the 'War Effort'.

Chapter 25 References

[1] I am grateful to Alan Denney for a copy of the engineer's notebook.
[2] LMWS Box 21-D2-BX3. Newmarket engine quote
[3] Anon. n.d. *The Works.* Twyford Waterworks Trust
[4] Pumping Engine at Kesteven Asylum, Lincolnshire. *Engineering.* 23rd January 1903.
[5] Hathorn, Davey. *Catalogue for Waterworks 1899.* pp.24 to 27
[6] White, A.J.L. n.d. Concerning the New Plymouth Graving Dock. A reprint from the *Great Western Railway Magazine*
[7] A single-stage centrifugal pump. *The Engineer.* 19th August 1910. p.208
[8] HL. 1940. *Hathorn, Davey and Company, Sun Foundry, Leeds.* p .8
[9] Patent 26913. Richard Clere Parsons. *Improvements in and relating to Centrifugal Pumps, Air Fans, Turbines and the like.* Filed 19th November 1910. Accepted 16th November 1911
[10] The Stereophagus Pump. *Engineering.* 5th April 1912. p.444

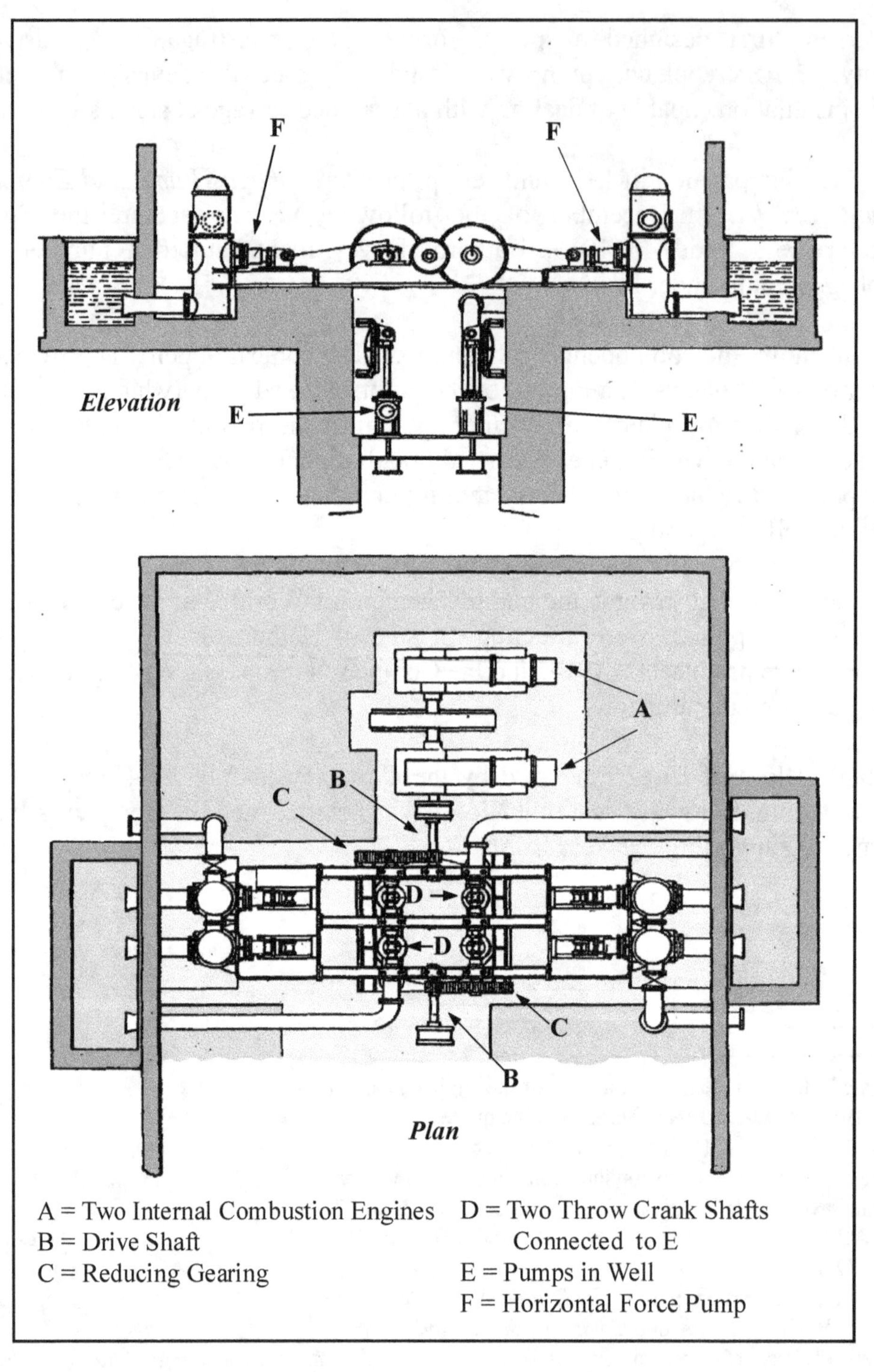

Figure 139. Davey's Patent for '*Improvements in and relating to Pumps*'
(British Patent Office. Henry Davey, Patent 12,224 accepted 26th November 1914)

26. Hathorn, Davey during World War I

There can be little doubt that 1914 was an *'Annus Horribilis'* for everyone concerned with the management of Hathorn Davey. Whilst the Company had made a small profit the previous year, partly attributed to the sale of the Belgian property, the actual manufacturing side of the business was realizing little profit.

Davey, who had been an advocate, and innovator, of steam power was now showing interest in other technologies. He applied for various patents after 1910, but not involving steam power. They were primarily concerned with pumping mechanisms actuated by internal combustion or gas engines although he did try to improve on some aspects of steam technology with a new design for an air pump. However, steam technology was probably in Davey's creative mind, a power of the past, perhaps exemplified by a series of papers he gave on the history of the Newcomen engine.[1] New technological developments, certainly were concentrating on the applications of the internal combustion engine, on both land and in air.

One of Davey's patent applications (Patent 8930, accepted 24th June 1909) was for a simple gas engine. It had four cylinders working one crank. There were two bearings on the crank and Davey considered the initial pressure on the crankshaft to be reduced, resulting in a better turning effect. The pistons were not connected directly to the crankshaft but to a rocking frame. This was connected to the crankshaft by a metal rod.

By 1914, Davey's interests had turned to waterpower. During the previous two years he had applied for several patents on new types of weirs, water gates and sluices. He was probably frustrated with his fellow Directors who wouldn't move with the times.

Davey did file patents utilising other power sources, that included pumps worked by the internal combustion engine (Figure 139). On the 26th November 1914, his patent (Patent 12,224) for *'Improvements in and relating to Pumps'* was accepted. His description states:

> *'It related to pumps actuated by cranks, particularly to a system in which water is first pumped by one set of pumps from a well, and then by another set of pumps to a further elevation.*
>
> *It has been usual hitherto in such installations to provide a crank shaft and gearing to each set of pumps, and if only two pumps were used in each set there would be considerable variation in torque on the gearing. To avoid this, it has been usual to provide three pumps in each set with cranks set 120 degrees apart. But this arrangement is complicated and costly as it involves the use of two three throw crank shafts, two sets of gearing, and six pumps.*

The object of the present invention is to secure similar results to those obtained by this complicated system, using simplified apparatus.

The invention consists of a pumping installation, comprising one set of gearing between the engine and a single two throw crank shaft, from which four pumps are operated, one pair of these pumps being placed at right angles to the other pair, and the pairs of pumps being arranged respectively to pump from the well and above the well to a further elevation.'

However, there were greater matters of concern. Perhaps it was the financial state of the company that inspired Davey to seek help from other quarters, to try to change the direction of the Company. It seemed apparent to him that the Company needed reorganising, with some new Directors.

The loss made by the Company for the fiscal year up to 31st July 1914, would have added to Davey's frustrations.[2] The Directors (Messrs. Davey, Meysey-Thompson and Lupton) elaborated on the situation in a draft financial report:

'They regret to state that after making all usual allowances a net loss of £3826-8s-9d has been incurred. As against the sum of £3678-18s-6d was, after payment of dividend carried forward from the previous year, and a further final profit of £178-19s-3d was realized on the sale of the Belgian property. These two sums together amount to £3857-17s-9d. There therefore remains after liquidation of the loss a credit balance of £31-9s-0d.

The Directors greatly regret the unfavourable result of the year. It is however fully accounted for by certain contracts entered into some years ago, and which owning to advance in material and wages and also to the present difficult financial position have turned out very unfavourably.'

A final sentence added, *'They do not however consider that there is any reason for anxiety as to the absolute financial soundness of the Company.'* However, this was crossed out, and presumably removed from the published report.

Whilst it is not recorded in the 1914 accounts what orders contributed to this predicament, later correspondence indicates that some of the blame might be attached to mining companies. Quite often such companies, certainly at this time, were ephemeral in nature often ordering additional parts after the initial contract was fulfilled, and then would go into liquidation with unpaid invoices.

Eventually, on the 30th November 1914, Davey wrote to Lupton with a suggestion that the Company should be reorganised, and that Lupton and Meysey-Thompson should retire from active management and that a salaried manager be appointed. He suggested that a Board meeting was held once a month, and that the ordinary Directors should be Lupton, Meysey-Thompson, Davey and E.G. Carey. Carey was a Civil Engineering

Consultant based at Uddingston, Glasgow, and it is unclear how he had reached this point to be considered for this high position in the Company. In all probability, Davey had made his acquaintance through his membership of learned Institutions, and may have even sold Carey some of his shares in the Company. Nevertheless, it is apparent from later correspondence that Carey had put some capital into Hathorn, Davey.[3]

Davey's letter prompted a Board Meeting in December. Lupton made various notes to review the finances of the Company, particularly the losses.

'The year resulted in a loss. As you will have noted after making all usual allowances for maintenance of plant and depreciation, and also an allowance for possible bad debts this loss amounts to £3,826. We believe that the allowance made, i.e. £2000 is adequate, but under present conditions [World War I] *it is of course impossible to be certain of this lead.*

The above unfavourable results is due principally to the fact that after certain large contracts had been booked, both materials and labour [costs] *rose seriously.'*

Lupton listed seven contracts where large losses had been made:

China Clay	£6,227	Booked	June 1911
Colne Valley	£8,342		November 1911
Bigrigg	£6,586		November 1911
Hants [Twyford]	£6,590		February 1912
Bristol	£6,631		April 1912
Crown Agents	£27,843		July 1912
Joint Dock	£8,486		June 1912

All seven contracts had been completed in the past year (year ending 31st July 1914).

In addition it was noted that the price of materials had risen sharply after quotations had been given. Manufacturing costs were compared for 1911 and 1913.

	1911		1913
Forging	£13-0s-0d	to	£18-10s-0d
Steel Plate	£6-16s-0d		£8-6s-3d
Cast iron	£2-17s-0d		£3-10s-0d
Steel Blooms	£6-5s-0d		£7-7s-0d
Copper	£61-0s-0d		£71-0s-0d
Firm	£150-0s-0d		£210-0s-0d

There were also four wage rises during this period - two in 1911, and one each in 1912 and 1913.

There was some optimism on the technical side. All the installed engines had worked well, mostly exceeding their specifications and expectations, and had even generated a bonus for the Company. The Hull Joint Docks project gave particular satisfaction as it had been inaugurated by the King, and had proved to be an unqualified success.

Lupton then concluded that the Company, under the circumstances was looking well, with ample orders and customers. It was apparent that he, and Meysey-Thompson, didn't think a reorganisation was needed.

Davey was not happy and wrote a further letter to Lupton on the 7th December 1914.

'I have been carefully through the Balance Sheet, and have come to the conclusion that it is useless to go on with the business as it is, and have determined to apply to the Courts for a winding up order.

I am sorry to have to do this, but you cannot fail to see that the business is going to the "dogs" and that we had better wind up while we have assets.'

Davey got support from Carey who also wrote to Lupton on the 9th December.

'I have received the report for the past year. Comment is needless.

It may be within your knowledge that any Shareholder may apply to the Courts for a compulsory wind-up order in any Company. This order can be made in the unfettered discretion of the judge. An expert opinion I recently took was fully confirmed verbatim by my Solicitors whom I consulted in London on Monday. I am further advised by them that in their opinion the grounds on which I shall present my petition are such as are likely to commend themselves to the favourable consideration of the Court. I therefore give you formal notice that unless one or other of the following resolutions is passed at the meeting on the 14th inst. [Company Meeting] I shall without further notice to you, or any notification other than the notifications required by law or by public advertisement, take immediate action.

The resolutions are as follows:-

A) The management of the Company shall be entirely reconstructed; by the appointment of a fourth Director, namely E.G. Carey (without salary); by the retirement of Mr. Meysey-Thompson and of Mr. Lupton from the active management, they retaining their seats on the board; the appointment of a general manager who shall not be a director; or

B) Failing this, that the Company be voluntarily wound up.

Unless I receive your assurance by midday on Saturday, the 12th inst. that you will support either of these resolutions in total, no good purpose is served by my attending the meeting and I shall instruct my Solicitors forthwith.

I cannot enter into any further argument or discussion, decision as above expressed being final.'

A reply from Lupton (9th December) rebuked Carey's resolutions as proper notice had not been given, and that they were not the views of the majority of shareholders. Lupton also asked on what grounds had Carey entered into by taking such action.

Lupton discussed the situation further and sought some legal clarity by consulting the Stockbroker and Member of Parliament, Sir Edwards Coates, for advice.

After reading Carey's proposals, Coates supported the Directors of Hathorn, Davey, but having seen the disappointing financial result, was of the opinion that there was some merit in bringing 'new blood' [new Directors] into the Company as it might have some advantages. The meeting with him on the 14th December covered a number of points.

It was pointed out to Coates that since the Limited Company had been formed, over £50,000 had been distributed, but that it was £21,000 less than the 'goodwill' that had been added at the time of the Company's formation.

In addition, no fresh capital had been raised, and all improvements were financed from revenue. Lupton blamed the losses on changes in trade, but the Company was doing it's best to overcome this difficulty.

There was some 'new blood' in the Company's management structure, as Maurice Lupton had taken up £6,000 worth of shares from retiring servants of the Company on the understanding that he was to take on a more active role in the management. He was regarded to be an *energetic, capable man.* The Company also had a sound technical man in charge of the Drawing Office. Lupton and Meysey-Thompson therefore felt that they had sufficient management advantages, and that any more might simply cause division.

Coates felt it was impossible for Carey to fill an additional role after his letter to Lupton, but considered that Henry Davey could nominate someone to represent him as a Director. It is apparent from later correspondence with Lupton, that Coates felt that the Company was very much a 'family business' and strongly advocated 'new blood' from outside, if it could be obtained.

A further letter from Coates to Hugh Lupton dated 4th January 1915 outlined the views of Coates on the composition of the Board of Directors.

'I have today received your letter of 2nd inst. for which I am obliged, and I gather from your letter that the suggestion I have made that the Board should be increased to five, Mr. Davey and his friends appointing one, and you and your friends appointing another Director, is not acceptable to Mr. Meysey-Thompson and yourself. I must regret this, as I thought this very possibly was a way to get over the

unpleasantness which has arisen and which will possibly increase in the future. Life and business is made up of compromise and personally I cannot see what harm it would do to the business to have two new Directors, and provided the Directors themselves are agreeable to their colleagues surely it would strengthen the business rather than weaken it.'

The letter then mentions that Coates was going to have a meeting with Davey and. Carey. A reply from Lupton the following day did express concern that Coates evidently did not agree with Lupton's point of view.

The situation rumbled on during the first quarter of 1915 with Davey asking for detailed information about the loss making orders, and the present orders in progress.

Hathorn, Davey's solicitors, Messrs Nelson, Eddisons and Lupton, met Coates on the 20th April and it is clear from the correspondence the situation had not been resolved. Clearly a take-over bid was in the offing, and a price for the Company was being estimated for the negotiations underway, apparently with Armstrongs. This was an engineering firm based at Newcastle upon Tyne, and who could be interested in taking a controlling interest. It was also apparent that the intention was to approach Davey to ask his views and determine at what price he would sell his shares in the Company.

On the 8th May 1915, Davey responded. He held 1,300 and his son, Norman held 1000 shares that they were both prepared to sell at £8 per share.

Henry Davey stated though, that he had retired from business and had no desire to return to it by forming a syndicate to buy the Hathorn, Davey Company, but if someone wanted to form such a syndicate Davey was more than happy to give advice, although he couldn't promise to invest in it.

However, Davey's suggested selling price was not acceptable, and further discussions were to be held with Armstrongs. It was agreed that Davey should be approached to ask if he would be prepared to sell at £7 per share to help to give Armstrongs a majority of the shares. Coates would also take a commission of 2½% on the sale of the shares.

Meysey-Thompson and the Lupton family were the principal shareholders and were reluctant to sell all their shares for the take over to happen, so if anything, the dispute was once again reaching a 'stale-mate' situation. *'Mr Davey has been the person who has raised the difficulty and he is the person who wishes to sell his shares.'* However, *'If the policy of selling a controlling interest to Messrs. Armstrong, Whitworth and Co. is eventually carried out, it cannot be carried out by the use of Mr. Davey's shares alone as they are only £23,000 out of £91,660.'*

'There would therefore be a further £27,000 to find - approximately a levy of 40% on the remainder of the Capital of the Company, and whilst it might be possible to obtain this amount by agreement with individual shareholders, it seemed to them that it would be fairer that the levy should at first be a general one, and that any sum obtained beyond the £7 per share should be divided amongst all the shareholders as a proportion of their holdings.'

It is evident that the ill-feeling between Davey and Messrs. Meysey-Thompson and Lupton continued, probably through the war years, but no take-over transpired and that the stale mate continued. Davey blamed the Company's Leeds management for the losses incurred. However, they blamed Davey for leaving Leeds and taking hardly any interest in the management of the Company, and leaving the management in their hands.

The continuance of World War I however, did seem to curtail the bickering. The War had officially started on the 28th July 1914 when Austria-Hungary declared war on Serbia. Britain declared war on Germany the following month. The supply of munitions for previous conflicts had been under control of the War Office. However, it soon became apparent that this conflict was going to be different to any war that had gone before it. It was a turning point in both tactics and weaponry. In early 1915, a shortage of shells brought recognition to the fact that a different approach to both the manufacture and supply of munitions was required, and in May 1915 the Ministry of Munitions was established to organise and expedite the supply chain, under the first Minister of Munitions, David Lloyd George.[4]

Hathorn, Davey did come under the Ministry's control, and manufactured machine tools for the production of munitions. Hugh Lupton records the war years in a short paragraph.[5] *'During the war years 1914/18, the firm supplied the Government with shell presses, accumulators and electrically driven plunger pumps; they also made a number of hydraulic presses for the production of electrodes and it may be mentioned that at this time Mr. Meysey-Thompson was a member of the Leeds War Engineering Committee, which was under the Chairmanship of the late Mr. - afterwards Sir John - McLaren, did very important work in the supply of munitions.'*

The Order Books support Lupton's statement. Some of the first orders in Books 15 to 18, that cover the War years were:

6720 & 6756	Armstrong - Whitworth	Three-throw and pressure pumps
6724/30/33/34	The Admiralty	Bearing brasses / castings / gudgeons
6713	New Explosives Ltd	Hydraulic accumulator
6753/61	National Shell Factory	Shell presses / Lathes
6740	Taylor and Challen Ltd	Cartridge Heading Presses

Figure 140. National Projectile Factory: Shell Presses
(National Projectile Factory. The Engineer, 21st July 1916. p.48 [GG])

Figure 141. National Projectile Factory: Lathes
(National Projectile Factory. The Engineer, 28th July 1916. Supplement [GG])

Hathorn, Davey also had to make some changes to fulfil the new type of orders. They manufactured an Acetylene Welding Plant for the Sun Foundry, for example. Orders did come in for engines and pumps from companies not involved with the 'War Effort'. Unfortunately, no engines were manufactured or delivered until hostilities had ceased.

The final years of the War remained very much the same for the Company, primarily making manufacturing equipment. Orders 6774 and 6839, for example were for twenty lathe beds (*Hunslet Engineering Company*), and two hundred punch noses (*Leeds Forge Company),* respectively. Other orders were again for the National Shell Factory, now renamed in the Order Books as the National Projectile Factory (Hunslet), and *Taylor and Challen.*

Figure 140 and 141 show the type of machinery being manufactured by Hathorn, Davey for the National Projectile Factory. Other engineering firms would also be manufacturing the same machine tools, but they were almost certainly made to the same specification.

By 1916, orders were being classified as 'A' or 'B'. This was presumably based on their strategic importance. An air pump body ordered (6807) on the 6th April 1916 by the *Cordoba Copper Company* for the San Rafael Shaft Vertical Compound Engine was class B, giving it a low priority. Hathorn, Davey also put their orders through the books, one of which (Order 6842) was for a new mess room, perhaps signifying that staff had increased.

The manufacturing companies of Leeds made a considerable contribution towards the War effort and orders for specific firms grew larger. Orders 6866/67/69/75/84 and 98 for the *Leeds Forge Company* included 256 castings, 300 cast iron punch noses, 100 shell noses and 12 piercing pots. The *Schoen Steel Wheel Company, Ltd.* of Newlay, Leeds also became a significant customer. Pumps, castings and other components were supplied to a whole range of local companies.

Elsewhere, the supply of raw materials, including coal and iron, was maintained. Contracts for pumps and engine components supplied to colliery and iron-mining companies were given priority. The Llanharry mines in the Vale of Glamorgan, South Wales that was operated by the *Cardiff Haematite Iron Ore Company, Ltd.* placed an order (6939) for the renewal of a Differential Gear.

Towards the end of the War, the *Alby United Carbide* factory also became a customer. The company based at Odda, Sweden was the largest factory of its kind at the time and had placed orders (7053 and 7083) for an electrode press and electrode moulds, respectively.

Financially, the war years were kind to Hathorn, Davey and it's contribution to the war almost certainly saved the firm from going out of business. The annual profits were:[6]

1915	£3784-8s-7d
1916	£3663-12s-6d
1917	£4837-13s-11d
1918	£4215-8s-1d

There was loss of life to Hathorn Davey employees in the conflict. The death of Maurice Lupton in 1915 near Lille, Northern France has already been mentioned. Other Hathorn Davey employees killed during the war include: Private Maurice Saxton [7] (2nd West Riding Regiment) December 1914; Bombardier Frederick Rothery [8] (Royal Field Artillery) October 1915; and Gunner J.T. Ashmell [9] (Royal Field Artillery) November 1915, almost certainly not a complete list.

It is not recorded how large the Hathorn, Davey workforce was during the War, or whether women were employed on any manual work, but it would seem that operations at the Sun Foundry carried on very much as usual, albeit producing a variety of products not associated with pumping technology.

It is possible that work at the Sun Foundry may have been suspended for a short period towards the end of the War. On the day that Armistice was signed (5:00am 11th November 1918) a notice appeared in the *Yorkshire Evening Post*.[10]

'HATHORN, DAVEY and CO'S WORKS will RE-OPEN TOMORROW (TUESDAY) at the usual hour.'

It was back to business, but would it ever be the same again?

Chapter 26 References

[1] Davey, H. 1903. The Newcomen Engine. Proceedings of the Institution of Mechanical Engineers. pp..655 to 704 and plates.

[2] LMWS. Box 21D2Bx1. 19865 Balance sheets 1902 to 1934.

[3] LMWS HD 19933 Unsorted Letters. Correspondence between Davey, Lupton and others.

[4] https://en.wikipedia.org/wiki/Minister_of_Munitions

[5] HL. 1940. *Hathorn, Davey and Company, Sun Foundry, Leeds.* p 6

[6] LMWS. Box 21D2Bx1. 19865 Balance sheets 1902 to 1934.

[7] Leeds Roll of Honour. *The Leeds Mercury*, 18th December 1914, p.2.

[8] Kirkstall Soldier's Funeral. *The Leeds Mercury*, 3rd October 1915, p.4.

[9] Casualties. *The Yorkshire Post*, 4th November 1915, p.5

[10] Notice. *Yorkshire Evening Post*. 11th November 1918. p.4

27. Hathorn, Davey in the 1920s

During the 1920s Hathorn, Davey continued the trend established in the war years, and overall it was probably a good period for the Company. However, the end of the decade saw the start of a period of uncertainty for many companies that culminated in the worldwide economic slump of the early 1930s. Hathorn, Davey generated modest profits, until 1927 when they made a loss.

The Company came out of the war making a profit of £3875-7s-6d in 1919. To this figure was added £950-15s-7d that was carried over from the previous year and totalled £4825-15s-7d. It enabled the Directors to recommend a dividend of 4%. The Company's assets at the time included cottages valued at £303-14s, land at £4345, and buildings at £15,277-5s-6d.

Year	Profit / (Loss)	Sundry Creditors	Sundry Debtors + Cash	Bank Overdraft
1920	£3,030-13s-0d	£28,127-4s-2d	£21,256-9s-5d	£6,702-11s-2d
1921	£3,125-11s-8d	£18,509-4s-2d	£59,800-19s-3d	£60,241-15s-2d
1922	£249-2s-0d	£17,822-9s-7d	£48,412-11s-7d	£48,539-9s-6d
1923	£1,859-12s-7d	£33,431-15s-3d	£49,502-19s-7d	£51,042-18s-11d
1924	£7,867-12s-7d	£24,804-9s-11d	£66,235-5s-3d	£69,325-11s-9d
1925	£8,379-8s-7d	£38,703-17s-1d	£95,634-8s-8d	£79,332-19s-1d
1926	£2,418-2s-9d	£21,671-19s-4d	£60,722-13s-6d	£40,657-13s-10d
1927	(£5,376-16s-4d)	£22,305-11s-8d	£65,671-11s-4d	£43,190-2s-9d
1928	(£4,207-1s-1d)	£11,835-12s-1d	£46,352-15s-1d	£35,677-16s-0d
1929	(£4,569-8s-3d)	£26,443-11s-6d	£20,182-18s-5d	£29,006-7s-4d

Figure 142. Hathorn, Davey: Financial Results 1920 to 1929
(Balance sheets 1902 to 1934. LMWS. 19865 Box 21D2Bx1)

Figure 142 shows the Profit and Loss, Creditors and Debtors, and the bank overdraft for the decade. The Profit and Loss figures show considerable variation. 1920 and 1921 are generally consistent with the levels of profit made in the war years. In 1922 the Company nearly made a loss, but the profits rose again to reach a peak in 1925.

The Creditors were mainly the companies providing raw materials, which was highest for 1925, when sales must have been at their peak. Debtors are predominantly purchasers and the figure includes unpaid, or partly unpaid orders that have been put through the Order Books. The bank overdraft is highest in the years of greatest profitability and mainly reflects the difference between Hathorn, Davey paying for purchased materials and receiving payment for orders.

Part of the costs incurred in 1920 may have been self-inflicted. Lupton mentions that after the War[1] *'A further considerable extension of the Works was taken in hand. Land, which had formed a part of the once celebrated Leeds Pottery, was bought from the Middleton Colliery Company and the new Foundry, with connections to the Colliery siding, built.'* This was probably land in the southeast corner, adjacent to Jack Lane.

Certainly Order Book 19 confirms that there was considerable expenditure on the Foundry and machinery between 1918 and 1920. Items listed include:

Order   Item and Date of order where known

7103   2 Lathes and 1 Planing machine for Pattern Shop (22nd October 1918)
          a) No.19 Double Ended 6" Centre Lathe
          b) No. 24 Lathe
          c) Combined Planer and Thickening Machine
7129   1 Second Hand Pearson type Kearns Boring, Milling and Surfacing Machine
7129A Foundations and Fixing (13th May 1919)
7130   Prepare Foundations, Pulleys, Shafting etc. fort 2 new Selson Shaping Machines (13th May 1919)
7144   Alteration of Key Seating Machine / Floor Plate Planing Machine
7144A Alterations to Steam Engine for use with above (27th June 1919)
7178   Repairs to Gearing of No. 49 Planing Machine (15th October 1919)
7192   New Foundry (Building)   }
7192A Plant (Materials and Labour)   }
7192B Crossing in Jack Lane   } Taken as completed 31st July 1922
7192C Incidentals   }
7192D Canteen   }
7195   1 - 4½" Spindle Vertical Milling Machine by Kendall and Gent to replace a Muir Machine (31st October 1919)
7196   1 Second hand 18" Slotting Machine by Muir and Co. (27th November 1919)
7197   1 Brass Finishers Lathe 9" x 6' - 0" by Alfred Herbert & Co. (12th January 1920)
7198   1 18" Muir Slotting Machine (12th January 1920)
7199   1 Radial Drilling, Boring, tapping, and Studding Machine and Box-work Table for the same (15th January 1920)
7205   1 Pneumatic Jarring Moulding Machine (4th February 1920)
7206   2½" x 30" Flat Turret Lathe (6th February 1920)
7207   1 Muir 18" x 23' (Between Centres) Taper Turning Lathe (7th February 1920)

The 1920s Accounts indicate that the Company had seven Directors: Arthur Herbert Meysey-Thompson, Hugh Lupton, Henry Davey, Henry Norman Davey, Arthur George Hopper, Hugh Ralph Lupton and Reginald Hamilton Lupton. It would seem that there had been a change of direction in the Company's management structure that Henry Davey had advocated for at the start of World War I, a fact that must have given Davey deep satisfaction.[2] The 'new blood' certainly had engineering skills. Hugh Lupton remarks, '*After the war Mr Arthur Hopper, who had been the manager of Messrs. Greenwood and Batley of Leeds and who was then Manager to Mr. Joseph Watson - afterwards Lord Manton - of his oil extraction works at Selby; Mr Reginald H. Lupton, who as before mentioned, had been the firm's representative in India; Mr. Hugh R. Lupton, who had obtained a first class in the Cambridge Mechanical Tripos and afterwards served in France with the 8th Leeds Rifles; Mr. Norman Davey, as representing his father's interest; Mr. Percey Gauvain, a London electrical engineer and with a long colonial experience; and later Mr Guy Meysey-Thompson, joined the Board of Directors.'* [3]

It would seem from this statement that Henry Davey, who still held shares in the Company, was in the process of considering complete retirement. Davey's wife, Elizabeth, had died on the 2nd March 1917, at their home, 'Conaways' at Ewell, Surrey. The house was a large property, standing in nearly 3 acres of grounds, just to the north of Ewell East Station, and would have required a lot of upkeep. The 1901 census shows that Davey and his wife employed a Domestic Housemaid, Maid and Cook.

Throughout the war years Davey had continued to file patents. He seemed to have now gone away from aspects of waterpower, and revisited steam and fuel economy. One Patent (115,582) he titled as, '*Improving the Fuel Economy of Multiple Effect Steam Engine and Steam Turbine Plants*' and was completely accepted on the 16th May 1918. It described a method to increase the efficiency of steam by heating the steam by means of heat from the furnace. Such a patent was typical of Davey throughout his life, always trying to attain a higher steam efficiency, and of course an aspect that gave Hathorn, Davey steam engines such a good reputation.

A year later he took the efficiency theme further and applied it to the Internal Combustion Engine, i.e. '*Increasing the Power and Efficiency of the Internal Combustion Engine.*' (Patent 139429 submitted 23rd September 1919 and accepted 4th March 1920).

Davey was now 77 years old and sometime in the first years of the 1920s, he moved to Tavistock, Devon, a location not too far from his birthplace. On the 8th November 1926, Davey applied for his final patent, fittingly titled, '*Improvements in or relating to steam engines,*' Patent 276531 was applied for on 22nd February 1927 and accepted on the 1st September 1927, eight months before his death. Henry Davey died on the 11th April 1928, at 1 Woburn Terrace, Tavistock, Devon, after a lifetime devoted to steam engineering.

Figure 143. The last known published photograph of Henry Davey
(Obituary: Henry Davey. The Engineer. 20th April 1928. p.43 [GG])

Davey's funeral was held on Monday 16th April 1928 in Tavistock. It is a testimony to how well known and liked Davey was when the following story appeared in many newspapers in the following days. It concerned Ramsay MacDonald the Labour Member of Parliament for Aberavon, who became the British Prime Minster the following year. [4]

'Mr MacDonald's Sad Experience.
Mr Ramsay MacDonald has had a sad experience during his walking tour in Cornwall with his daughter, Miss Ishbel. He went by 'bus from Liskeard to Tavistock to pay a surprise visit to a friend, Mr Henry Davey, a retired civil engineer and inventor. On his arrival, however, he found that Mr Davey was dead, and that his funeral was to take place that afternoon. It is understood that the sole purpose of Mr MacDonald's visit to Tavistock was to renew his earlier acquaintance, with Davey who had received many medals from exhibitions and societies for his inventions. Davey was 85.'

Davey left to his son, and to his nephew Harold William Davey, as joint tenants, all his interests in a patent for improvements in steam-jacketing and super-heating, and models and papers connected therewith. £50 to Nurse Trethewey, £15 to his Chauffeur Webb, £1,000 to Kate Gertrude Gauvain, £1000 to Mary Florence Gauvain, and the residue of the property to his son Henry Norman Davey.[5]

It is interesting to note Davey's connection to the Gauvain family. Kate and Mary Gauvain were the daughters of Percey Gauvain who had become a Director of Hathorn, Davey. The Gauvain's originated from Alderney and are believed to be relatives of Davey's wife, Elizabeth. Percey Gauvain was actually involved with banking and was also the Managing Director of the *Antrim Electricity Supply Company, Ltd.* as well as being the Chairman of several other similar Companies.[6]

There were many obituaries for Davey in both the press and professional journals. Perhaps the conclusion to his one page obituary, with his photograph (Figure 143), was in *The Engineer*. It sums up the contribution he made to the steam age.[7]

'He had an excellent memory, a vast experience, and an innate feeling for what was mechanically right, and he would chat delightfully for hours to sympathetic listeners, particularly if the subject was the Cornish pumping engine. As he had an intellect far above the common and was of a very loveable disposition, such conversation brought around him, not only his contemporaries, but young engineers and often those who were not engineers at all but who were fascinated by his reminiscences and by his clear descriptions of mechanical devices. Wherever he lived he had his private workshop, and although it is scarcely right to speak of it as his hobby, for it was there that he worked out his inventions or tested principles and ideas that came to him in his professional work, yet it was, as with so many of us, a place in which he found recreation and rest. He was sometimes assisted by one or two mechanics; but being an admirable craftsman, he did not a little of the work himself.

By his death we lose very literally a link with the past, not only because he started his engineering career nearly seventy years ago, but because, whilst keeping abreast of the present, he retained a rare devotion to that wonderful period in the development of the steam engine which was only coming to an end when he was a boy and which saw its rise in the mines of Cornwall and Devon amongst which he lived. To him and to the great firm [Hathorn, Davey] which he founded and directed for so long the world owes more in the progress and improvement of reciprocating steam and hydraulic pumping plant than can easily be recorded.'

There is no doubt that the age of the stationary steam engine really was slowly coming to an end, and Davey's death was part of that process.

The Great War had been a showcase for the versatility of the internal combustion engine with the introduction of the tank and other armoured vehicles on the ground, and a variety of flying machines. Certainly the 1920s had seen the technology expand. It was similar for electricity, which was slowly being introduced into the domestic and industrial environment, although some thought that it's spread was too slow.

As far as the Directors of Hathorn, Davey were concerned, they themselves must have realised that they were fighting a losing battle in the promotion of steam power. Not a person to give in easily, in 1924 Hugh Ralph Lupton wrote a letter, that was published in *The Engineer.*[8] Entitled *'Steam v Electricity in Pumping Stations'*, it might be regarded as a desperate tactic, or an expression of frustration.

'Sir,- The British manufacturer is always believed by the politicians to be an extremely stupid individual. Your readers will have observed a paragraph in the speech which the Chancellor of the Exchequer delivered in the House of Common on July 30th, in which, dealing with unemployment scheme and particularly with the electrical industry, he says, "Another reason why electrical development was backward in this country was the failure of the industrialists themselves to appreciate the value of it," and then he goes on to tell of a foreigner who made an offer to a number of our large manufacturers to re-equip their works with electrical machinery on the principle of payment by results, but who could not get a single English firm to accept the offer.

It seems to be an axiom or politicians that power in the form of electrical energy is superior to all others, and in many instances, owing to its convenience, this is no doubt the case, as the rapid growth of electrical generators during the past ten or a dozen years has shown, but after all every case must be considered on its merits. Dealing with the particular product with which our own firm is concerned, i.e., pumping machinery, and supposing that, as is usual in this country, coal is the source of power, it appears to be the reverse of the fact.'

Lupton then cited various examples and concluded: 'The best electrically-driven pump of which we have trial results gave an overall efficiency pump delivery as compared with electrical input of just over 84 per cent., but this was a high-pressure pump, and for ordinary work an efficiency of 75 per cent. water delivery compared with electrical input may be taken as an excellent figure. On this basis coal per water horse-power would be practically the same as coal per electrical unit...

Even if by the elimination of the smaller and worse-equipped stations greater economy is obtained, it is unlikely that the fuel consumption per actual horse-power in water delivered can be reduced to less than the consumption of a good pumping station, while, and this is the really important point, power at the pumping station is generated where it is wanted, while that produced at the central electricity station has to bear the very heavy cost of transmission and distribution mains which absorb a large capital and involve correspondingly heavy annual charges.

There seems therefore, no doubt that for main pumping station work the general adoption of central station electric drive would not result in any saving of the nation's fuel stores and would involve a large addition to the money cost of the work done.

Another obvious consideration to which the present threatened electrician's strike very forcibly draws attention, is the danger of putting too many eggs in one basket, but this is outside the scope of the present letter.

For Hathorn, Davey and Co., Limited'

Nevertheless, the Directorship Hathorn, Davey eventually concentrated much of its effort on modifications to centrifugal and hydraulic pumps. Hugh Ralph Lupton who by now was the Technical Director, had became well-known for his work on the mixed flow centrifugal pump impeller, and had applied for a number of patents relating to pumps during the decade. His patents included:

255271 Improvements in or relating to hydraulic pumps   22nd July 1926
256420 Improvements in or relating to hydraulic pumps   12th August 1926
         They were both designs to materially lessen the stresses on the crankshaft of high-pressure three throw pumps.
257111 Improvements in or relating to centrifugal pumps   26th August 1926
         This was a small, more compact type of centrifugal pump capable of being placed in a small suction well.
265275 Improvements in or relating to valves for ram pumps   2nd February 1927
         A power saving valve that was applicable to any type of ram pump that would automatically cut off the delivery if the pump exceeded a predetermined pressure.
317660 Improvements in or relating to multi-stage turbo pumps   22nd August 1929
         Changes to the internal design of turbo pumps.

By 1930, Hugh Ralph Lupton had also made improvements to the pumping methods used for sewage disposal at sea. A report in the *Yorkshire Post (1930)* referred to the pump as a 'New Disintegrator.'[9]

'The second recent development which merits special attention is the "Gargantua" disintegrator. The object of this disintegrator is to deal with the solid matter contained in crude sewage; not merely to pump it, as was the purpose the "Stereophagus" pump—another Hathorn, Davey speciality—but break it up into small pieces not exceeding a specified size. The importance of this service will be readily appreciated by visitors to some of our seaside resorts. With the growth of these resorts the accepted method of discharging sewage some distance out to sea, which was at one time entirely satisfactory has in some cases seriously interfered with the amenities of the beach. The mechanical disintegration of solid matter is doing a great deal to restore these amenities and the attractions of sea-bathing, and there must obviously be

a pressing demand for a satisfactory disintegrator in localities which desire at once to become popular, and to remain health resorts.

The "Gargantua" disintegrator may be regarded as a two-stage scissor pump, although pumping is quite a secondary function. The first cutting stage takes place between fixed steel knife-blade and a series of revolving blades, and the second between these revolving blades and another set of fixed blades beyond them. The effectiveness of the cutting action is maintained by adjustments provided in the design, and all cutting edges can easily be skimmed up when necessary. A disintegrator of this type has shown the strength its digestive organs by dealing, to order, with such objects as a lined flannel jacket, a boot, a scrubbing brush, and a canvas satchel. When objects like this can be masticated there is little fear that softer organic sewage matter will able to survive, and the disappearance of expectant flocks of seagulls from the outfalls of some of our larger seaside towns is clear proof that the practical results also of this disintegrator have been entirely satisfactory to every one except the birds.'

The latter comments are probably a reference to Hathorn, Davey's sales pamphlet for the Gargantua pump. There are two photographs before, and after that show the view around a sewer outfall pipe, with and without a circling flock of seagulls!

Figure 144. Horizontal version of the Gargantua Disintegrator
(Gargantua Disintegrator Pamphlet. Courtesy of Mike Beevers)

The pamphlet must have been published about the same time as the newspaper article (1930). It would be reasonable to assume that both the horizontal (Figure 144) and vertical (Figure 145) versions of the Gargantua Disintegrator were certainly in production during the previous year. The pamphlet also shows that Hathorn, Davey was now using a Company logo that is shown at the bottom of Figure 145, a complete page from the pamphlet.

It is noted that it was essential that the Horizontal Disintegrator should run continuously, either by (i) arranging the speed of the machinery involving a variable speed drive, noted as always being a difficulty with an alternating current electricity supply or by, (ii) arranging it to always deal with its maximum flow. Any lesser quantity would have to be augmented by the return flow through a by-pass.

With the Vertical Disintegrator, it's speed was controlled by a float in the suction pump. If the flow exceeded the maximum capacity when operating at full speed, the surplus was arranged to flow out to sea as a screened liquid, through an emergency screen.

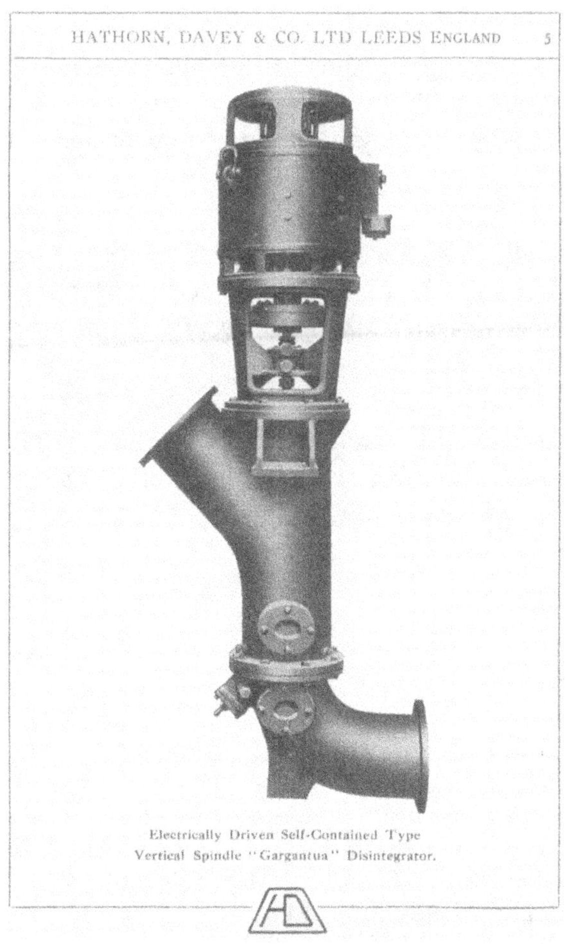

Figure 145. Vertical version of the Gargantua Disintegrator
(Gargantua Disintegrator Pamphlet. Courtesy of Mike Beevers)

Hugh Ralph Lupton did not patent the designs for a modified Gargantua Disintegrator until the mid 1930s.

Despite the Gargantua being an improvement on the Stereophagus pump, some orders were placed. Stereophagus pumps were ordered for the new sewage works at Bognor, which at the time were the largest in existence. The pumps were arranged at the bottom of pump wells and driven through vertical shafting by electric motors supplied by *Lawrence, Scott and Company, Limited* of Norwich.[10]

At this time the Company also produced a variety of other pumps that were considered ideal for being worked by electricity, and suitable for underground use in mines. They also produced an improved three-throw ram pump, as well as the multi-stage turbine pump.

Hathorn, Davey seemed to recognise some of the inadequacies of the centrifugal pump and had started to replace it with an Axial-Flow pump for low-lift work. Apparently they had spent some time and money in its development and it became referred to as the Helivane pump.

'Complete hydraulic supply pumps were also manufactured at the Sun Foundry, which is specially adapted to make the very long vertical castings required for hydraulic accumulators. For hydraulic water supply, a special balanced three-throw pump has been evolved.'

The Company did of course still manufacture the Vertical Triple Expansion Pumping Engine, and in the 1920s, a number were manufactured. One Horizontal Compound Engine was also ordered, which is still working in Staffordshire.

Chapter 27 References

[1] HL. 1940. *Hathorn, Davey and Company, Sun Foundry, Leeds.* p.7

[2] LMWS. Box 21D2Bx1 19865 Balance sheets 1902 to 1934.

[3] HL. 1940. *Hathorn, Davey and Company, Sun Foundry, Leeds.* p.6

[4] Mr MacDonald's Sad Experience. *Western Daily Press*, 17th April 1928. p. 12

[5] No Flowers or Mourning. *The Western Times.* 25th May 1928. p.10

[6] Bath Wills. *Western Daily Press.* 13th January 1943. p.2

[7] Obituary: Henry Davey. *The Engineer.* 20th April 1928. p.437

[8] Steam v Electricity in Pumping Stations. *The Engineer.* 22nd August 1924. p.215

[9] 'New Disintegrator'. *The Yorkshire Post.* 14th March 1930. p.3

[10] New Sewage Pumping Plant at Bognor. *The Engineer.* 20th April 1928. p.425

28. The Staff and Workforce

Hathorn, Davey and Company, Limited, had the reputation for building reliable and efficient engines and this must partly be attributed to the Company's loyal workforce. This is exemplified by the fact that there were often several generations from one family employed by the Company. Skills were passed down from father to son. Industrial relations seem to have been good, and accidents, when they occurred, were few. Strikes and similar industrial disputes were rare, if at all, and as the engineers' strike shows, initiated by outside elements. It would seem that Hathorn, Davey made every attempt to treat their workforce with kindness and respect. Worker involvement in decision-making may also have been a key in promoting good industrial relations.

After World War I, loyalty incentives for the workforce were examined. One possibility that was considered was to give staff and workmen the option to buy shares in the company, via a Trust.

At an Extraordinary General Meeting of the Company on the 2nd September 1920, a resolution was discussed to issue 1,200 participating preference shares of £10 each.[1] However, the first batch of 49 shares may have been offered to Miss Lupton (presumed to be Hugh's daughter).

The holder of the Participating Preference Shares would be, *'entitled to receive out of the profits of the Company in each year a non-cumulative preferential dividend at the rate of 7% per annum on the capital for the time being paid up on such shares respectively and such dividends shall run from the date on which the instalments in respect of such shares are paid.'* In addition, if there was still a surplus of profits after the 7% was paid, the shareholder would be entitled to receive a portion of that as well. Should the company be wound up, the shareholder would also be entitled to receive a percentage of the surplus assets after the capital was paid up on each share.

A year later on the 19th September 1921, in response to a question asked by Charlie Lupton, the Company's Solicitor of the firm Nelson, Eddisons and Lupton, Hugh Lupton wrote:[2]

'We are considering the issue of preference shares to our workmen at the same rate of interest as those already taken up by some of our other shareholders i.e. 7% non-cumulative participating preference shares of £10 each'

'The total amount of these shares authorized was £12,000 of which £8,050 has been taken up, leaving a balance of £3,950, and it is this balance which we propose in the first instance, to issue to the men. In order that they may have security for their capital and also that we may be able to buy back the shares when any of the men who hold them leave, we propose that the money which they subscribe it be placed in the

hands of trustees, in trust to invest the capital subscribe in War ..?.. or other trustee stock, and to pay the interest after deducting the expenditure of Hathorn, Davey and Company.'

'Hathorn, Davey and Company will not therefore really receive the money paid for the shares, but they will receive interest upon it and will pay the owners of the shares interest at the same rate as a percentage as the other holders of preferential stock.' Exceptions were men who did not own a complete £10 share.

'A representative of the men taking shares is to have the right of attending all general meetings of the Company, but not a right to vote excepting in cases where all preferential shareholders we have such rights, i.e. on propositions for the sale of the Company's undertakings, or the altering of the regulations that might interfere with the rights and privileges of the preference shareholders.

I would like to know generally whether you think there would be any difficulty in drawing up a trust fund of the kind indicated?'

Another year passed and finally on the 14th August 1922, a draft Engrossment of Deed of Trust was forwarded by Charles Lupton.[3] Hugh Lupton and his son Hugh Ralph Lupton were named as the funds Trustees. Membership of the Trust was to holders of preferential shares in the Company and it would cost £5 to join.

Certificates denoting membership of the Trust would only be issued to those workmen who held the equivalent amount of fully paid up Participating Preference Shares. There were other clauses denoting requirements for membership including that the Certificate holder must be an employee of Hathorn, Davey and Company. On leaving the service of the Company, or death, the Certificate must be surrendered and amount repaid.

If the money was required because of hardship, the employee first had to get permission from the Managing Director of Hathorn, Davey and Company for the value of the certificates to be repaid.

The Trust document seems to be fair to both management and the employees and that perhaps signifies that there was a good working relationship within the Company.

Shortly after, a *'Summary of Scheme for the Issue of Participating Preference Certificates, Secured as Trustee Securities, to Employees of Messrs. Hathorn, Davey and Company, Limited'* was published for employees, announcing the scheme.[4]

The principle objective of the scheme was, *'to enable the employees of Messrs. Hathorn, Davey and Company, Limited to become financially interested in the prosperity of the Firm.'*

The scheme was described as, *'a most advantageous one which it is hoped will have the close consideration of everyone employed by the firm.'*

Considering that most employees were probably earning relatively small wages, it is not known how many actually joined the scheme.

Hugh Lupton knew his employees well, and when he produced his notes on the history of the Company in 1940, he could name many of the senior people employed at the Sun Foundry, and their predecessors.[5]

From the time he had become apprenticed to the Company in 1886, Lupton named three cashiers: John Telford, the son of the John Telford, who was a partner in Carrett and Marshall; Walter Barrus, who was the son of a local artist, and Robert Howcroft, who had joined the firm in 1882 as an office boy earning 4s per week. However, Hugh Lupton believed that Frederick (Fred) Pawson had superseded Howcroft in the 1930s.

According to Lupton, the Pawson family were an outstanding example of the longstanding connections some families had with Hathorn, Davey. Fred Pawson's grandfather Charles Pawson joined Carrett and Marshall in 1857 and continued working at the Sun Foundry until 1884. Fred's father, Robert Pawson began work with Hathorn, Davis and Campbell in 1873, while Fred started working for Hathorn, Davey in 1903, with an absence of five years while on active service.

It is presumed that in the early years of the Partnership that Henry Davey designed and drew up many of the plans for his engines. Alfred Towler, who was later promoted to Works Manager, took on the role of Leading Draughtsman. It is unclear who filled this role after Towler was promoted, as the next person to be mentioned by Hugh Lupton was Francis George (Frank) Heseldin. He only held the position for a short time before he was drowned swimming whilst on holiday at Robin Hood's Bay on the 5th August 1913.[6] Towler had already retired from Hathorn, Davey in 1911.

John Anderson who Lupton described as, *'a real dry Scotchman who had been a marine engineer,'* replaced Heseldin. Richard H. Smith was the fourth leading draughtsman and stayed with the Company when it was taken over by *Sulzer Brothers* in 1936.

Lupton had high regard for another draughtsman, Henry A. Sims. *'The whole of his working life was passed in the service of the Company. His particular charge was to replace jobs, spares etc. for pumps and engines supplied by the Company, of which he had an encyclopaedic knowledge. A most devoted man - the difficulty was not to get him to come to the office, but to leave it at night, in fact he lived for his work and nothing else.'*

The various shop foremen at the Sun Foundry were also mentioned in some detail. *'In shop foremen there were more changes, but the outstanding ones during the writer's experience* [Hugh Lupton] *were first Jim Hargreaves. In his time in the eighties of the last century* [19th] *a good deal more was entrusted to the foreman than is now the case. There were of course, no blueprints, which now make duplication of drawing office work easy, and the foreman used to prepare lists of bolts and sundries for engines erecting in the shops, without troubling that office.'* [James William Hargreaves aged 64 died in February 1912. After a career with Hathorn, Davey, he became a Director of the *Leeds Engineering and Hydraulic Company, Limited.* He also held various patents for pumps and valves. [7]]

Many of the plans retained in the Hathorn, Davey Collection at the West Yorkshire Joint Services Archives (Leeds) are blueprints. [Sir John Herschel invented the process in 1842 and it was essentially a contact print process using light-sensitive sheets of paper that allowed rapid and accurate reproduction of a drawing. A blueprint was a negative of the original drawing that was typically reproduced as white lines on a blue background, hence a blueprint.[8]]

Other shop foremen included: Joe Barafirth, who had kept the underground engines running in the Collieries owned by Lord Durham [Northeast Coalfield] during the Durham Coal strike in the early 1890s.

Harry Briggs, who went to Australia on behalf of the Company to supervise engine erection. When he left the Company in November 1898 he was presented with a gold watch and a walking stick. He became landlord of the Parkfield Public House.[9] [The Parkfield Hotel was located at the junction of Jack Lane and the Dewsbury Road]

Arthur Nicholson died in the Company's service.

Tom Mallinson held the post of foreman for many years. He, *'was never known to be a minute late even on the coldest morning, in days when the shop opened at 6:00 am.'*

'Harry Mallinson, a nephew of Tom and son of the firm's old cost clerk Joe, was a good engineer and it was a matter of very great regret that he was lamed in an accident to the firm's motor wagon when on his way to an erection job.'

'In addition to shop foremen, the firm, having always a number of erecting jobs on hand, employed a so-called outforeman. For many years this position was occupied by J.S. Milner, a man with a quite wonderful gift for the diagnosis of engine troubles. Later this work was in the hands of Tom Metcalf, a most all round able man. Whilst on erecting work in India he was taken over by the Karachi Municipality as their Municipal Engineer, while their permanent engineer was on furlough.'

Lupton was proud of the fact that employees who left the firm often went on to fill good positions in other organisations, or set up their own business, for example he mentions:

Benny Nicholson (brother of Arthur) was in charge of one of the *London Metropolitan Water Board's* principal pumping stations.

Reginald Young, who began his career as an office boy, became an Indian coal owner.

Ernest Newell, who was apprenticed to the firm, established his own works for the manufacture of cement mixing machinery.

George Melton, who for a number of years worked in the drawing office, became Station Construction Superintendent to the *Yorkshire Electric Power Company.*

Fred Temple, originally a draughtsman, became assistant engineer to the *South Staffordshire Waterworks Company.*

Harry Wilks who had also come from Messrs. *Tannett, Walkers* was with the Company from a number of years, whilst his son Bernard had left the firm and become a partner in *Tangyes.*

During the whole of Lupton's time at the Sun Foundry, he records that there were only two Pattern Shop Foremen. Jonas and Robert Wilkinson. The latter had come originally from Messrs. *Tannett, Walkers* works in Hunslet and proved to be an excellent foreman.

'During the same period there were three foundry foreman, that included *Stewart, a Scotchman. At that time, the 1880's and 90's there were a great number of Irish labourers in Leeds'* that were presumably employed in the engineering sector.

Other employees named by Lupton included Robert and Martin Aldridge.

Finally Lupton mentioned the two leading smiths, who presumably worked in the foundry. They were George Brow, who he described as a burley Yorkshireman, and Jim Stewart, a far more refined personality.

In another section of Lupton's notes, he included several pages of comments about the standard wages for various jobs in the Leeds area.

'In 1887 the wages of a fitter or skilled turner in Leeds were 28s per week, and for an Engineer's Labourer 18s and the standard week was 54 hours works beginning at 6 a.m.'

'In 1897 the Firms workmen, in common with other members of the A.S.E. [Amalgamated Society of Engineers] *struck for shorter hours, but after having been out of work for 6 months....the men were defeated by the Employer's Federation, and returned to work practically on the same terms.'* The working week according to Lupton was shortened by 1 hour.

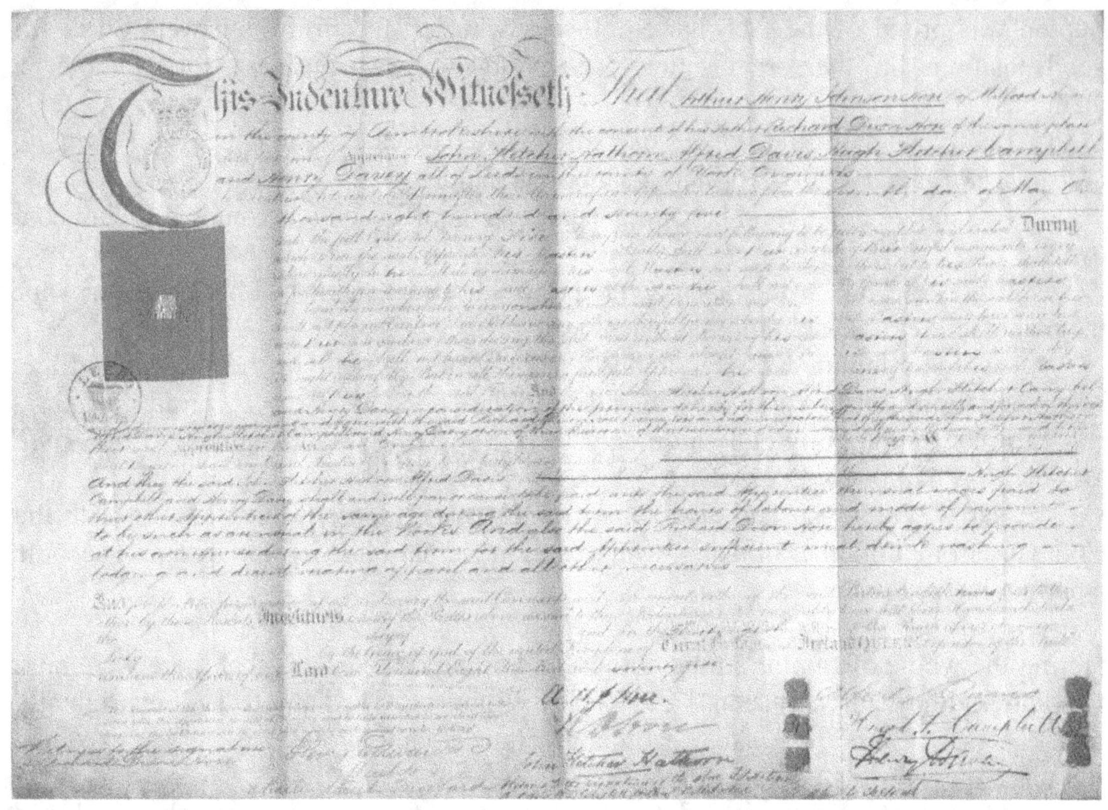

Figure 146. Indenture of Apprenticeship for Arthur Henry Johnson Hore, dated 1875
(Courtesy of the LMWS. Box 21D2BX2. Item 20234)

The 53 hours week is verified in an advert placed by Hathorn, Davey in 1900 for non-Society Patternmakers with a weekly wage of 37s.[10]

'*After this strike wages and hours remained nearly constant until after the 1914-18 war.*' In 1914, just prior to the outbreak of that war, a labourer's wage in Leeds was 20s and that of a fitter or skilled turner 37s.

Following a brief boom after the war, by 1922 wages for fitters and turners had risen to 57s per week, and labourers 42s-6d, while a 47 hours working week was standard. Work began at 7:30 a.m. but a half-hour that had previously been introduced for breakfast was abolished. Prior to World War II fitters and turners were getting 67s per week, and labourers 52s.

Lupton also remarked that prior to the Education Act coming into force in 1870, the workmen were much less literate. On out-erection jobs for example, the filling in of timesheets and other correspondence was generally left to the Premium Apprentice, one of whom was commonly sent out with the out-erection squad.

The apprenticeship scheme would usually bind a person to a craftsman for a period of up to 9 years, to learn a trade. Premium Apprentices were those where a premium or fee was paid by the apprentice's family, and was usually legally binding in the form of a signed Indenture. Figure 146 is an example of a Non-Premium Indenture dated 7th May 1875 between Arthur Henry Johnson Hore [the Apprentice] and his father, Richard Dixon Hore, both of Milford Town, Pembrokeshire, and Messrs. Hathorn, Davis, Campbell and Davey.[11] In this example, the apprenticeship period is 5 years. The Partnership also, '*shall and will pay or cause to be paid unto the said Apprentice the normal wages paid to their other Apprentices the same age during the said term the hours of labour and mode of payment and to be such as are normal in the works And also the said Richard Dixon Hore hereby agrees to provide at his own expense during the said term for the said Apprentice sufficient meat, drink washing and lodging and decent wearing apparel and all other necessaries.*'

Although in this instance the Apprentice's father did not pay a premium to the Partnership, he did have to pay for his son's total upkeep during the five-year term. Normally apprentices were generally taken from the surrounding area. In this instance young Arthur Hore was probably from a family that Davey may have encountered when he was involved with the Milford Pier contract a few years earlier.

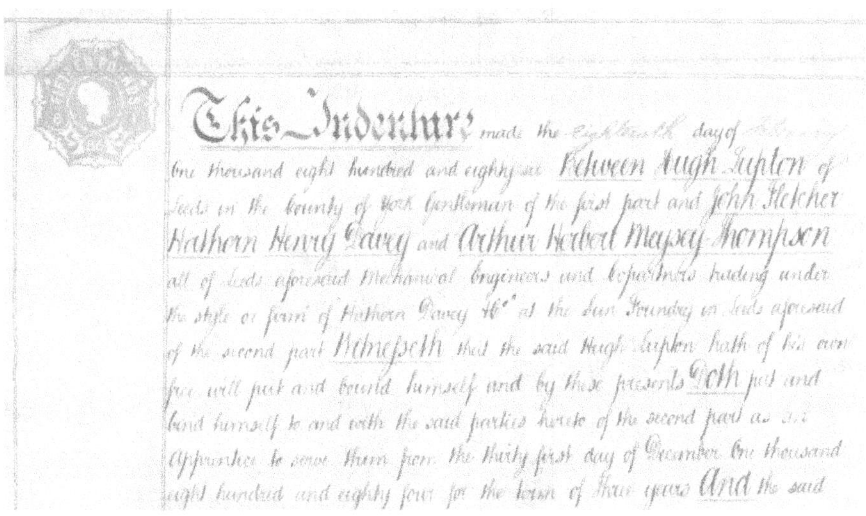

Figure 147. Portion of Hugh Lupton's Indenture of Apprenticeship, dated 1886
(Courtesy of Sulzer (UK) Holdings, Ltd., Leeds)

Hugh Lupton, of course joined Hathorn, Davey as an Apprentice. His Indenture (Figure 147) with the Partnerships is dated 18th February 1886 and was for a period of three years, but he was rapidly elevated to a Partner in 1887. In the Indenture, Lupton agreed to obey his master's lawful commands, keep Company secrets, and not to be absent, etc.

Name	£	s	d	Occupation
R Wilkinson	5	7	0	Foreman - Pattern make
A. Saville	5	0	0	Foreman - Machine Shops
W. Nicholson	6	0	0	Foreman - Moulder
T.E Metcalf	5	0	0	Foreman - Outdoor
H.A. Pollard	5	0	0	Estimating Clerk
F. Pawson	4	15	0	Cashier and Purchasing Department
A. Crowther	3	13	6	Forwarding Department
J. Farrow	3	3	0	Invoices and Work Ordered Out
F.E. Parsons	2	15	0	Wages, Insurance and Orders to H.D. & Co.
C. Drake	2	5	0	General Clerk
Mrs. Norfolk	2	2	0	Costing
Miss J. Halliday	1	12	0	Costing
Harold Heaton		12	0	Office Boy
Miss Lofthouse	2	5	0	Typist
Miss Lucas	1	15	0	Typist
H. Hobson	3	13	6	Piece Work Clerk
C.K. Booth	2	15	0	Shop Costing Department
A. Seamen	3	1	0	Time keeper
P. Henderson	4	0	6	Store Keeper
Miss Nora Longley		15	0	Typing and General
G. Fenman	6	2	0	Draughtsman
C.H. Hutchinson	5	9	6	Draughtsman
A.B. Hustvil	4	16	0	Draughtsman
C.N.White	4	10	0	Draughtsman
E.A. Beevers	4	4	0	Draughtsman
C.W. Ellis	4	0	0	Draughtsman
R. Coultas	3	5	0	Draughtsman
S.A. Yeadon	3	12	0	Draughtsman
G. Milner	3	10	0	Draughtsman
S. Couch	2	15	0	Draughtsman
C. Smith	2	0	0	Draughtsman
W. Rodgers	2	0	0	Draughtsman
H. Campbell	1	1	0	Tracer
A. Wilkinson	1	1	0	Apprenticed in Drawing Office
J. Burley		15	0	Apprenticed in Drawing Office
C. North	2	5	0	Blue Print Taker etc.
Miss D. Winder	1	18	0	Tracer
Miss I Garnett	1	15	6	Tracer
Miss C. Bottom		15	0	Tracer

Figure 148. Salaried Staff below Works Manager, January 1932
(LMWS 21D2Bx1. Item 19871.6)

In 1940, Lupton remarked that, '*The apprentice system was more in vogue then* [1870s] *than it now is and there were usually several in the shops - A contemporary of the writers* [Lupton] *in this capacity was the late Mr. Roger Smith, afterwards Electrical Engineer to the Great Western* railway.'

As with many firms of this period, the typewriter was a major innovation that saved time. Prior to its introduction, '*letters were as a rule confined to necessary matters and customers generally contented themselves with stating their requirements, leaving the company to advise as to the best means of fulfilling them.*'

Figure 148 lists the staff employed by Hathorn, Davey in January 1932 below the level of Works Manager, together with their weekly salaries. [12]

There is a Foreman in charge of each of the main shops or places, although it is unclear who was in charge of the Erecting Shop as it isn't listed. Perhaps that duty was the responsibility of either the Machine or Outdoor Foreman?

Sixteen people are involved with the administration (costing, orders, wages, timekeeping, stores, etc.).

By far the largest department was the drawing office employing nineteen people. There were twelve draughtsmen with seven support staff that included two apprentices, four tracers, and one person employed to make blueprints.

It is almost certain that there would have been a similar number employed up to the point prior to the Company's takeover by *Sulzer Brothers* four years later.

Chapter 28 References

[1] LMWS Box 21D2BX2. Item 19929.6 Notice to approve issuing of preference shares.

[2] LMWS Box 21D2BX2. Item 19929.3 Issue of preference shares to workmen.

[3] LMWS Box 21D2BX2. Item 19929.4 Deed of Trust.

[4] LMWS Box 21D2BX2. Item 19929.5 Workmen - financial involvement.

[5] HL. 1940. *Hathorn, Davey and Company, Sun Foundry, Leeds.* pp.9 to 12

[6] Beach Tragedy at Robin Hood's Bay. *Leeds Mercury.* 6th August 1913. p.5.

[7] Death of a Leeds Engineer. *Newcastle Daily Chronicle.* 8th February 1912. p.3

[8] https://en.wikipedia.org/wiki/Blueprint

[9] Leeds and District. *Leeds Times.* 5th November 1898. p.7.

[10] Clerks, Assistants and Co. Wanted. *Derby Daily Telegraph*, 3rd May 1900. p.1

[11] LMWS Box 21D2BX2. Item 20234 Indenture of Apprenticeship. Dated 7th May 1875

[12] LMWS Box 21D2Bx1. Item 19871.6 List of Office Staff 1932

Figure 149. The Triple Expansion Engine (12"+23"+34") at Ahmedabad, India
(Courtesy of the Lupton family)

Figure 150. The Pump House at the Mount Crosby Waterworks, Brisbane, Queensland
(Courtesy of the Lupton family)

29. Engine Sales in the 1920s

In the post-war years, orders for steam engines never achieved the levels previously seen. In addition to a requirement to fulfil orders placed just before, and during the war, there were engine orders from Uruguay, Australia and India, and a few from the larger British waterworks companies.

The appointment of Reginald Hamilton Lupton as the firm's representative to India after World War I was an effort to expand into areas where there was still a need for steam power. Consequently it is not surprising that he made a number of journeys there, and to Malaysia and Australia, during the 1920s that did result in some sales. The decision to concentrate on those countries was probably due to their status in the British Empire, as well as having good accessibility by ship, together with the relative lack of electric power, and there were significant sources of fuel in those countries.

The vastness of India, with its extensive coal reserves, presented an excellent market for steam engines. In the past Hathorn, Davey had sold engines to Indian colliery and railway companies. Even by 1920, only the major Indian cities, and a few industries, had limited access to electric power. The companies on the Kolar Gold Field that had previously bought a number of Compound Pumping Engines from Hathorn, Davey now had their very own hydroelectric power supply.

During the 1920s, Reginald Lupton made at least five trips to India and beyond, and certainly by the end of the decade it is believed that Hugh Ralph Lupton had made one visit to India.[1] Most of the engines sold during this period were Triple Expansion, but a few Compounds were still manufactured.

It is fortunate that during his travels Reginald Lupton also carried a camera. His photographic albums, that are now deposited in the Leeds Record Office, courtesy of the Lupton family, chart his travels. In addition to showing the places he visited and the people he met on the voyages, there are also a small number of photographs showing Hathorn, Davey engines.

Reginald's first journey started from Liverpool on the 20th December 1919, bound for Bombay.[2] He certainly visited the cities of Ahmedabad (northwest India) and Lahore (now northern Pakistan) and photographed several existing Hathorn, Davey engines.

Both cities had previously bought the same size Vertical Triple Expansion Engines (12"+23"+34"). Ahmedabad Municipality was order 6493, dated 2nd January 1911 and cost £5,819 (Figure 149), and the one for Lahore Municipal Waterworks was order 6469, dated 6th June 1910. Reginald was no doubt pleased to see both engines still working, especially as Lahore had only just ordered a second engine to the same specification (Order 7174, date 16th September 1919).

Figure 151. Singapore: The two Triple Expansion Engines being installed
(Courtesy of the Lupton family)

Reginald Lupton departed from London on the 13th April 1923 on his second business journey. The trip was longer than the first and took him to Australia, where he arrived at Sydney on the 18th May 1923.

He seems to have spent a large portion of his time in Queensland where he visited the working Hathorn, Davey engines supplying water to Charters Towers. At Brisbane he initiated a contract to supply engines for the Mount Crosby Pumping Station (Figure 150) that supplied water to Brisbane. According to Hugh Lupton, the contract was valued at £100,000, and administered by Reginald.[3] Certainly the photograph album for this period indicates that he had meetings with the waterwork's engineers. No specific details about the engines are recorded. The Hathorn, Davey Order Books are missing.

On the return journey to England, he had a short stay in India where he revisited Ahmedabad and photographed the Hathorn, Davey engine again. Perhaps his persistence paid off, because a few years later the Municipality ordered, through Agents (*Turner, Hoare and Co. Ltd*), two Inverted Vertical Triple Expansion Engines for the Dudheshwar Pumping Station (Order 7635, dated 27th January 1927). Once again the engines had 12"+23"+34" cylinders.

Reginald's third (1925) and fourth (1926) business trips also took him to Australia. Photographs indicate that he stopped off at Singapore to see the installation of two Triple Expansion Engines (Figure 151), and also visited Malaysia.

The Singapore contract would have been listed in the missing Order Books. 'Singapore' is stencilled on the engine as well as the various components being numbered, to match numbering on the drawings.

A note in one of the photographic albums indicates that by the middle of May 1926 two engines for the Mount Crosby Waterworks, Brisbane had been installed. On the 29th and 30th May 1926, Reginald Lupton represented the Company at the 2nd Unit Pumping Test. In actual fact the contract seems to have been for three engines, as two years later, an article in the *Brisbane Daily Standard* (12th March 1928) proclaimed:[4]

'Pumping Engines Tested: A final test of the third unit of pumping engines at the Mount Crosby pumping station was made yesterday. The test started at midnight on Saturday, and the engines were run for 12 hours, with satisfactory results. Mr. J. Henderson, chief inspector of machinery, was appointed by the Metropolitan Water Supply and Sewage Board to make the test. Mr. J. Peart (engineer for water supply) represented the board, and the contractors (Hathorn Davey, Ltd of Leeds, England) were also represented. The engines have been running for 12 months, and they will now be taken over by the board.'

Figure 152. Installation of a Triple Expansion Engines at Mount Crosby, Brisbane
(Courtesy of the Lupton family)

Reginald Lupton may have made a fifth journey to Sydney in 1927 but it is unlikely that he witnessed the test as he arrived back in London on the 6th April 1928. A voyage from Australia to Britain usually took about a month.

The Mount Crosby Pumping Station had been constructed before the turn of the century and it is believed that the Hathorn Davey engines were replacing an earlier pumping system. There is an unlabelled photograph in the albums that may possibly show one of the engines being installed. Certainly the building shows signs of having had a previous use. (See Figure 152)

There were other contracts in India. The Karachi Municipality (now Pakistan) placed an important order (Orders 7642 and 3) on the 22nd February 1927 for four Inverted Vertical Triple Expansion Engines (10"+19"+29", two with 2', and two with 2' 6" strokes).

The last Horizontal Compound Engines manufactured by Hathorn, Davey were probably two rotary engines ordered by the agents *J. Birch and Company, Limited* for Ranikhet, India (Order 7803, dated 16th February 1929). Ranikhet nestles in the foothills of the Himalayas. It was a popular destination for those wanting to avoid the Indian summer heat.

Hugh Ralph Lupton also visited India in 1924 to supervise the installation and testing of electrically driven high-pressure plunger pumps and accumulators that supplied water to Simla (now Shimla). The town, located north of New Delhi and in the foothills of the Himalayas, was the summer location of the Raj Government. The pumping scheme delivered water from the Valley of the North River in one 4000 feet lift.[5]

Hugh Ralph Lupton was Technical Director at this time, and there is no doubt that it was his role to manage the various large contracts that the Company had been awarded in Britain. One particularly big contract was with the *London Metropolitan Water Board*, for three large pumping engines. Whilst, the actual Order Book entries are missing, there is sufficient information about the engines in various Journals. The three *Metropolitan Water Board* orders were for:

i) Lea Bridge (Valley) Pumping Station. It is fortunate that Hugh Ralph Lupton presented a paper about the engine to the Institution of Civil Engineers.[6] The engine was designed to pump 12 to 14mgd against a head of 180 to 200 feet, by pumps located directly beneath the steam cylinders. It was a 24"+44"+66" engine with a 4' stroke, operating at a speed of 30.5 revolutions per minute. Typically the valves were Corliss operated by the Craig valve gear.[7]

ii) Walton on Thames Pumping Station.[8] The engine was built to the designs of the Water Board's engineers. It was referred to as a majestic engine because the height from the pump room floor to the tops of the cylinder was 52 feet. The order number is again not known but it was a 25"+47"+70" cylinder engine, with a 5' stroke (See Figure 153).

The engine and pumps were designed to pump against different heads at different times. '*The rams at Walton consist of an inner plunger, 28" in diameter, which is encircled by an annular plunger having an outside diameter of 36". For pumping 24 million gallons per day against the lower head of 139 feet, the two plungers will be bolted together, and will move as one piece, whereas for pumping smaller quantities against greater heads, the two portions will be disconnected, and the annular portion bolted to the pump casing, so that only the smaller-diameter portion reciprocates.*' The operation to switch the plunger to different diameters took 2 to 3 hours.

iii) Deptford Pumping Station. This was listed as Order 7721, dated 6th March 1928 for one Triple Expansion Pumping Engine (23½"+46"+70", 5' stroke) as well as three turbines and centrifugal pumps, a water turbine, three 22½" well pumps.

Figure 153. Walton on Thames Triple Expansion Engine (25"+47"+70") and Pumps
(The Engineer, 13th August 1926, p. 172[GG])

In all three contracts it is almost certain that the Company provided more than just the engine and pumps. An article in the *Yorkshire Post* in 1930 entitled '*Economical Plant*' is probably a description of the Deptford installation.[9]

'Among work of a more traditional type, Messrs. Hathorn, Davey have just completed an exceptionally large triple-expansion steam pumping set for the Metropolitan Water Board, in which a single engine drives no less than four sets of three-throw pumps, each providing for a different service. By this arrangement, although each service only demands a moderate horse-power, it possible to secure the economies provided by a large engine. Moreover, the cost of the whole plant is very much less than would be that of four separate units. In order to obtain flexibility, two of the pumps have provision for stroke-variation. The fact that all four services were to driven by one engine made a complete standby plant desirable. Since this standby plant will only run on rare occasions, efficiency is quite secondary consideration; reliability and low overhead costs are of more primary importance.

For the three force-pump services, which pump water from the surface into service mains, the standby units take the form of centrifugal pumps driven by steam turbines, while for the well-pump unit, which raises water to the surface from 108 feet below ground level, a water turbine drive to a centrifugal standby is provided, power water being obtained from the delivery side of one the steam driven units just mentioned. It is of some interest to note that the Metropolitan Water Board still use steam as a rule for their largest new units, particularly for those whose duty is to be continuous. For medium-sized units, oil engines are often used, and for small units electricity. This last is, is of course, particularly well suited where automatic working required.'

The *Staffordshire Potteries Water Works Company* also placed orders with Hathorn, Davey during this period.

A Horizontal Compound Engine, which is still working, was installed at the Mill Meece Pumping Station located just to the north of Eccleshall, Staffordshire. (See Figure 154) In the same engine house, there is another working Horizontal Compound Engine manufactured by Ashton, Frost and Company, Blackburn, Lancashire.

The Mill Meece engine house was originally constructed by the *Staffordshire Potteries Water Works Company* to house two engines, but by 1914, only the Ashton Frost engine had been installed. It had been the intention to install a second Ashton Frost engine at a later date, but World War I interrupted this process. Ultimately, in the 1920s when the decision was made to order a second engine, it was found that Ashton Frost was going through the liquidation process and so the engine was ordered from Hathorn, Davey. Although in the 1920s the Triple Expansion Engine was *de rigueur*, the design of the engine house committed the Company to install a Horizontal Compound Engine.

Figure 154. The Hathorn, Davey Compound Engine (27"+52") at Mill Meece
(Author)

Order 7557 (dated 12th January 1926) was placed in 1926 for a Horizontal Tandem Compound Engine (27"+52", stroke 60"), to operate a double-acting ram pump (15¾") with the two cylinders placed end to end, and well pumps in external boreholes, operated by two surface pumping quadrants. (See Figure 155) Elsewhere this pumping arrangement is referred to as the 'Pitman', and is the only known working example of this type of pumping method by a Hathorn, Davey Compound Engine.

The air pump and condenser arrangement is also different to that previously seen on earlier engines. The air pump operating mechanism is connected to the piston rod between the low-pressure cylinder and the ram pump. The condenser is located in a water chamber beneath the engine house. The steam having condensed to water is then passed through an oil separator into the feed water tank and is returned to the boiler via an economiser.

Figure 155. Mill Meece: The two pumping quadrants are still worked by the engine
(Author)

The pumps lifted the water from the boreholes 120 feet to an impounding chamber that was located beneath the engine house, and had the capacity to hold 26,000 gallons. The ram pumps then pumped the water a further four miles to a reservoir at Hanchurch. The specification required the engine to pump 2 million gallons per day, but in actual fact it exceeded this amount. The total cost of the order was £11,230.

The Hathorn, Davey engine was installed in 1927 and worked at 19rpm. A Marshall steam barring engine initially turned the engine. This had a pinion that engaged into the toothed inner edge of the 19' 6" diameter flywheel.

After a long period of manufacture, the engine was delivered on the 31st December 1927, and was started on the 17th February 1928. By the 5th May 1928 it was operating equipment. The engine finished pumping on the 22nd December 1979.

The *Mill Meece Pumping Station Preservation Trust* that operates the engine has an interesting account of the engine's installation on their Internet site.[10]

Harry Cusworth, the Hathorn Davey installer, had the help of two farm labourers to erect the engine. The parts were delivered to Standon Station, a few miles to the north of Mill Meece and brought to the site on a horse-drawn trolley in the correct assembly order. Harry established a base line for aligning the engines by driving rods into the two end walls of the engine house, and everything was aligned to a wire stretched between the rods. Harry was called away to another contract so the engine and pumps were actually commissioned by Harry Greenfield, who had recently completed his apprenticeship with Hathorn, Davey. Some 60 years later Harry Greenfield returned to Mill Meece and was pleased to see the working engine. He described its running as, 'Smooth as oiled silk.' Clearly an engineering term!

The *Staffordshire Potteries Water Board* also installed (Order 7801, dated 21st December 1928) two Triple Expansion Pumping Engines (17"+29"+44", 3' stroke) at their Cresswell Pumping Station.

Other examples of engines ordered in the 1920s include a second engine (16"+28"+46", stroke 3') for the Umberumberka Pumping Station, Broken Hill, New South Wales, Australia (Order 7599, dated 30th July 1926) that still survives *in situ*.

In Britain, order 7819, dated 27th May 1929 was for one Triple Expansion Pumping Engine (24"+43"+66", 3' stroke) for *South Staffordshire Waterworks Company* (Slade Heath Pumping Station).

Chapter 29 References

[1] HL. 1940. *Hathorn, Davey and Company, Sun Foundry, Leeds.* p.7

[2] Passenger list for SS Circassia, Anchor Steamship Line, sailing from Liverpool to Bombay on the 20th December 1919. Ticket 943. R.H. Lupton. Age 34, Engineer.

[3] HL. 1940. *Hathorn, Davey and Company, Sun Foundry, Leeds.* p 7

[4] Pumping Engines Tested. *Daily Standard* (Queensland) 12th March 1928, p.6

[5] High Head Pumping Plant for Simla Water Supply. *The Engineer.* 16th February 1925. pp.162 and 167

[6] Lupton, H.R., 1926. The Triple-Expansion Steam Engine as Applied to the Pumping of Water, with Special Reference to the Plant at Lea Bridge. *Minutes of the Proceedings of the Institution of Civil Engineers.* v.222, p.212

[7] An Efficient Waterworks Pumping Plant. *The Engineer.* 16th April 1926. p.430

[8] New Filtration Plant at Walton-on-Thames. *The Engineer.* 13th August 1926. pp. 161 to 164, 172.

[9] Engineering Notes. *The Yorkshire Post.* 14th March 1930. p.3.

[10] http://www.millmeecepumpingstation.co.uk/

30. The 1930s: Lack of Orders

Hathorn, Davey's three years of financial loss (1927 to 1929) could not have happened at a worse time because the 1930s commenced with a period of economic downturn in Britain. It was a worldwide depression and Britain's world trade halved during the period 1929 to 1933. The output from heavy industry fell by one third. It was the heavily industrialized north of England that felt the repercussions the most, and that included firms like Hathorn, Davey.

	Unemployed		Temporary out of work	
	Men	Women	Men	Women
9th January 1930[1]	10,299	1,631	7,158	2,622
7th January 1931[2]	16,196	4,138	4,056	4,187

Figure 156. Table showing the number of unemployed in the Leeds area: 1930-1931
*(Sheffield Daily Telegraph - Thursday 09 January 1930[1], and
Yorkshire Evening Post - Wednesday 07 January 1931[2])*

The unemployment figures for Leeds show the effect of the economic slump on the city. At the start of 1930 over 10,000 men were without work, but this rose to above 16,000 at the start of 1931 (See Figure 156). The number of men temporary out of work showed a decrease for the same period, but it is presumed that a good proportion of those had moved into the unemployed category. Overall the figures for women of an employable age showed an increase in both categories. It is assumed that the situation remained roughly the same throughout 1931.

Lupton seems to confirm this:[3] *'The slump which occurred in 1931, followed by continued adverse conditions, particularly in the Dominions and Colonies, with whom about half the firm's business had been done, coupled with the failure of one of their debtors for a large sum, placed the firm in difficulties.'* He doesn't state just how the slump affected the numbers employed at the Sun Foundry. Nevertheless, a newspaper report in March 1930 painted a very optimistic picture of the firm's future.[4]

'Engineering Notes. Progress in Pumping Machinery. (From Our Engineering Correspondent.) Among the traditional industries of Leeds hydraulic engineering demands a high place, and there can few responsible waterworks or sewage engineers

in the Empire to whom the name of Hathorn, Davey, for example, is not synonymous with all that is best pumping machinery. A visit to the Sun Foundry is always an attractive feature among engineering societies, for the technical man is sure at any time to find a variety of plant passing through the shops, all of considerable interest and much of original and novel design.

The traditional reputation of Messrs. Hathorn, Davey and Co. was founded largely on pumping engines and on waterworks, sewage, mine and other high duty pumps. But the firm has not allowed its respect for the past to dim its eye for the present and future, and during the last few years at least two important developments have emanated from the Sun Foundry. In times when Continental hydraulic engineers seem to have been voted a monopoly in brains and enterprise, and when the employment of foreign designers or foreign designs has become almost the rule in water power plant and pumping machinery, it specially pleasing to find proof of British inventive thought backed by a progressive policy in a British hydraulic undertaking.

The axial-flow "turbine" pump possesses in principle certain clear advantages where large quantities of water have to be dealt with at comparatively low heads, particularly when the head is liable to vary down to the vanishing point. For such purposes as irrigation, fen drainage, and the emptying of docks this typo of pump has an assured future, and the " Helivane " pump, which created so much interest among engineers at the Leeds meeting the British Association, has a long lead in this field. A similar conception is applied in the "Contrawhirl " axial-flow pump, a still newer H.D. development, of which a batch of five dozen are at present on their way through the shops for the circulation of oil through the windings of large transformers under the national electrification scheme. These pumps are designed in such a way that when the load on the transformer is insufficient to necessitate their running they offer but little resistance to natural thermo-syphon circulation'.

The remainder of the article mentioned other notable achievements by the Company, including the 'Stereophagus' pump, and its successor, the 'Gargantua,' for the disposal of sewage. The final paragraphs referred to the triple-expansion-steam pumping engine recently completed for the *Metropolitan Water Board, London*.

Both the Company Accounts for 1930 and 1931 recorded profits, on paper at least, of £5,436-10s-4d and £5,908-7s-0d, respectively.[5] In the previous decade, the Company's accounts had listed Sundry Creditors, Sundry Debtors and a Bank Overdraft. The successful years of the Company had also enabled them to have a Reserve Fund, but three years of losses had eliminated that from the accounts. In addition, the accounts for both years show that the Bank Overdraft was now secured by Mortgage, which was presumably held on the Sun Foundry.

'Stock in Hand' and 'Work in Progress' was included in the Company's assets, and this figure also showed a reduction from £34,384-6s-9d to £11,064-4s-9d. Certainly the problem seemed to be one of cash flow, at this stage in the Company's affairs. A photograph (Figure 157) of the Erecting Shop in the 1930s indicates it was a busy place.

Figure 157. The Sun Foundry Erecting Shop probably taken in the early 1930s. Visible are two Triple Expansion Engines, pumping quadrants, a stack of connecting rods in the foreground, and a variety of pumps including a three-throw pump, in the distant background.
(Courtesy of Mike Beevers)

In January 1932 Hugh Ralph Lupton wrote a short article in the *Yorkshire Post* summarising the state of the economic situation as he saw it.[6]

'PUMPING AND HYDRAULIC MACHINERY.
By Mr. H. R. Lupton, of Hathorn, Davey and Co., Ltd.
In the domain of pumping and hydraulic engineering none of the local firms has anything of outstanding interest to report. With most of them, trade has been very quiet,

a large proportion having again been on behalf or public bodies—not, perhaps, a very healthy sign.

The main cause of slackness is, of course, the world-wide trade depression, but there are two subsidiary causes, specially noticeable where public bodies are concerned, which it is not wholly beyond our power to influence. The first is an undue exaggeration of the cult of economy. To achieve a saving in initial expenditure—on British labour and materials in a period of unparalleled stress—at the risk of inefficiency, increased running costs and shorter "life," is the falsest of economy. Yet there is an undoubted tendency to such "achievement," often not unconnected with party politics. Undoubtedly, this has diminished the volume of trade in certain directions.

The other subsidiary cause is uncertainty of political action. Hope of financial assistance from, say, the Unemployment Grants Fund has caused many a necessary scheme to be held up just as it has caused many unnecessary schemes to be begun. Uncertainty regarding tariffs and their effect on costs and prices makes tendering, especially for future work or schemes of considerable duration, practically impossible. Every effort must, therefore, must be made to stabilise political action at the earliest possible moment.'

'Partly owing to the magnitude of many of its undertakings, and partly because its products are themselves mainly ancillary to the production of goods for other industries, the engineering industry as a whole is subject to a considerable lag, and cannot feel the benefit for some time to come. Hydraulic engineering, of which Leeds is a notable centre, has been very slow.'

'Waterworks machinery, both for home and abroad, has maintained a fair demand and an exceptionally large number of Leeds-made plants have just been, or are about to be, started. The drive is shared by steam, oil and electricity. A specially fine steam-driven plant for the Metropolitan Water Board, started in August, consists of a high efficiency unit with a less economical, but very much cheaper, stand-by [A reference to the Deptford Pumping Station contract]. The savings thus effected in combined capital and running costs are obvious. It is also interesting to note that a body which applies the most careful scrutiny to every item of expenditure has adopted steam-drive for its latest large pumping stations.

Requirements for pumping-plant sewage-works, mainly electrically-driven, have also been fairly good, chiefly, however, for the home market. Disintegrators, also, for the prevention of beach-pollution have been installed on the South Coast, and their field seems likely to extend.

Industrial and mining pumping plant has been in very poor demand, except in the case of the electrical trade, where it has been moderate. Land-drainage schemes also have been few and far between, Government grants for this purpose having been much curtailed.

Regarding prospects for 1932, it would be rash to prophesy in these days. Perhaps, however, the hope that trade in the coming year may not, at any rate, be worse than in 1931 would not be regarded as extravagant.

However, the financial fall for 1932 was even lower. Not surprisingly, the Company recorded a loss of £1,760-18s-2d in 1932.

Other firms in the immediate area of the Sun Foundry were also seeing the effects of the slump. One firm was *Joshua Buckton and Company*, manufacturers of testing machinery, engines and wagons. They owned the Well House Foundry on the west side of the Dewsbury Road that faced the northern end of the Sun Foundry.

In April 1932, Bucktons had discretely put some of their premises up for sale. Hugh Lupton showed an interest to purchase and had met their Agents, *J.W. Watson and Sons*. After the meeting, Watsons sent a letter to Lupton with a suggestion for an offer, to which Lupton must have tentatively agreed. The offer was for 12958 square yards at 27s-6d per square yard (£17,817), a sizeable amount considering Hathorn, Davey's financial situation.[7]

However, the possible expansion of the works may have been associated with a deal Hugh Lupton was trying to arrange with *Sulzer Brothers* of Switzerland. Sulzers also manufactured pumping machinery and had in the past often been in competition with Hathorn, Davey when tendering for work. A letter from Hugh Lupton to Meysey-Thompson dated 26th April 1932 indicated that J.F. Schubeler, the Managing Director of *Sulzer Bros (London), Limited* was visiting the Sun Foundry on the 28th.[8] The letter related to a proposal that Hathorn, Davey and Sulzer should collaborate on a joint project to manufacture machinery for Sulzer, subject to shareholders approval. Clearly Sulzer wanted some security as part of the deal. The points raised were:

i)	Sulzer would supply any machinery needed, valued at £10,000.
ii)	Hathorn, Davey to issue preference shares to Sulzer to cover the value of the machinery. The letter indicates that if the deal went ahead then further preference shares would have to be issued, as the current limit was £12,000
iii)	Hathorn, Davey would sell to Sulzer 50% ordinary shares, which they valued at 5s per share, but this was clearly open to discussion. The total capital of the ordinary shares was valued at £99560, but with depreciation Hathorn, Davey would sell 50% to Sulzer for £12,445.
iv)	Provision to be made for Hugh and Reginald Lupton to be retained as managers.

The situation became clearer in a letter[9] from Hugh Ralph Lupton to Sulzer dated 1st June 1932:

<u>'Machining Costs of Centrifugal Pumps</u>

Following upon our recent discussions, we now have pleasure in reporting as to the estimated machining times of the typical horizontal turbine pump and vertical spindle borehole pump, of which you kindly sent us the particulars.

In the first place we would apologise for the delay in sending you the figures. You will readily understand that we did not wish to put the job through our departments in the usual way. We therefore handed over the drawings to our Works Manager and requested him to get out the machining times. This he did by reference to his actual records of corresponding parts of our own pumps - a far more lengthy process than if it had been carried out by our estimators as a routine job. Having done this the matter was, we fear, further kept back in order to have the consideration of the writer [Hugh Ralph Lupton], who has unfortunately been much engaged lately.

As mentioned above, our Works Manager got out figures entirely independently of yours, and moreover his figures assume the use of our existing machines only. Doubtless the provision of additional machines specially adapted for the work in question would considerably accelerate the work.'

Figure 158. Three Throw Ram Pump driven by an Electric Motor
(*From the 1933 Hathorn, Davey Catalogue. Courtesy of Mike Beevers*)

Other correspondence refers to specific drawings and procedures for manufacturing the pumps. How far this deal progressed is not known, but clearly it didn't go ahead at this time. There isn't for example, reference in the Annual Accounts to additional work, or any change to share holdings.

In 1933, the Company produced a new catalogue. The Triple Expansion Engine featured strongly in it. The work carried out for the *Metropolitan Water Board* being used as an example. Various pumps were also included and some were advertised as capable of being driven by electricity (Figure 158) or an oil-engine (Figure 159).

Figure 159. Set of Three Throw Ram Pumps
(From the 1933 Hathorn, Davey Catalogue. Courtesy of Mike Beevers)

As an aside, it is interesting to note that on the 13th May 1932, Hathorn, Davey submitted a petition to the High Court of Justice for the winding up of the firm of *Davey Paxman and Company, Limited*, of Colchester. The Company specialised in portable steam engines and other types of engine that included Compound engines. Although no other evidence has been found, it is probable that Davey Paxman owed money to Hathorn, Davey. Neither the Company, nor the Davey in the Company name, were connected with Hathorn, Davey. Paxman's were restructured later that year.[10,11]

Whilst the Hathorn, Davey Order Books for this period are missing, it is possible to identify some of the Contracts that were ongoing in the early 1930s that may have contributed to the Company's financial difficulties. Lupton listed them, whilst trying to assess outstanding amounts, and whether any surety had been placed against the contract. In many instances only part payment had been received, which was presumably due to stage payments agreed in the contracts. Probably the final payment would have been made once the contract was commissioned.

Staffordshire Potteries Water Board (Cresswell Pumping Station)
    Order 7801 dated 21st December 1928. Two Triple Expansion Pumping Engines etc. (17"+29"+44", 3' stroke)

Wolverhampton Corporation (Dimmingsdale Water Works).
    Order 7789 dated 5th March 1929. One Triple Expansion Pumping Engine etc. (21"+37"+54", 4' stroke)

South Staffordshire Water Works Company (Slade Heath Pumping Station)
    Order 7819 dated 27th May 1929. One Triple Expansion Pumping Engine etc. (24"+43"+66", 3' stroke)

Stourbridge and District Main Drainage Board (Enville Street Pumping Station)
    Order 7827 dated 13th September 1930. One Triple Expansion Pumping Engine etc. (10½"+19½"+29", 2'6" stroke)

Poole Corporation (Corfe Mullen Pumping Station)
    Order 7992 (date of order not known, but installed 1932) One Triple Expansion Pumping Engine

Lupton lists other orders, of which very little is known, that includes: Barbados (8023); Cape Town, South Africa; Eggborough Pumping Station (Pontefract and Goole Waterworks) (8035);[12] Folkestone Water Works, possibly Dover; Cheltenham Corporation; Clacton Urban District Council, probably for the Hertfordshire and Essex Waterworks Company, Limited; Exmouth Urban District Council; Ipswich; Paignton; and Aldershot.

Figure 160 The Sutton Poyntz Triple Expansion Engine. The engine is now a working exhibit at the Internal Fire Museum of Power, near Cardigan, Wales.
(Courtesy of Chris Allen)

One engine from this period that Lupton doesn't list was Order 8040 (dated 14th May 1934) that was originally installed at the Sutton Poyntz Pumping Station, near Weymouth. It was for a relatively small Inverted Vertical Triple Expansion engine (12"+19"+29", stroke 24"). After it finished working, it was taken from Dorset to West

Yorkshire with the intention of re-erecting it in the Leeds Industrial Museum. However there was a fire at the Museum, so the engine was never fully assembled. The engine was removed and taken back to Sutton Poyntz. Unfortunately, it couldn't be erected there, so it was eventually taken to the Internal Fire Museum of Power, Castell Pridd, Tanygroes, Ceredigion, West Wales were it has been assembled, and is now back in working order. (See Figure 160)

The Corfe Mullen Engine also survives at Chippenham, Wiltshire. It is currently dismantled, but will eventually be restored to working order.

Chapter 30 References

[1] Leeds Unemployment. *Sheffield Daily Telegraph.* 9th January 1930. p.4

[2] Leeds Unemployment. *Yorkshire Evening Post.* 7th January 1931. p.9

[3] HL. 1940. *Hathorn, Davey and Company, Sun Foundry, Leeds.* p.8

[4] Engineering Notes. Progress in Pumping Machinery. *Yorkshire Post.* 14th March 1930. p.3

[5] LMWS Box 21D2Bx1 Item 19865 Balance sheets 1902 to 1934

[6] Engineering in the North of England (continued). *Yorkshire Post and Leeds Intelligencer.* 14th January 1932. p.30

[7] LMWS 19933 Unsorted Letters. 28th April, 1932 Letter re purchase of part of Well House Foundry.

[8] LMWS 19933 Unsorted Letters. 26th April, 1932 Correspondence with Sulzer

[9] LMWS 19933 Unsorted Letters. 31st June, 1932 Manufacturing work for Sulzer

[10] Davey Paxman and Company Limited. *London Gazette.* 20th May, 1932. p. 3295

[11] https://www.gracesguide.co.uk/Davey,_Paxman_and_Co

[12] Eggborough was for Pontefract and Goole Waterworks. George Watkins photographed it in 1962 and states there were a pair of engines. He gives the sizes as 12", 18" & 28" x 2' stroke. Dated as 1933. They drove well and force pumps. See: *Stationary Steam Engines of Great Britain*, Vol 1, Landmark Collector's Library, Picture No. 144. (Information courtesy of Chris Allen)

31. Receivership, and a Possible Take-Over

Hathorn, Davey suffered further financial losses of £7,499-5s-9d and £438-13s-1d in 1933 and 1934, respectively. For the first time 'Bad Debts' of £5,212-12s-0d and £2,783-12-10, were listed under losses for the two years. It was inevitable that with both bad debtors and financial losses, that the mortgage provider would have to be informed of the financial difficulties in which the Company found itself.

On the 3rd July 1934, just three days after Hathorn, Davey's financial results were declared, their Accountants, Messrs. *Smithson, Blackburn and Company*, wrote to the Manager of the National Provincial Bank, Leeds.[1]

'We have been instructed by Messrs. Hathorn, Davey and Company, Limited to report to you as to our opinion as to their present position and future prospects, especially in connection with the Bank Overdraft and the financing of two proposed contracts.'

Following the statement, Hathorn, Davey attempted to demonstrate to the Bank that it could cover the overdraft amount with two contracts that they hoped to be awarded in the coming twelve months, and that the orders would be obtained without competition. The two estimated contract values were:

South Staffordshire Water Works Company.	£33,786
Hertfordshire and Essex Waterworks Company, Limited	£17,316
A total of	£51,102

No doubt they were two dependable Water Companies that would reliably pay their invoices quickly. The Accountant's letter to the bank continues to state, *'If these two orders are executed it appears to us that on completion and receipt of payment the Bank Balance would be reduced by a very substantial profit obtained during the coming financial year of the Company, as well as by the amount of Depreciation written off the fixed assets but expended (say £1,800).*

'It would therefore appear to us to be as much to the benefit of the Bank as to our clients, that these orders if received, should be executed, which are apparently by far the most favourable our clients have had the chance to secure for many years.'

The letter then went on to show how the Company would invoice the Water Companies at different stages of the work. It addition, an enclosed statement of terms of payment for the previous 4½ years showed overhead charges and other expenses. As a footnote, there was a comment that, *'A 10% reduction of staff wages and salaries took place, commencing for the week ending April 2nd, 1932 and also a reduction of the number of employees.'*

How many people were involved is not recorded. However, the Accountants did mention that an additional overdraft of £25,000 would be required to complete the two contracts.

Clearly the Bank was not impressed with the proposal because during October 1934, a Receiver was appointed.[2]

'MESSRS. HATHORN, DAVEY AND CO
Receiver Appointed

The difficulties which British engineering firms have encountered in recent years are well known to have been serious. Loss of trade has led in several instances to the contraction of credit, followed by necessary steps for the protection of special interests, and the latest instance is that of Messrs. Hathorn, Davey and Company, of Leeds, whose position, we learn, is the moment briefly this:-

Owing to adverse conditions in the Dominions and Colonies, to which formerly nearly half their products were sent, and to the slump the engineering trade generally, which has prevailed ever since 1926, orders have been difficult to obtain, prices low and customers' conditions exigent.

The failure of one of their own debtors for a large amount added to these difficulties, and the bankers, who are their principal creditors, have appointed a receiver. Mr. K S. Morrison, of Peat Marwick, Mitchell and Co., who is carrying on the business.

In point of fact it is understood that conditions are more encouraging than they have been for some years.'

Certainly there were some elements of optimism in a confidential report that Hugh Lupton had presumably prepared for the Receiver.[3] Lupton listed the type of work that Hathorn, Davey carried out.

During the past ten years the products consisted mainly of waterworks pumping plant driven by steam, oil-engines and electricity, and sewage and drainage pumping plant driven mainly by electricity. These were supplied to public utility bodies both at home and abroad. Other products included a whole variety of pumps, for mines drainage, hydraulic power, etc.

Other types of work included overhauls, renewals and repairs to their installed systems both home and abroad as well as heavy presswork, both mechanical and hydraulic. This latter work was based on considerable experience gained during World War I. Hathorn,

Davey also specialised in low-lift pumps for dock-work and large valves, especially for sewage plant. This was a fair assessment by Lupton as the 1933 Catalogue shows a number of items that were not manufactured by them before the War, including 750-ton presses. (See Figure 161)

Figure 161. 750 ton Hydraulic Forging Press
(From the 1933 Hathorn, Davey Catalogue. Courtesy of Mike Beevers)

Lupton blamed Government policy for a reduction in the manufacture of engines. *'Owing to the Government's economic measures of 1931, coupled with the difficulties of export trade,'* that sector had shrunk dramatically. *'The large schemes especially require a period to pass through their different stages (Parliamentary sanctions, etc.) but the present cheapness of money should favour the high-class plant.'*

The Company was trying to foresee future demand and had bought the rights to manufacture a high-speed piston pump and had developed a large sewage disintegrator

305

for which there was much worldwide interest. There were also a number of pending waterworks and drainage schemes, including an enquiry for two Triple Expansion Engines driving both well and force pumps valued at about £30,000 from Dover.

After listing other potential contracts, Lupton concluded, '*I would draw your attention to the reputation of the firm, which, until its present distress was, I think, higher than it had ever been. This should prove no mean asset in the event of a reconstruction, and, particularly in the development of new lines (e.g. the taking up of a Schoene's* [high speed piston pump] *and of turbine-type borehole pumps) it would be of immense assistance. It might also be not out of place to mention the extent to which any of the firm's customers treat it in an advisory capacity, the goodwill thus engendered being a material asset in procuring work at a later date.*'

Hathorn, Davey was invited to tender for the extension to the Dover waterworks. However, the operations at the Sun Foundry had been slowly reduced to meet a demand set by the Receiver that no additional contracts would be taken on after 30th June 1935. To continue with a new contract, permission was required from the Bank.[4] The Company was unable to fulfil the contract even if it was awarded to them. Lupton, with it seems the blessing of the Receiver, looked for available engineering companies to try and broker a collaboration deal, or even a take-over, to complete the tendered contract.

The first company he approached was Messrs, *Glenfield and Kennedy, Limited* an engineering company. They were based at Kilmarnock, Scotland and Hugh Lupton paid them a visit in August 1935. A second company, *Crossley Brothers*, of Manchester was also contacted, but neither gave enthusiastic responses.[5]

A third company, *Crossley-Premier Engines*, Sandiacre, Nottingham gave their reasons why they couldn't strike a deal to work with Hathorn, Davey.[6] On the 28th August they wrote a letter to K.S. Morrison, the receiver.

'*After considering the matter from every point of view, we have reluctantly decided that we are not able to go further with the negotiations.*

The greater part of Messrs. Hathorn, Davey's productions are quite outside our scope inasmuch as we have no knowledge and experience in that line of work.'

There was definitely a positive response from a meeting with Messrs. *Markham and Company*, Chesterfield. At least they contacted the National Provincial Bank on the 19th September 1935 to determine how they would respond to any deal. Naturally the Bank was only interested in a cash settlement, and confirmed that they would prefer an offer of £30,000 for Hathorn, Davey. For that amount, the purchasers would acquire the whole of the assets of the business, free of the Bank's Debenture.

If that deal had gone ahead, then the purchaser would have been free to take any profits from current work. In addition, if this had transpired Hugh Lupton would have paid the Pre-Receivers Creditors account personally. This would have cleared all debt, and the suggestion was that Hathorn, Davey could be established as a new company, for example *Hathorn, Davey and Company (1935), Limited*. There would be no break in the continuity of the Company. Another suggestion was that with the debts cleared, the Company could continue as normal. There was some pressure to complete a deal by the end of the month, as it was the deadline for submitting tenders for the Dover contract.

Fortunately, the pressure to close down the Sun Foundry as soon as practically possible was thrown a lifeline, and it seemed to have the Receivers support. On the 3rd September 1935, K.S. Morrison (The Receiver) wrote to the bank seeking a further extension to keep the shops working.[7] *'Owing largely to delay by one of the customers in taking delivery of machinery on a pre-Receivership Contract, it will, I fear, be possible to be in a position to close the Shops, if desired, for some months yet, and the big contract which I took with the Bank's authority will still require some two months or so in the Shops.'*

Clearly business was picking up again, as Morrison continues, 'I have at the moment a considerable number of Tenders out, all for work which can be completed fairly quickly, with the exception of two of the value of some £6,000 which, if orders were received before a decision were to come to about the Works, would, I estimate, take three to four months to complete in the Shops.'

'It is, however, becoming increasingly difficult to deal with any enquiries coming through, and at the moment they are pretty numerous, owing to the doubt as to how long the Shops will remain open.'

'This is very forcibly brought before one in the case of an enquiry at present out for Dover Corporation for big new pumping machinery. I have instructed the Company's Officials to go on preparing the estimates for the enquiry as if they were going to tender, but not to submit any tender without the approval of the Bank. The estimates are not yet completed, but rough figures lead one to believe that the tender will be in the region of £36,000, which we estimate to show a profit of some £5,000.'

Morrison estimated that the Dover contract would take 15 to 18 months to complete, and that another year would be needed for total completion of installation etc. He thought that being awarded the Dover contract would be advantageous for a number of reasons:

i) It would give more time to find a scheme for the continuance of this very old established company.
ii) It would definitely ease the position with regards to the seeking and carrying out of smaller contracts.

iii) It should bring a profit to the business.

iv) It would enable any maintenance in connection with current work to be carried out in the Company's own shops instead of outside.

v) If a scheme was arranged, it would avoid the Bank realising its security on a break-up basis.

The disadvantages of not extending the period were that it postponed the Bank realising its security, and additional monies would be required to finance the Dover Contract.

Morrison then remarked that under his receivership the Company had made an operating profit of approximately £2,000, without considering depreciation, contingencies and interest, and any other expenditure on Pre-Receivership Contracts.

He concluded the letter by indicating to the bank that there were several parties interested in buying Hathorn, Davey.

With regard to the Dover tender, there was certainly a letter drafted to indicate that Hathorn, Davey would like an option to withdraw if awarded the contract. It would also seem that an arrangement was made for Messrs, *Glenfield and Kennedy, Limited* to complete the contract to Hathorn, Davey's design and specification, if Hathorn, Davey couldn't fulfil the contract.

However, all this negotiation was irrelevant, as Hathorn, Davey was not awarded the contract. It was given to *Worthington-Simpson*, of Newark, Nottinghamshire.[8]

Chapter 31 References

[1] LMWS 19933. Unsorted Letters. Correspondence with Bank 3rd July 1934

[2] Messrs. Hathorn Davey and Co. Receiver Appointed. *Yorkshire Post and Leeds Intelligencer.*25th October, 1934, p.9

[3] LMWS 19933. Unsorted Letters. List of type of work undertaken by Hathorn, Davey dated 25th June 1935

[4] LMWS 19933. Unsorted Letters. Letter to National Provincial Bank 3rd September 1935

[5] LMWS 19933. Unsorted Letters. Letter Crossleys and Dover contract 14th August 1935

[6] LMWS 19933. Unsorted Letters. Possible buyer of Company - Crossley-Premier 25th August 1935

[7] LMWS 19933. Unsorted Letters. Letter from Receiver to Bank 3rd September 1935.

[8] https://www.forncettsteammuseum.co.uk/the-dover-engine.html

32. Sale to Sulzer Brothers

In October 1935, Hathorn, Davey was still operating under the direction of a Receiver, with very little progress being made to secure the Company's future on a reliable footing. It was left to Hugh Ralph Lupton to try and end this deadlock, so he wrote a letter on the 3rd October 1935 to a possible rescuer, J.F Schubeler, the Managing Director of *Sulzer Brothers (London), Limited.*[1] Fortunately, the draft letter survives.[2]

' Dear Mr Schubeler,

You will remember the discussions you had with this Company some three years ago with regard to a possible arrangement for the manufacture of your products in this Country.

You will also have noticed from the Press that this business has for the last 12 months or so been carried on by a Receiver appointed by the Debenture holders.

The Receiver is now engaged in efforts to reconstruct the Company, and it seems not unlikely that some scheme may very shortly mature. I heard recently that you were probably making certain changes in your arrangements for manufacture in the Country, and it immediately occurred to me that you might very possibly be interested in such scheme. I am therefore writing to you to suggest that I call at an early date to discuss the possibility.

I realise that in 1932 the terms we were able to offer were not sufficiently attractive to your Company: in our present position, however, I think you might well find the matter quite otherwise. Should you think anything of the suggestion I would bring such figures as would probably be required, or, better still, our Receiver would call and put the whole matter before you. Owing to the imminence of the arrangements now in train, the matter is one of considerable urgency, and I will ring you up tomorrow morning to make an appointment if possible. Tuesday next would be convenient to me.

Yours Very Truly
Hugh R Lupton'

The draft letter was almost certainly sent to Schubeler, because a simple footnote states,

'Saturday Morning
Rang up - Schubeler very genial, regretted could not manage Tuesday, so fixed Monday at 3:00pm.'

It is also fortunate that Hugh Ralph Lupton retained the notes that he made from the meeting that he held with *Sulzer Brothers (London)* on the 14th October 1935. Also present was Morrison, the Receiver, and various officials from Sulzer that included; Mr Robert Sulzer, one of the Company's partners, responsible for technical development,

J.F. Schubeler, Mr Paul Hurliman, M. Staff (Sulzers Company Secretary) and F. Oerdelin.[3,4]

'Mr Schubeler confirmed that they are now looking for a place to manufacture in this country and that our letter seemed to them to come just at an opportune moment. Our works and machinery however were not suitable for their equipment and they would have to build a further bay, or bays, and put in new machinery. Schubeler also said that our idea of accuracy and methods of manufacture were old fashioned and that a complete reorganisation would be needed. Our machinery he thought old. [The latter is not an unsurprising statement from Schubeler, as it is probable that much of the equipment was installed before 1920]

Mr. Morrison explained the financial position generally along the lines of the letter written to Macarthey. [Not known] *He also described the three causes of HD & Co's trouble - Government and economy, D-P bad debt* [D-P Not known], *Guarantee penalties. Sulzers expressed doubt as to whether there would be any more vert-triple* [Reference to the steam driven vertical triple-expansion engine, which was now largely superseded by electric motors], *but Morrison mentioned that Dover was imminent and explained that we thought our chance, at a profit of £4,000 / £5,000, very good. They agreed that it would be a great mistake, not to take such work, which still available. We explained that we had made efforts to obtain orders for centrifugal and axial flow pumps and that, particularly in the sewage field we had done moderately well. Mr* [Robert] *Sulzer mentioned the Stereophagus pump a useful branch of manufacture and we told him that we make the larger sizes. Also the Gargantua Disintegrator. Sulzer said that sewage pumping was a field they had not sufficiently exploited. Lupton mentioned the oil-circulating pumps* [used with electrical transformers] *as an instance of our efforts to fit centrifugal orders.*

Sulzer's Company Secretary was particularly keen on turn-over figures, which Mr. Morrison gave him for the past four years. We made it clear, however, that they should be much greater. They also wanted to know the number of men employed now (about 150) and normally (about 300).

Sulzer were of the opinion that it was better not to run own iron foundry. We explained to them the terms of our let to Glover and Wood and gave the capacity of the foundry at 50 tons / week, though not so much as that is going through now. Single castings up to 30 tons.

Sulzer were interested as to the extra land available for new shops, particularly to enable them to manufacture their oil-engine. We told them Buckton's works, across Dewsbury Road, were to a large extent still available (NB - held by National Provincial Bank, Agents J. and W. Watson).

Sulzer enquired regarding personnel and staff. Lupton said he would, provisionally, be willing to stay on if the firm could be reconstituted. He thought he [Lupton] *would stay for a short time if required. Sufficient existing staff could easily be retained to carry on existing products.*

We mentioned that Dixon had a friend who would probably be willing to put in £6,000 / £8,000 should a reconstitution with Sulzer be brought about. Mr. Morrison mentioned our friend (Thornton) who wants a place for his son / son-in-law, and might put in money under such terms. Sulzers expressed themselves why keen on having a few prominent British engineers on their Board.

We explained our possible cooperation with Premier. This did not appeal to Sulzer at all, as their ultimate idea is to develop their oil-engine manufacture in this country. So far this only oil engine made in this country are the [?] engines.

We also mentioned Schoene, but they were not much attracted, having a natural prejudice in favour of the Centrifugal Pump. However, they by no means turned it down.'

The meeting closed after Sulzer's had expressed a desire to visit the Sun Foundry. Messrs. Schubeler, Oerdelin and Hurliman made the visit on the 18th October 1935. Mr Schubeler, also sent for both a Mr. Kubler from the *Armstrong-Whitworth's*, Scotswood works at Newcastle-on-Tyne, which was manufacturing machinery for Sulzers, and also for Mr Staffs, to meet him on the 19th. Schubeler also had a meeting with a representative from the Leeds Chamber of Commerce.

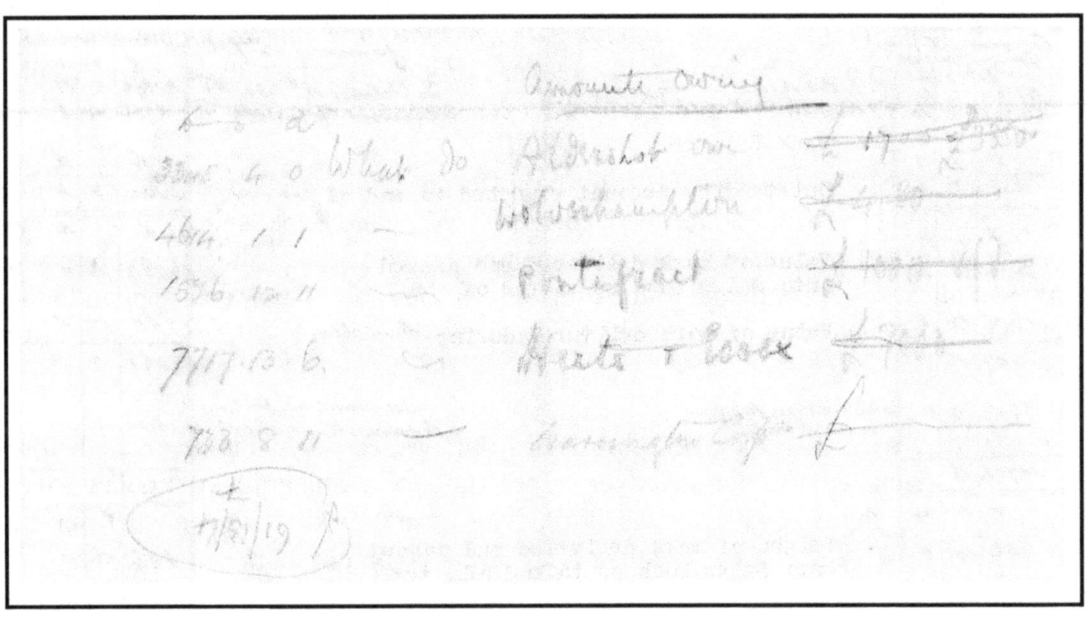

Figure 162. The five organisations that owed money to Hathorn, Davey.
Pontefract and Wolverhampton probably relate to Eggborough and Dimmingsdale.
(LMWS HD Box 21D2BX2. Item 19872.11)

Mr Staffs held a meeting with Mr. Matthews to go through various figures, and specific Hathorn, Davey contracts were discussed that included; Wolverhampton, Pontefract, Warrington, Aldershot and Shanghai. The discussion would have centred on amounts owed to Hathorn, Davey. A pencil note (Figure 162) on the reverse of accounts dated 22nd January 1936 identified five debtors. The debts totalled £17,937.[5]

The machinery supplied by Hathorn, Davey to those organisations is not recorded, but it is more likely to be pumps for water or sewage schemes rather than engines. For example, the debt of £3,345-4s incurred by Aldershot Town Council probably related to a joint sewage scheme with Guildford Rural District Council.[6] A new sewage treatment works was constructed at the outfall in the mid-1930s and it is probable that Hathorn, Davey supplied the appropriate pumps. Other debts probably arose from the sale of pumps for waterworks extensions. Information relating to specific contracts has not been found.

A few weeks later there was some optimism regarding the Dover Contract, and Lupton wrote to Schubeler.[7] *'I have now a further bit of news which will be of interest to you in connection with Dover, in that I have arranged with Messrs. Glenfield and Kennedy, that assuming I obtain the order for Dover they will take over the contract should I request them to do so. We are supplying the designs and any available patterns.'* Lupton felt certain that if he wrote to the Dover Borough Engineer, the engineer would accept a re-submission of Hathorn, Davey's tender for the contract.

On 10th January 1936, Sulzer made a formal offer to purchase Hathorn, Davey. They were prepared to offer £30,000 for the Debenture held by the Bank, and £550 for all the issued shares. The Debenture and Shares to be transferred to Sulzer free of all costs and stamp and other duties incidental to the transfer. Various conditions were applied to the offer that included a safeguard should the business be valued at less than the offer price.[8]

However, there were still a number of obstacles Hathorn, Davey had to negotiate before the deal was clinched. For the takeover to be completed, Sulzers had stipulated that they held all the shares. One potential difficulty was getting total agreement from the shareholders to sell the shares. In most cases it was a simple matter of accepting the sale of the shares to Sulzers, as the majority of the shares were held by members of the Lupton family and Meysey-Thompson. However, there were a few anomalies.

Despite an early career in engineering, and positions of responsibility within Hathorn, Davey, Norman Davey had pursued an alternative profession as a novelist in the 1920s. He lived with his second wife in the south of France and had to be contacted via his bank, in Monte Carlo. He received very much the standard letter sent to all the shareholders by the Receiver. In Norman's case, he may not have been *au fait* with the details of the takeover by Sulzer. The letter outlined the present state of the negotiations; the details of the offer from Sulzer; and the consequences if the offer wasn't accepted. On the 20th January 1936 the Receiver sent a letter to Hathorn, Davey's Company Secretary outlining the proposal that would ultimately be put to all shareholders.[9]

'As I think you are aware, I have for some time been in negotiation with a wellknown Engineering concern as to the taking over of this business, and I have now received an offer from them which, subject to the settlement of a few minor points, I propose to accept.

The amount of the offer will not enable the Debenture Holders [National Provincial Bank] to be paid in full; they will, in fact, have to write off approximately $33^1/_3\%$ of their secured debt, and in the ordinary course of events there would, I regret to say, be nothing available for shareholders.

Mr H. Lupton was most unwilling that any Creditor of a Company of Hathorn, Davey's standing should not be paid, and, as you may be aware, Mr. Meysey Thompson and Mr. Lupton have paid the Trade Creditors out of their own pockets. Recognising this the proposed Purchasers make a contingent offer of a certain allowance to these gentlemen, not exceeding the amount they have paid, if certain benefits accrue to them as a result of the scheme. Failing this, they have no chance of being repaid any of the amount they have expended.

As a result of their action the continuance of the present Company becomes possible, and the proposed Purchasers, in addition to their offer to the Debenture Holders, make an offer of 1/- [One Shilling] per share (free of Stamp Duty and Costs of Transfer) for each of the Preference and Ordinary Shares of the Company, this offer being contingent upon the whole of the Shares being transferred to them.

If any Shareholder is not prepared to accept 1/- per Share for his holding in the Company, the scheme falls through, subject to the provisions of Section 155 Companies Act 1929, whereby a power is given, in certain circumstances, to acquire the Shares of dissentient Shareholders in accordance with the scheme approved by the majority.

Should the scheme fall through, I shall be faced with one or two alternatives:-

(a) Endeavour to arrange with the proposed Purchasers a fresh offer for the Assets of the Company, or

(b) Close the business down and realise the Assets of the Company.

In either event I think you will agree that there is no prospect of any Shareholder receiving anything, and that the Shares are, therefore valueless. I trust, therefore, that every Shareholder will accept the offer.

I would add that I am raising with the proposed Purchasers the possibility of some amendment to their offer for Shares which may result in slightly better terms for the Shareholders, but I cannot, of course, give any guarantee on this point.

I would, however, suggest that your Directors should consider the problem, and if the approve the offer, should immediately circularise their Shareholders and enclose a blank transfer form. The Transferer, I would suggest, should leave the consideration blank, so that if I am able to get a better offer for the Shares the amended figure can be included. I, of course, would undertake not to use a transfer for any consideration less than 1/- per Share.

Yours faithfully,
K.S. Morrison.'

A Statement (Figure 163) dated 22nd January 1936 summarised the state of Hathorn, Davey's finances and showed an improvement in the business. It must have pleased all concerned. The value of work delivered and passed into the Sales Book up to the end of November 1934 had risen from £6,837-3s-10d to £16,553-3s-0d for the same period in 1935, but during December it had nearly halved, almost certainly due to the actions of the Receiver.

Also on the 22nd, January 1936 a letter went out to all Shareholders from the Company Secretary with a copy of the Receiver's letter. He confirmed that the Directors held some 60% of the shares and were recommending accepting the offer.

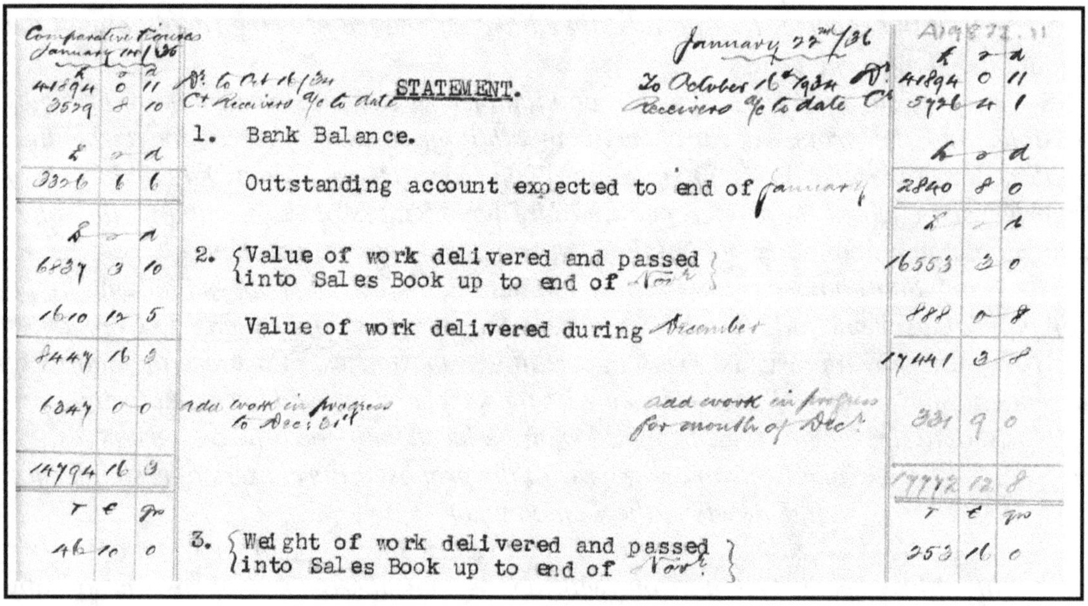

Comparative Figures January 22nd/36 — January 22nd/36 — A19872.11

STATEMENT.

£	s	d		£	s	d
41894	0	11	Dr. to Oct 16/34 / To October 16th 1934 Dr.	41894	0	11
3579	8	10	Cr Receivers a/c to date / Receivers a/c to date Cr	5926	4	1
			1. Bank Balance.			
3326	6	6	Outstanding account expected to end of January	2840	8	0
6837	3	10	2. Value of work delivered and passed into Sales Book up to end of Novr	16553	3	0
1610	12	5	Value of work delivered during December	888	0	8
8447	16	3		17441	3	8
6347	0	0	add work in progress to Dec 31st / add work in progress for month of Dec	331	9	0
14794	16	3		17772	12	8
T	C	Qr		T	C	Qr
46	10	0	3. Weight of work delivered and passed into Sales Book up to end of Novr	253	16	0

Figure 163. Statement showing Hathon, Davey's basic finances in 1935 and 1936.
(LMWS HD Box 21D2BX2. Item 19872.11)

Another minor difficulty from Sulzer's point of view is that they did have an arrangement with *Armstrong-Whitworth* for some items of machinery to be manufactured at the Scotswood Works, Newcastle. It was considered that it would be a relatively easy Agreement from which to be released. In fact, Armstrong's engineers visited the Sun Foundry to determine where items of the machinery could be sited. It also became apparent that Hathorn, Davey was about to tender for other work at Blackpool which if successful would require manufacturing in the transition period. There was concern from Sulzer on who would take responsibility should the machinery not have the required efficiency. [10]

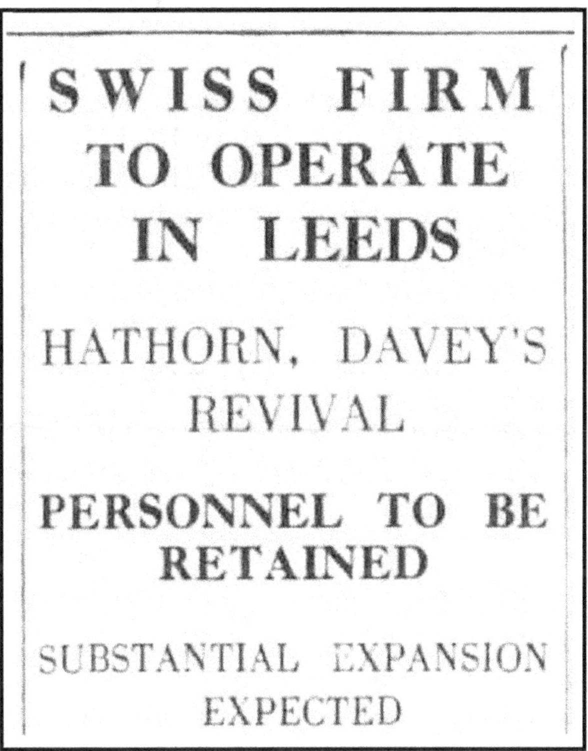

Figure 164. Announcement of the Take-Over of Hathorn, Davey by Sulzer
(Yorkshire Post. 2nd May 1936. p.13)

By May 1936, all the formalities had been completed, and on Saturday 2nd May 1936 a brief article was published in the *Yorkshire Post*, proclaiming the event.[11] (See Figure 164) After briefly outlining why the firm had gone into Receivership, the article extolled Mr. Morrison's belief in the Firm.

'Mr. Morrison quickly realised that the firm, whose fame is world-wide numbered among its asserts an excellent reputation and a valuable goodwill, and his efforts

therefore, were directed towards a reconstitution such as would preserve the firms identity intact.

> *We are glad to record that Mr. Morrison has been successful, and that the firm of Hathorn, Davey and Co. has been acquired as a going concern by Sulzer Bros. (London) Ltd., the English branch of the Swiss engineering firm. Messrs. Sulzer's intention is to extend and adapt the works to cover the manufacture, in the first instance, of centrifugal pumps, and later of the whole range of their other products; at the same time they intend Hathorn, Davey and Co. to continue to produce their own specialities. Further to preserve the entity of the Leeds firm, they intend to retain, as far as possible, the present personnel, which will form the nucleus for the Anticipated expansion.'*

Lupton records in his history that Mr. M. Hurliman was appointed Manager of the Leeds Works. But Mr. Reginald H. Lupton and Mr Hugh R. Lupton were retained in executive positions and together with the writer [Hugh Lupton], remained on the reconstructed Board of Directors until 1938, when as they had no direct financial stake in the Company, thought it better to resign.[12]

The Luptons should be congratulated for this remarkable achievement that could only have been brought about by their gentlemanly management qualities that were so evident in the way they conducted their business.

Chapter 32 References

[1] Johann Jakob and Salomon Sulzer at Winterthur, Switzerland established Sulzer Brother in 1834. Initially the company produced cast-iron, but then expanded into the field of heavy engineering, specialising in the manufacture of pumping equipment.

[2] LMWS 19933 Unsorted Letters. Draft Letter by H.R. Lupton to Sulzer 11th October 1935

[3] LMWS 19933 Unsorted Letters. Notes on a meeting held with Sulzer 14th October 1935

[4] Obituary for Robert Sulzer. https://www.gracesguide.co.uk/Robert_Sulzer

[5] LMWS 21D2BX2. Item 19872.11 Statement of accounts and owed amounts 22nd January 1936

[6] Aldershot and Ash Sewage. *Surrey Advertiser*. 9th September 1933. p.9

[7] LMWS 19933 Unsorted Letters. Correspondence re Dover Contract November 1935

[8] LMWS Box 21D2BX2. Item 19872.13 Completion of Sulzer takeover 10th January 1936

[9] LMWS 19933 Unsorted Letters. Details about takeover by Sulzer 20th January 1936

[10] LMWS 19933 Unsorted Letters. Notes regarding current commitments 24th January 1936

[11] Swiss Firm to Operate in Leeds. *Yorkshire Post*. 2nd May 1936. p.13

[12] HL. 1940. *Hathorn, Davey and Company, Sun Foundry, Leeds.* pp. 8 and 9

33. Hathorn, Davey: Part of Sulzer Brothers of Switzerland

Although there are no Order Book references to confirm that this is the case, the general consensus of opinion tends to agree that the last steam engine to bear the Hathorn, Davey name was manufactured in 1937/38, for the *Constantinople Waterworks Company*, for installation at the Terkos Pumping Station.

The first hint of the order can be found in an article in the *Yorkshire Post*.[1]

'Leeds Engineers' £29,000 Order
Waterworks Plant for Constantinople

In the face of competition both in this country, and also in Germany and France, the Leeds engineering firm of Hathorn, Davey and Co. Ltd., have secured an important order from the Constantinople Waterworks Company.

The value of the order is slightly in excess of £29,000, the work consisting of a triple expansion steam engine driving directly a set of vertical 3-throw ram pumps of a total horsepower of about 480. It also includes the necessary steam-raising plant, namely boilers, superheaters, economiser, steam piping, instruments, and all the accessories necessary to initiate and maintain running at a very high efficiency. This of course is a feature of this type of plant, the economy of which is unequalled by any other.

With regard to the financial arrangements, these comply with the new Anglo-Turkish Trade Agreement, according to which the customer deposits the whole of the payment in Turkish currency immediately on delivery at Constantinople, payment being made in sterling in this country as soon as the exchange permits, debts being discharged in strict chronological order.

This is the first considerable purchase made in this country by the Constantinople Waterworks Company, and the first part of the order is to be shipped September next. The remainder of the work will follow about the end of the year.'

Frank Mallinson described the Terkos engine.[2] He was the Hathorn, Davey erection engineer responsible for installing the engine. Unfortunately whilst carrying out this contract World War II broke out and Frank remained trapped in Istanbul until the War ended.

The Inverted Triple Expansion Engine (23"+43"+62", stroke 36") (Figure 165) was steam jacketed with two reheater receivers typically located between the cylinders. The ram pumps ($17^5/_8$" diameter) were located beneath the engine, and auxiliary pumps were attached to the main ram crossheads. The engine would run at 40rpm. Superheated steam was applied at 200psi and at 182°F at the stop valve, supplied by two 30' long Lancashire boilers. The engine and pumps were specified to be capable of pumping 3,666gpm against a total suction and delivery head of 430 feet.

The lack of an adequate water supply to Istanbul at the time meant that the engine was put into service before the plant was fully installed. This meant that the engine could not be regarded as 'new' when the required trial took place. The engine worked until 1967.[3]

That was the end of Hathorn, Davey's contribution to the age of steam.

Figure 165. Terkos, Turkey: The last steam engine manufactured by Hathorn, Davey
(Photograph courtesy of Chris Hodrien)

At the time of the Sulzer take over, Hugh Lupton had generously paid off outstanding debt that was owed customers to the Company, and the wages of salaried staff. In 1937 an agreement was reached with both Hathorn, Davey and Sulzer, that should tax savings be made in the subsequent years, Lupton would be paid back, and a payment also made to shareholders of the old company of 1s per share.

It must have been a pleasant surprise to Lupton, when that agreement was honoured in 1941, and he received part-payment from Sulzer via the Receiver, Kenneth Morrison.

The letter to him from the Receivers detailing the reimbursement concluded, '*May I say how very glad I am that you are, in part at least, refunded what you so generously and following such high principle paid to the old creditors, I repeat that I hope next year you will be reimbursed in full.*'[4]

By 1941, Hugh Lupton had been in retirement for three years. Together with Hugh Ralph Lupton and Reginald Hamilton Lupton, Hugh Lupton had stayed with Hathorn, Davey in an executive position. All three remained members of the Board of Directors until 1938 when they resigned and left the running of the Company totally under Sulzer's management. [5]

By 1939, Hathorn, Davey mainly manufactured pumps, and perhaps some machine tools. The only involvement with steam engines appears to have been the servicing of working engines, or manufacturing spare parts.

At the start of World War II, Major Alan Moncrieff, a veteran from the previous world war, and with wide engineering experience in the Leeds area, held a senior position in Hathorn, Davey. During the war the Company came under the control of the Ministry of Supply, and Moncrieff was promoted to the position of Assistant Regional Controller within the Ministry. All the skylights at the Sun Foundry were painted black, and for probably for the first time, all the work had to be done under electric lighting. One of the employees at this time was Geoffrey Luty who had started as an Apprentice with Hathorn, Davey. He had worked his way up to the drawing office but served the Company on the shop floor during the war. In addition to studying at night school, Luty shared a machine with a lady engineer in shifts. '*He was always proud of her as she kept the lathe free of swarf, and well-oiled, ready for use.*' The ratio of women workers replacing men's jobs is not known.

The austere period after World War II was a time when manufacturing companies were trying to re-establish their old work patterns, and markets. It meant nationalisation for some industries, for example coal and steel producers. Perhaps because of its links to Switzerland, Hathorn, Davey and Sulzer escaped this mania for mergers and remained a private concern. As always seemed to be the case, the Hathorn, Davey work force remained relatively settled and content. In the fifties and sixties the Company hosted children's Christmas parties, trips to the pantomime and dinner dances. Even a Company Art's Show was arranged in 1948.[6]

Leeds educationalists visited an exhibition of Arts and Crafts staged by employees of Hathorn, Davey and Co. Ltd., the well-known Yorkshire firm of hydraulic engineers, at their Dewsbury Road works on Saturday.

There was a fine show of 200 exhibits, ranging from an embroidered picture of Durham Cathedral by Miss J. Lucas, a typist, to a toy theatre, complete with stage lighting and figures, by Mr. S. Spofforth, who is in charge of the stores. Mr. R.M. Atkinson, managing director, submitted an original oil painting.

The exhibition was arranged by Mr. T. Mitchell, the works manager, and a trio of enthusiasts from the social and welfare section - Mr. A. G. Alldrit, Mr. A.H. Wilkinson and Mr. K. Smith.'

Figure 166. Hathorn, Davey: Drawing office employees in October 1947. Geoffrey Luty on the right. The gentleman in the centre of photograph is probably Mr Fennor, who may have been in charge of the drawing office? Other people in the photograph include: Archie Guy, Jock Malcolm, Tom Malcolm, Bob Mitchell, Steve Donague, Jack Taylor, Bob Lee, Frank Dyson, and (?) Morris.
(Courtesy of Mrs Christine Hart (née Luty))

The Company staff and workforce at this time enjoyed other leisure activities that included an Angling Club.

There was clearly a strong bond and loyalty between the Hathorn, Davey workforce at this time. Perhaps this is exemplified by an interview with Jim Stuart and his workmate John Healey that appeared in the *Yorkshire Evening Post* in 1948.[7] Lupton earlier described Jim Stuart (p.277), as having a *'refined personality.'*

'Jim (74) is an "artist in metal," but has never made a horseshoe.
As you walk through the busy, extensive workshops of Hathorn, Davey and Sulzer. Ltd., Jack Lane., Leeds, you come to an old part of the building housing furnaces, where the red glow of the heaped cinders throws a ruddy glare on to two grey-haired men stooping over their anvil. Jim Stuart, holding the hot iron, is 75 next month. His striker, John Healey, is 70, and they have worked together for 45 years. "I daren't think what's going to happen when they decide to retire," said Mr. C. Martin, under works manager. "There are few apprentices these days - they think the work's too hard."

"Sweating" for health - Jim, a Leeds Loiner of Harlech Mount, Hunslet, started as a smith as a lad of 13 before the days of electric welding when he learned to sweat it out with a hammer. "Sweating's kept me healthy," Jim said. "Nothing like it. In the old days we were given nitre** to make us sweat but I never really needed it." He went to Hathorn, Davey and Sulzer's 50 years ago this week and is now a foreman forger.*

"When I first came here there were 14 men and boys in the smithy," he said, "but now there's only myself and the two strikers.' That shows how the trade is slowly going, what with modern machinery and the lads of today who the work's too hard for them' I've made forgings weighing seven hundredweight. John and I have done some intricate jobs in our time." "He is an artist in metal on the forge," said Mr. Martin.

10-hour day - "But do you know I've never made horseshoe my life," added Jim with a grin. He is a grandfather who celebrated his golden wedding anniversary last month and his only handicap a slight deafness, is due, he thinks, to 62 year's among the noise of hammers and machinery. He doesn't smoke or drink, occupies his spare time gardening and prefers the Isle of Man for his holidays. He works a 10-hour day on staggered hours. "They played Hamlet with me at home for working till nine at night." As I left the old forge Jim and John turned back to straightening some chequer plates "Come on young 'un," said Jim and the clack clang of the hammer was soon drowned by the roar of the lathes as I walked back through the machine shop.'

*   An old term for a native of Leeds.
**   19th century prescriptions indicate that nitre was prescribed to induce sweating.

Hugh Lupton, died on the 18th February 1947 at the age of 85, and was buried in Saint John the Evangelist Churchyard, Roundhay, Leeds. Arthur Herbert Meysey-Thompson survived him by a few years and died on the 5th February 1950 aged 98.

It is not proposed to examine accounts in detail for this time. It is sufficient to mention that in the 1950s the Company was going through a period of growth. The portion of the Company's assets that included stock of materials, jigs, gauges, tools, and work in progress rose. In 1949 and 1950 the Directors valued them at £152,354 and £196,771, respectively.[8]

The War and its aftermath had ensured that the old social order was fragmenting in many of the traditional manufacturing industries. No longer was it common for a skill to be handed down from father to son. Most companies now usually required employees to have appropriate qualifications and that included Hathorn, Davey. They advertised widely for qualified staff, including in Scotland, and areas where such industries were starting to go into decline. It seems likely that the drawing office (Figure 166) was still the largest department and they regularly advertised for draughtsmen.

Figure 167. Hathorn, Davey Advertisement for a Jig and Tool Draughtsman
(Newcastle Evening Chronicle.24th July 1965. p.3)

The advertisement (Figure 167) for a Jig and Tools Draughtsman placed in the *Newcastle Evening Chronicle* states that there was a Company Pension Scheme in place, and a set pattern of daily work where an employee could enjoy a canteen break.

The main business at the time was manufacturing pumps. There is no doubt that pumps have survived from this period like the one shown in Figure 168. It can be seen in the grounds of the *Museum of Technology*, Cheddars Lane, Cambridge. The brass maker's plate carries a wealth of information. The order was 878541/1 dated 1964 and it was specified to pump 8500gpm, from a total head of 50 feet turning at 980 rpm. It was rated at 153 horsepower.

In the 1960s, the Company was awarded some major contracts. For example Hathorn, Davey placed an order worth more than £500,000 with English Electric in the summer 1965. It was for sixteen 8,850 horsepower electric motors, and their controllers for installation at the *Central Electricity Generating Board*'s new power stations at Pembroke, South Wales and Didcot, Berkshire. It is presumed that the motors were to drive water pumps.[9]

Figure 168. A Hathorn, Davey pump on display at
Museum of Technology, Cheddars Lane, Cambridge
(Author, 2019)

In 1972, a new office block and machine shop was added to the Sun Foundry.

However, this was found to be inadequate, and by 1980/81 Sulzer had moved their factory to the Millshaw area of Leeds, about 1 kilometre to the southwest of the Sun Foundry. The Sun Foundry remained in a derelict state and in the early 1980s the old Hathorn, Davey offices were used as a film set for the television programme, 'Harry's Game' set during the troubles in Northern Ireland.[10]

During the 1980s the whole of the Jacks Lane area that is now adjacent to junction 4 on the M621 was redeveloped. What remained of the Sun Foundry was demolished, but perhaps the new offices constructed by Sulzer on the corner of Jacks Lane and the Dewsbury Road was reused.

It was probably sometime during the 1980s / early 1990s that Hathorn, Davey was categorised as a dormant company on the Company's House Register, London. Annual returns were minimal. Sulzers did show their office was the Registered Office of Hathorn, Davey and Company, Limited, with an appropriate plaque on their entrance.

Hugh Ralph Lupton, the last surviving Director of the pre-Sulzer era of the Company died on the 8th March 1983. He had a long career in engineering. In 1939, he had been appointed the Mechanical Engineer to the Metropolitan Water Board and during World War II had been responsible for maintaining London's water supply. On leaving the Board he became a consultant on pumping machinery in 1951 and was made President of the Institution of Water Engineers and Scientists in 1955.[11]

Hugh had survived Reginald Hamilton Lupton by eight years. Reginald died in retirement at Budleigh Salterton, Devon 27th April 1975.[12]

Figure 169. A brass plate at Twyford bearing the manufacturers name - Hathorn, Davey
(Author, 2020)

During 2015, a decision to wind up Hathorn, Davey was made by Sulzers and a liquidator was appointed. On the 16th September 2016 the final liquidators meeting was held. The Company was struck off the Company's Register in December 2016, ending 144 years of innovation, enterprise and success.[13]

Thanks to the many ideas of Henry Davey, Hathorn, Davey was one of a relatively few companies that filled the technological gap between the slow decline in steam power, and the introduction of newer forms of motive energy, i.e. electric motors, gas, diesel and petrol engines.

Hathorn, Davey engines still survive in many parts of the world, and some still display the manufacturers name, either on a brass plate (Figure 169) or in the base castings. The pair of Inverted Compound Engines, ordered in 1888 and 1895, on the banks of the Burdekin River, Charters Towers, Queensland, Australia, (Figure 170) may well be the oldest surviving Hathorn, Davey steam engines.

Such engines are testaments to the reputation that British engineering had in the 19th and early 20th centuries, together with the associated qualities of robust construction, efficiency and reliability. The engines are also monuments to the loyal workforce that manufactured them, and who put Hathorn, Davey at the forefront of what was once the manufacturing district of Hunslet, Leeds, West Yorkshire.

That is quite an accolade for a Company founded on the aspiration of a retired Army Officer wishing to find a worthwhile career for his half-brother.

Chapter 33 References

[1] Leeds Engineers' £29,000 Order. *The Yorkshire Post*. 7th April 1937. p.7
[2] Mallinson, F. 2009. The Istanbul Water Administration with Especial Reference to the New Plant at the Main Pumping Station at Lake Terkos. Bull. of the International Stationary Steam Engine Society.31:2.
[3] http://tayakadinli.blogspot.com/2019/01/ilk-yap-islet-devret-modeli-dersaadet.html
[4] LMSW Box 21D2BX2
[5] HL. 1940. *Hathorn, Davey and Company, Sun Foundry, Leeds.* p.9
[6] Leeds Firm's Art Show. *The Yorkshire Post*. 2nd December 1946. p.1
[7] Jim (74) is an "artist in metal," but has never made a horseshoe. *Yorkshire Evening Post*. 6th March 1948 p.5
[8] Hathorn, Davey and Company Ltd. (Company 71739) Annual Report and Statement of Accounts for year ending 31st December 1949. Companies House Records, London.
[9] £½ million order for English Electric. *Coventry Evening Telegraph*. 12th August 1965. p.17
[10] http://www.leodis.net/display.aspx?resourceIdentifier=1298&
[11] Hugh Ralph Lupton: Obituary. *Transactions of the Newcomen Society*. v.55. 1983. p.251.
[12] Probate Office Records
[13] Hathorn, Davey and Company, Limited. (Company Number 00071739). *The London Gazette*. 13th December 2016, Supplement: 61787 p..75

Figure 170. The two Inverted Compound Pumping Engines at
Charters Towers, Queensland, Australia
They were manufactured in 1888 and 1895
They are probably the oldest surviving Hathorn, Davey steam engines
(Author, 2014)

Appendix 1.   Hathorn, Davey Order Books

Order Books deposited at: WYAS, Leeds, West Yorkshire Joint Services
Nepshaw Lane South, Morley
Leeds LS27 7JQ

Order Book Number	Years Range	Order Number Range
?	1852 to 1868	2 to 1734
?	1868 to 1872	1735 to 2108
1	1872 to 1879	2057 to 2750*
2	1879 to 1883	2751 to 3709
3	1883 to 1886	3710 to 4409
4	1886 to 1890	4410 to 4802
5	1891 to 1893	4803 to 5111
6	1893 to 1896	5112 to 5345
7	1896 to 1899	5346 to 5696
8	1899 to 1902	5698 to 5889
9	1902 to 1904	5890 to 6111
10	1904 to 1906	6112 to 6228
11	1906 to 1908	6229 to 6361
12	1908 to 1910	6362 to 6470
13	1910 to 1912	6471 to 6576
14	1912 to 1914	6577 to 6702
15	1914 to 1915	6703 to 6772
16	1915 to 1916	6773 to 6863
?	1915 to 1916	6774 to 6899
17	1916 to 1917	6864 to 6965
18	1917 to 1918	6966 to 7098
19	1918 to 1920	7099 to 7219

The first two Order Books contain orders placed with Carrett and Marshall.

*Order 2112 was placed at about the time that the Partnership of Hathorn, Campbell and Davis took control of the Sun Foundry.

Missing Order Books

Order Book number 20, and onwards, are missing.

Order Books 23 to 26: examined by Brian Hillsdon at the Sun Foundry in the 1970s.

Order Book Number	Years Range	Order Number Range
23	1926 to 1927	7546 to 7635
24	1927 to 1928	7636 to 7743
25	1928 to 1929	7744 to 7842
26	1929 to 1930	7843 to 7927

<table>
<tr><td colspan="2">Appendix 2.</td><td>Henry Davey's Patents</td></tr>
</table>

1860s

GB 3139/1864	Improvements in Steam engines
GB 1008/1867	Obtaining motive power and using combustion products of liquid and other fuels
GB 1936/1867	Centrifugal governor for steam and other engines
GB 2430/1867	Supplying water to steam boilers
GB 1033/1868	Improvement to steam engines, part applicable to steam pumps
GB 2953/1868	Improvements to steam engines, part applicable to steam pumps
GB 48/1869	Raising and injecting water for feeding steam boilers
GB 971/1869	Actuating hydraulic machinery, partly applicable to steam engines

1870s

GB 277/1871	Steam and pumping engines
GB 277/1871	(Amendment dated 1876) - Steam and pumping engines
GB 3208/1872	Evacuating condensers of steam engines
GB 1943/1874	Working valves of steam or water pressure engines
GB 323/1875	Working by hydraulic pressure (regulating speed)
GB 2152/1875	Improvements in the method of working the slides of steam or water pressure engines and apparatus for that purpose
GB 2156/1876	Compressing and forcing air, water, etc.
GB 2814/1876	Compound pumping engines
GB 4160/1876	Hydraulic engines
GB 280/1877	Steam engines
GB 346/1879	Water meters, water pressure engines
GB 844/1879	Motive power engines and generators

1880s

GB 3070/1880	Differential valve gear of steam engines
GB 5489/1880	Steam engines
GB 1028/1882	Indicating and registering apparatus for pumping engines
GB 2527/1882	Production of flammable gas and applying its combustion for the production of motive power
GB 3785/1882	Regulating apparatus for steam engines
GB 3787/1882	Apparatus for generating elastic fluid under pressure for working engines
GB 4273/1883	Pumping engines
GB 3833/1884	A new or improved construction of low pressure steam motor
GB 15185/1884	A new or improved construction of low pressure steam motor

1880s (continued)

GB 1981/1885	A construction of low pressure steam motor
GB 3312/1885	Double acting pump
GB 2930/1886	An improvement in steam generators
GB 5607/1886	Improvements in steam motors
GB 7713/1886	Connecting steam engines to deep well pumps
GB 9364/1886	A governor and expansion valve for engine worked by steam or other elastic fluid
GB 9356/1886	An improvement in escapement for clocks
GB 11194/1886	An electric motor escapements for clocks
GB 14895/1886	Improvements in clocks
GB 2859/1887	Improvements in pumping engines
GB 2860/1887	Method and means of automatically raising sewage
GB 8113/1887	Improvements in pumping engines
GB 10535/1887	Improvements in air compressing pumps
GB 13269/1887	Improved valve gear for pumping engines
GB 13270/1887	Improvements in pumping engines
GB 1160/1888	Improvements in the construction of low pressure steam boilers
GB 14056/1889	Automatic water pressure sewage pump
GB 19197/1889	Improved pumping engine

1890s

GB 15/1890	Improvements in and connected with games
GB 5625/1891	Improved apparatus for feeding steam boilers
GB 8317/1892	Improved construction of water pressure pump
GB 8318/1892	Improvements in pumping engines
GB 15620/1892	Improved pumping engine
GB 18257/1892	Improved triple expansion steam engine
GB 17452/1893	Improved triple expansion steam engine
GB 17453/1893	Improved compound steam engine
GB 17492/1893	Improved triple expansion steam engine
GB 17493/1893	Improved compound steam engine
GB 17975/1893	Non-rotative steam pumping engines
GB 3772/1895	Water pressure engines
GB 16320/1895	Improved steam engine
GB 10935/1897	Steam engines.
GB 11995/1897	Improvements in apparatus for raising and lowering pumps in mine and other shafts
GB 29632/1897	An improvement in compound pumping engines (marine ?)
GB 22283/1900	A method of, and apparatus for, controlling pump valves

1900s (continued)

GB 23028/1900	An improvement in pump valves
GB 23905/1901	Improved pumping engine
GB 23906/1901	Improved valve gear for pumping engines
GB 26046/1901	Improved valve gear for steam engines
GB 26047/1901	Pumps
GB 4343/1902	Improvements in pumping engines
GB 6509/1902	Improved means for regulating and indicating the water level in steam boilers, applicable also for regulating the supply of liquid fuel, draining steam pipes and like purposes
GB 9234/1902	Condenser or cooler for motor cars
GB 16060/1902	Steam pump
GB 16061/1902	Improvements in hydraulic engines
GB 17370/1902	Differential governing gear for brake wheels
GB 20729/1902	Improvements in steam pumps
GB 21708/1902	Pumping engines
GB 23245/1903	Marine engines
GB 23713/1903	Improvements in steam turbines
GB 24371/1903	Improvements in gas engines
GB 26226/1903	Improvements in steam turbines
GB 5219/1905	Steam turbines
GB 10556/1905	Steam turbines
GB 15581/1905	Centrifugal pumps
GB 356/1907	Steam superheater and boiler feed water heater
GB 6822/1907	Steam turbines
GB 24085/1907	Combined hot air and steam turbine
GB 24573/1907	Ammonia gas engines
GB 24574/1907	Turbines
GB 26580/1907	Steam and combustion product generators for turbines
GB 5644/1908	Steam superheaters
GB 23296/1908	Universal joint or coupling for transmitting rotary motion
GB 25909/1908	Centrifugal pumps
GB 8930/1910	Gas engines
GB 7529/1910	Improved method of raising or forcing liquids and apparatus there of

1910s

GB 2935/1911	Floodgate
GB 6207/1911	Turbine
GB 6208/1911	Air compressors

1910s (continued)

GB 17577/1911	Weirs and floodgates
GB 21326/1911	Automatic sluice or floodgate
GB 11445/1912	Improvements in and relating to sluices and water gates
GB 29236/1912	Improvements in and relating to weirs, sluices and the like
GB 1485/1913	An improved air pump
GB 25064/1913	Improvements in pumping engines
GB 25065/1913	An improved hydraulic compressor

World War I Years and Post-War

GB 2319/1914	Improvements in pumping engines
GB 8466/1914	Improvements and relating to pumping plants
GB 12121/1914	Improvements in pumping engines
GB 12224/1914	Improvements in, and relating to pumps
GB 14186/1915	Gas burners
GB 101673 (1916)	Improvements in compound condensing steam engines
GB 102217 (1916)	An improved compound steam engine
GB 115582 (1917)	Improving the fuel economy of multiple effect steam engine and steam turbine plants
GB 112726 (1818)	An improved method of operating steam engines and steam turbines and means of carrying the method into effect
GB 139429 (1920)	Increasing the power and efficiency of internal combustion engines
GB 276531 (1927)	Improvements in, or relating to, steam engines

The list has been compiled from various sources, but the Author is indebted to John Hole for providing a list of his own research work about Henry Davey's patents.

Index

Abell (Derby) 39
Aberavon 266
Aberdare 54, 66
Accident 193
A.D. Mining Co 81
Admiralty 25, 27, 259
Aguas Corrientes (Uruguay) 233
Ahmedabad 240, 282, 283, 285
Albaret, Auguste 119
Alby United Carbide Factory 261
Alderney 125, 267
Aldershot 300, 312, 316
Aldridge family 277
Almadén (Spain) 105, 113
Almería 108
Altham Colliery 91
Amalgamated Society of Engineers, Machinists, Smiths, Millwrights and Patternmakers 194
Anglian Water Authority 243
Anglo-Turkish 317
Antrim Electricity Supply Co 267
Apprentice 47, 128, 130, 187, 193, 203-06, 275, 277-81, 319, 321
apprenticeship 32, 127, 131, 205, 278, 279, 281, 292
Argentina 232
Armstrong-Whitworth 85, 258, 259, 311, 314, 315
Ashmell, J.T. 262
Ashton Frost and Co 289
Atkinson 320
Aubrey Street 157, 164
Australia 71, 104, 145, 162, 170, 206, 229, 231, 240, 276, 283, 285, 286, 292, 325, 326
Austral-Otis 230

Babcock 210, 232
Badami Bagh (Bombay) 240
bagatelle 148
Balaghat Mysore Mines 219

Barafirth 276
Barbados 145, 300
Barbenson Le Ber, Elizabeth 125
Bargoed 226
Barnstaple 145
Barrus, Walter 275
Basset Mines 179, 180, 186
Beaconsfield 76, 78, 219-24, 227
Beckermet Mining Co 245
Bedford Foundry 32
Belalcázar 52, 105, 107, 113
Belalcázar Silver Lead Co 107
Belgian 196, 218, 253, 254
Belgium 196, 197, 201, 203, 207, 218
bell-crank 150, 154
Bennis' Sprinkler Stoker 159
Beyer and Peacock 85
Bigrigg Mining Co 245, 255
Bilston 132-34, 138
Birch, J. and Co 286
Birkenhead 96
Bloemfontein (South Africa) 240
Bognor 272
Bombardier 262
Bombay 240, 283, 292
Bonham-Carter, Reginald 181
Boulton, Matthew 3
Boulton and Watt 3, 105
Bover, José 108
Bower, George 35, 37, 39, 40
Bradley 132, 134-140, 214
Brazil 152-155
Bridgtown (Barbados) 145
Briggs 276
Brillador and San Carlos Copper Mine (Chile) 102, 103
Brindley Bank 227
Brooks, Shoobridge and Co 27
Buarcos (Portugal) 105, 113
Buckton 87, 88, 297, 310
Buenas Aires 232
Bull 6, 71, 204
Burma 122

Burnt Fen 89-91
Burton on Trent 94
Butterknowle 71

Calyx drill 230
Campbell, Hugh Fletcher 22-25, 53, 58, 65-67, 85, 130, 192
Cambridge 51, 173, 175, 205-07, 227, 265, 323
Cambridge, Duchess of 207
Cambridgeshire 88
Cape Town (South Africa) 300
Cardiff 149, 151, 261
Cardiff Haematite Iron Ore Co 227, 261
Cardigan 301
Carey E.G. 254-256, 258
Carrett (and Marshall) 14-20, 22-25, 27, 47, 77, 87, 192, 212, 275, 327
Castlemaine (Australia) 230
Castlewigg 23
Central Electricity Generating Board 323
centrifugal 33, 87-89, 96, 137, 216, 217, 248-251, 269, 272, 287, 289, 298, 310, 311, 316, 329, 331
Ceredigion 302
Cerro Muriano (Spain) 180, 181, 186
Champion Reef Gold Mining Co 219
Charters Towers 145, 147, 170-72, 285, 325, 326
Cheltenham Corporation 300
Chester 94
Cheddars Lane 44, 51, 173-75, 191, 323
Chicago 119, 124
Chichester 46, 47, 52
China and Japan Trading Co 102
China Clay 255
Chinese Engineering and Mining Co 98
Chiswick 67-73, 75, 94
Chives and Fraser 108
Chumka (Odessa) 213
cistern 32, 54, 55, 117
clack 56, 76, 78, 82, 98, 223, 321
clack-boxes 214

Clacton 300
Clay Cross 24, 26, 42
Clayton 32
Coal Pit Heath (Bristol) 112
Coates, Sir Edward 257, 258
Colchester 300
Colne Valley 255
Compound Engine 47-52
Compton, Reverend Lord A, 23
Comstock Lode (Nevada) 52, 87
Conaways (Ewell, Surrey) 265
Constantinople Waterworks Co 317
Contrawhirl 294
Coquimbo (Chile) 102
Córdoba 50-52, 105, 113, 181, 186, 261
Cordova 181
Corfe Mullen 300, 302
Corliss 153, 155, 157, 159, 160, 209, 229, 230, 232, 241, 287
Coseley 138
Cosford 145, 151
Craig Trip (Valve) Gear 209, 232, 287
Cresswell 200
Crofton 3
Crosby, Mount 206, 282, 285, 286
cross-compound 47, 248, 249
Crossley-Premier 306, 308
Crown Agents for the Colonies 145, 234, 255
Croydon 58, 63, 64, 66, 73
Crystal Palace 84
Cumberland 245
Cusworth, H. 292

Dan, Takuma 167, 175, 235, 238
Darlaston 138, 186
Darlington 24, 71
Davey 1-8, 24, 26, 29-49, 51-54, 56-58, 61-63, 65, 70-75, 77, 78, 80-82, 84–87, 89, 94, 102, 104, 109, 111, 113, 115, 116, 118-29, 134-36, 139, 141, 142, 145-49, 151, 153-57, 163, 164, 166, 168, 169-77, 179, 180, 185-87, 190, 193, 197-99, 201, 202, 209, 211,

212, 224, 238, 246, 251-256, 258, 259, 265-67, 279, 325, 270, 293, 297, 303, 306, 308, 312-14, 318, 329, 332
Davey, Norman 125, 148, 212, 258, 265, 267, 312
Davey Paxman's 300
Davey Safety Motor (see Domestic Motor)
Davy Brothers (Sheffield) 33
Davis 20–25, 30, 47, 53, 56, 58, 63, 65-67, 71-75, 192, 193, 275, 279, 327
Deepfield 138, 186
Delft (Netherlands) 145
Deptford 287, 289, 296
Devon 31, 37, 105, 206, 265-267, 324
Dick-Lauder, Charlotte 62, 130
Didcot 323
Differential Gear 5, 7, 29, 33, 37-39, 41-46, 51, 56, 57, 70, 73, 91, 104, 111-13, 130, 134, 135, 150, 154, 163, 172, 177, 179, 209, 226, 246, 261, 329, 331
Dimmingsdale 300
dip-working 77, 80, 214
Domestic Motor 114-24, 130
Dowlais 250
Dover 300, 306, 312
draughtsman 32, 192, 193, 275, 277, 280, 281, 322, 323
drift-heading 81
Dudheshwar 285
Dunwell Ings, Hunslet 9
Durham 24, 26, 53, 73, 78, 79, 84, 85, 215, 247, 276, 320

East London (see Waltham Abbey)
Ebbw Vale 165, 166, 226
Ebbw Vale Steel, Iron and Coal Co 166
Eggborough 300
El Bole Mining Negotiations Co 104
electrode 259, 261
engine-recorder 87
Enville Street 300
Erin Colliery (Germany) 56, 57, 62, 66, 80
Etherley 91

Exhibition 15, 20, 30, 31, 71-75, 108, 119, 120, 124, 147, 266, 320
Exmouth 300

Fenman 280
Findley and Co (Motherwell) 180, 183
Fishguard 249
Fleam Dyke 227
floodgate 331, 332
Folkestone 300
Fondón 108, 113
Fordingbridge 32
Fraser and Chalmers 165
Fuveau 109

Gainsborough 94
Garforth 91, 201
Gargantua 269-72, 294, 310
gasworks 35, 189, 192, 212
Gauvain 265, 267
Glasgow 32, 105, 255
Glenfield and Kennedy 306, 308, 312
Gold Medal (Paris Exhibition) 74
Goliaths (Travelling Cranes) 187
Graces Guides 8
Graving Dock 216, 218, 248, 249, 251
Great Western Railways 32, 216, 249, 281
Green's Economiser 159, 210, 239
Greenwood and Batley 265
Griff Colliery 80
Grubbs Shaft 219, 221-23
Guibal Fan 24, 27, 99
Gunnerside 82
Gunnislake 32

Haematite 227, 245, 261
half-brother 23, 325
Hansa Colliery (Germany) 56, 57
Hardcastle, T. and Son (Bolton) 122
Hargreaves 276
Harrogate 85
Harrow 23

Hart, J.W. 100
Harts Shaft 219
Harvey and Co 1, 47, 183
Harvey and West Valve 76, 78
Headingley 158-160, 229
Heath, Wakefield 9
Helivane 272, 294
Henderson 183, 280, 285
Herberts Park 185
Hertfordshire and Essex Water Works Co
300, 303
Heseldin 275
Hewitt, George (Bristol) 112
Higgins 10, 11, 192, 195
hippopotamus 78
Hopper, Arthur 265
Howcroft 275
Hunslet 9, 11, 193, 195, 261, 277, 321, 325
Hunslet Engineering Co 261
Hurliman, P 310, 311, 316
Hutton Henry Coal Co 84

Ilford 244
impeller 249, 251, 269
Indenture 75, 127, 131, 202, 203, 207, 278,
279, 281
India 23, 30, 104, 105, 145, 170, 206, 213,
219, 231, 234, 240, 265, 276, 282, 283, 285-
87
Indian State Railways 25
Ingram 32, 40
Internal Fire Museum of Power 301, 302
Ipswich 300
Isochronal Governor 33
Istanbul 317, 318, 325

jacketed 54, 67, 157, 160, 162, 232, 317
jackets 153, 155, 161, 162
Japan 17, 52, 97, 101, 102, 104, 105, 121,
153, 167-69, 175, 209, 210, 213, 224-26,
235, 238-40
Jersey 37
John Taylor and Sons 103, 104, 145, 181,

219
Joy (Organ Blower) 16, 77, 84

Kachidachi 168
Kai ping 97-100
Karachi 276, 286
Katsudachi 168
Kendall and Gent 264
Kesteven (Lincolnshire) 245-47, 251
Kettering Type 247
Kippax Colliery 17
Knostrop (Leeds) 249
Kolar 145, 219, 283
Kubler 311
Kyūshū 102, 167, 210, 224, 238, 239

Lagos 234
Lahore 283
Lambert 102
Lambton Colliery 53, 56
Las Piedras 233
Lea Bridge
Leather, James 9
Leather (George and William) 15
Leeds Corporation 189, 249
Leeds Forge Co 261
Leeds General Infirmary 194
Leeds Pottery 264
Leeds University 249
Leeds Workpeople's Hospital Fund 194
Lewtrenchard 31
Lhuillier 109-111
Liancourt-Rantigny 119
Liege 196, 197
Linares 47, 105, 112, 181
Lincoln 71
Lincolnshire 245, 251
Little Ouseburn 85
Liverpool 96, 98, 157, 164, 192, 283, 292
Llanharry 227, 261
Llanover Colliery 226
Lock and Warrington 17
London Museum of Water and Steam 243

London Office 128, 141
London Tower 146, 147, 151
London Tower Co 147
Louis, Professor Henry 215
Lumpsey 227
Lupton's 65, 128, 202, 203, 205, 212, 258, 259, 277, 279
Lupton, Charles 274
Lupton, Geoffrey Henry 204
Lupton, Hugh 8, 11, 41, 65, 85, 86, 113, 124, 127-30, 138, 140, 153, 187, 196-99, 201-205, 209, 210, 212-214, 218, 250, 254-59, 264, 265, 268, 274-79, 281, 285, 293, 297, 300, 301, 304, 306, 307, 310, 312, 313, 316, 318, 319, 321, 322, 325
Lupton, Hugh Ralph 205-7, 265, 268, 269, 272, 274, 283, 287, 295, 298, 309, 319, 324, 325
Lupton, Isabella 202, 203, 207
Lupton, Maurice 204, 207, 262
Lupton, Olive 207
Lupton, Reginald Hamilton 206
Luton 93, 94
Luty, G. 319, 320

MacDonald, Ramsay 266
mahogany 107
Malaysia 283, 285
Mallinson 276, 317, 325
Malta 145
Manda Shaft 213, 224, 225
man-engine 102, 103
manganese 215, 249
Mansell, Walter 153
Mansfeld 80, 87
Manton 265
Marine Colliery 165, 166, 226
Markham and Co. (Chesterfield) 306
Marriott 's Shaft 179, 180
Marseilles 84, 109
Mathews, Joseph 32
McOnie and Co 183
Meiji Period (Japan) 102, 240

Melbourne 229, 230, 240
Melton Mowbray 64
Mersey Railway Tunnel 96
meter 88, 89, 115
Metropolitan Water Board 287, 294, 296, 299, 324
Meysey-Thompson, Herbert 85, 86, 104, 126-28, 138, 140, 196, 197, 201-3, 207, 210, 213, 214, 218, 254, 256-59, 265, 297, 312, 322
Middleton, Richard 207
Middleton Colliery Co 189, 212, 264
Miike 102, 167, 169, 213, 224, 225, 235-37
Milford Haven 35-37, 40, 77, 279
Mill Meece 175, 227, 289-92
Milnes, Fryston Hall 9, 10
Millom and Askham Haematite Iron Co 245
Ministry of Munitions 259
Mitsui 167, 168, 213, 235, 238
Miyanohara 168-170
Miyanoura 235, 238
Moat 132, 134, 137-139, 214
Moncrieff, Major A. 319
Montevideo 233
Montreal 161, 217
Montreal Iron Works (Whitehaven) 71,
Moore, Joseph (San Francisco) 52, 118
Mortar Mills 69
Morrison, K.S. 304, 307, 309-11, 315, 316, 319
Mountain Ash 54
Mount Morgan Gold Mine 162
Muir and Co 264
Mulvany 56, 57, 66, 87
Mysore 145, 219

Nagasaki 17, 101
Nakama City 210, 238, 239
National Shell (Projectile) Factory 259, 260
Navigation Colliery 54, 55, 66
Nelson 108, 129, 258, 273
Netherley 149
Newcomen 1-3, 8, 140, 207, 253, 262, 325

New Disintegrator (see Gargantua)
New Explosives Ltd 259
New Hartley Pit 28-30
New Joint Dock (Hull) 255, 256
Newmarket 241-243, 251
Newstead Colliery 71
Newton Gap 24
New Zealand 6, 122, 177, 178, 182-84, 186, 213, 219
Nicholson and Mathews 32
Nicholson, John 23
nitre 321
Northampton 23
North of Ireland Iron Ore Co 23
Northumberland 24, 29, 206
Norwich 272
Nuneaton 80

Odda (Sweden) 261
Odessa 213, 240
Oerdelin 310, 311
Onga, River 210, 238, 239
Ooregum Gold Mining Co 219
organ 17, 23, 24, 30, 77, 84
Organ-blower 16, 17, 22, 24, 77
Oriental Gold Mining Co 219
Osnabrück 205
Oxford 116, 127, 141, 203, 207
Ouro Preto Mining Co 103

Paddington 249
Paignton 300
Pakistan 283, 286
Parallel motion 70
Paris 15, 71-75, 119
Parral Silver Mine (Mexico) 86
Parsons R.C. 251, 280
Patent 8, 14-16, 22, 24, 32-35, 37, 39-43, 52, 54, 61, 62, 64, 72, 73, 116, 117, 121, 124, 144, 145, 148, 164, 169, 170, 177, 179, 183, 185, 186, 209, 215, 250-53, 265, 267, 272
Patternmakers 194

Pawson family 275, 280
Peacock Brothers (Montreal) 161, 216, 217
Peart, J. 285
Pemberton 10
Pembroke 323
Perran Iron Ore Mines (Cornwall) 66
Pitman 243
Plymouth 216, 218, 248, 249, 251
Pontefract 10, 300, 311
Poole Corporation 300
Portland 25, 27
Port of Spain (Trinidad) 240
Portugal 105, 113, 119
Powell Duffryn Steam Coal Co 167
Potternewton 205
Preece, William 120
Preston Show 130
Prussian 56
Public Health Act 62, 63

Queensland 71, 145, 162, 170, 172, 282, 285, 292, 325, 326

Rand Water Board 213, 234
Ranikhet (India) 286
Rhur 41, 56, 57, 66
Rio Tinto 109
Ritchie, Robert 9
rocking-frame 154
Rosario 231, 232, 240
Rothery, Frederick 262
Roundhay 203, 322
Rowbarton 131
Royal Windsor Show 121
Russia 213, 240

Saint Helens 65, 73
Saint Neots 8, 35, 37, 39, 40
San Rafael Shaft 181, 261
Savery, Thomas 1, 8
Saxton, Maurice 262
Sayers 10
Schoen Steel Wheel Co 261

Schoene 306, 311
Schubeler, J.F. 297, 309-12
Schulze, Edmund 24
Science Museum (London) 72
Seaton Delaval 24, 29, 30, 47
Shanghai 101, 104
Shepherd 13
Sheppard 32, 40
Shimla (Simla) 287
Shuttleworth 32
Sims 5, 47
Singapore 284, 285
Sir Francis Level (Swaledale) 81-84
Slade Heath 300
Smith, Roger 281
Smithson 9, 303
Société Anonyme de Charbonages, Bouche de Rhone 109
Société Anonyme des Charbonages des Kessales 196
Société Anonyme Miniere de la Province de Murcia 108
Société Anonyme Hathorn, Davey Belgium 196
Solana Mine 50, 51, 105, 108
Solana Mining Co 107
South Durham Colliery Co 73, 79
South Field (Newmarket) 241
South Frances Mine 179
South Staffordshire Mines Drainage Commission 131-40, 185, 214
South Staffordshire Waterworks Co 227, 292, 300, 303
Spotswood 228-230, 240
Staffordshire 53, 93, 131-33, 136, 138-40, 163, 164, 175, 186, 191, 214, 227, 272, 277, 289, 292, 300, 303
Staffordshire Potteries Water Works Co 289, 292
steam-jackets 141
Stereophagus 250, 251, 269, 272, 294, 310
Stereophagus Pump and Engineering Co 251
Stobart, H. and Co 91

Stourbridge and District Main Drainage Board 300
Stowheath 132, 134, 137, 138, 186
strike 184, 194-200, 269, 273, 276, 278, 306
Stuart, J. 277, 321
Sudbury 37-40, 63
Sugden, John 11
Sulzer 7, 131, 206, 275, 279, 281, 297, 298, 302, 309-319, 321, 324
Sun Foundry 9-13, 15, 18, 20, 22-25, 30, 52, 58, 65-67, 72, 73, 77, 87, 92, 96, 99, 108, 113, 124, 126, 128, 130, 131, 141, 164, 184, 187-89, 192-95, 197, 199-203, 207, 212, 213, 218, 224, 251, 261, 262, 264, 272, 275-77, 281, 292-95, 297, 302, 306, 307, 311, 315, 316, 319, 324, 325, 327, 328
superheated 153, 209, 210, 232, 317
Sussex 246, 247
Sutton Poyntz 301, 302
Swaledale 81, 82, 84
Sweden 261
Swedish 23, 30
Sydney 240

Tannett, Walkers 277
Tanygroes 302
Tasmania 76, 78, 219-23, 226, 227
Tasmanian Gold Mining Co 219, 223
Tasmanian Gold Mining and Quartz Crushing Co 219
Tavistock 31, 32, 47, 265, 266
Taylor and Challen Ltd 259, 261
Telford 15, 17, 18, 22, 24, 67, 85, 192, 193, 197-99, 202, 212, 275
Terkos 317, 318, 325
Thames 25
Thames Engineering Co 183
Thompson's 230
Tipton 132-34, 138-40
Todd, Charles 13
Toro Mining Co. 107
Tower (see London Tower)
Towler, James Alfred 157, 158, 161, 164,

193, 197-99, 202, 205, 209, 275
Trent Valley 163, 164, 191
Trethewey (Nurse) 267
Townsville (Australia) 145, 170
Trevithick 3, 5, 8, 47
Trinidad 240
Trinity House 130
Turner, Hoare and Co 285
Twyford 240, 243-45, 251, 255, 324

Ullcoats Mining Co 245
Umberumberka (Australia) 292
undercroft 29, 166, 186, 226, 241
UNESCO 240
United Mines 50
Unwin, Professor William 159, 175
Uruguay 233, 240, 283

Venezuela 145
Ventnor (Isle of Wight) 94
Vienna 30
Vulcan Foundry 35

Waihi 6, 177, 178, 182-84, 186, 213, 219
Waihi Gold Mining Co 177, 182
Wakefield 9, 210, 218
Wales 54, 165-67, 226, 227, 250, 261, 292, 301, 302, 323
Waltham Abbey 156, 161
Walton on Thames 287, 288
Watkin, Sir Edward 147
Watt 3, 48, 70, 105, 113, 156, 207
Watt's Medal 212
Webb 267
weir 212, 253, 332
Westhausen 113
Weston-super-Mare 143, 144, 151
Westphalia 56, 66, 73, 87
Weymouth 301
West Sussex 247
Wheatley Wood Colliery 92
Wheldale Coal Co 27
Wilkinson family 277

Wilks, Harry 277
Wingate 84
Whitehaven 71
Whithorn 23, 62
Whitworth 258, 259
Widnes 145, 149
Wilkinson 277, 280, 320
Willard, Charles P. (Chicago) 118, 119, 124
Williams' Perran Foundry 1
Wiltshire 3, 302
windbores 98
Winterthur 316
Wipperman 112
Wolverhampton 132, 138, 143-45, 300, 311
Woolf 47
Woolley Colliery 91, 92, 96
Woolwich 156
Worcester 23
Worthington-Simpson 165, 308

Yacatecas 104
Yangpu 100
Yarlside 94-96
Yate Colliery 91
Yawata Steel Works 238
Yeadon 94, 280
Yorkshire Electric Power Company 277

Zollern Colliery (Germany) 56
Zwaartkopjes 234